PRINCIPLES OF APPLIED MATHEMATICS

PRINCIPLES OF APPLIED MATHEMATICS

Maria Catherine Borres

ARCLER
PRESS

www.arclerpress.com

Principles of Applied Mathematics

Maria Catherine Borres

Arcler Press
2010 Winston Park Drive,
2nd Floor
Oakville, ON L6H 5R7
Canada
www.arclerpress.com
Tel: 001-289-291-7705
001-905-616-2116
Fax: 001-289-291-7601
Email: orders@arclereducation.com

ISBN: 978-1-77361-374-1 (Hardcover)

ABOUT THE AUTHOR

Catherine is currently taking up Master of Arts in Education Major in Mathematics in Philippine Normal University – Manila. She is currently working as a Content Developer for Mathematics at the Affordable Private Education Center (APEC Schools).

TABLE OF CONTENTS

PREFACE

In many respects, this book is for beginners. It combines a decoration-hearted approach, well-rounded explanations and much of practice opportunities. It is an ideal friend for those students who are commencing a course in applied mathematics and differential equation at a seminary or for the undergraduate layer in Mathematics, Engineering or Physics.

The study of differential equations is a wide domain in pure and applied mathematics, physics, and engineering. All of these disciplines are anxious with the properties of differential equations of different types. Pure quantity emphasis on the existence and singularity of solutions, while applied mathematics emphasizes the rigorous occasion of the medium for approximating solutions. Differential equations are an important senate in modeling virtually every physical, technical, or biological process, from celestial motion, to bridge design, to interactions between neurons. Differential equations such as those used to solve real-life problems may not necessarily be directly solvable, i.e., do not have closed form solutions. Instead, answer tins be approximated using numerical methods.

Many fundamental customs of physics and suitability can be formulated as differential equations. In biology and economics, differential equations are used to mold the dowry of complex systems. The mathematical opinion of differential equations first developed together with the proficiency where the equations had originated and where the results found the application. However, diverse problems, sometimes originating in quite distinct scientific fields, may give rise to identical differential equations. Whenever this happens, mathematical opinion seats the equations can be viewed as a unifying outlook incubation diverse phenomenon. As an example, consider assembly of enlightenment and sound in the atmosphere, and of the shaft on the surface of a pond.

Similarly, different sections of Mechanics are of much importance. They recreations important role in mathematical problems. From different realms, we tin easily model a configuration and then apply different methods to get some unique solution which suits that model.

CHAPTER 1

BASIC CONCEPTS OF ANALYTIC GEOMETRY

CONTENTS

1.1. INTRODUCTION

In analytic geometry of two dimensions the underlying space is $R \times R = R^2$ and each point of R^2 is an ordered pair (x, y) of real numbers. In analytic geometry of three dimensions we consider the space $R \times R \times R$ or simply R^3. Any element of R^3 is an ordered triple (x, y, z), where x, y, z are real numbers. As in the case of element of R^2, two ordered triples $(x, y, z) and (a, b, c)$ belonging to R^3 are equal if, and only if, their corresponding components are equal i.e., if, and only if $x = a, y = b and z = c$. A coordinate system in R^3 is defined as under.

Definition: We select a fixed point O for the origin. The x and y axes are drawn mutually perpendicular through O as in the case of plane analytic geometry. A line drawn through O and perpendicular to both the x and y axes is taken as the z-axis. The direction of $z-axis$ is positive if the $xy-plane$ viewed from a point on the positive $x-axis$ to the positive $y-axis$ is in the counter- clockwise direction. The three axes taken in pairs determine three mutually perpendicular planes. The plane determined by the x and $y-$ axes is the horizontal plane and is called the $xy-plane.$ Similarly, the plane determined by the $y and z-axes$ is a vertical plane and is called the $yz-plane.$ The $xz-plane$ is also vertical and is determined by the z and $x-axes.$ These three planes are referred to as the coordinate planes.

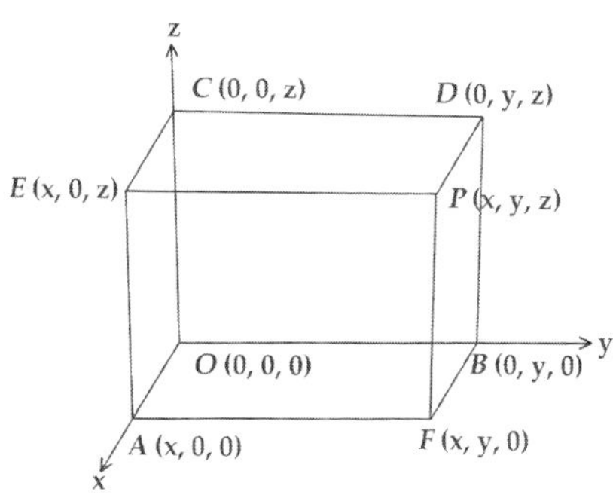

Let P be any point in space. We draw through P three planes parallel to the coordinate planes. These three planes together with the coordinate planes. These three planes together with the coordinate planes from a box (a parallelepiped) as shown in above Figure. The plane $PEAF$ parallel to the yz *plane* meets the $x-axis$ at A. Similarly, the other two planes through P meet the $y\,and\,the\,z-axes$ at B and C, respectively. If the directed distance OA, OB and OC are $x, y\,and\,z$, respectively then the point P is said to have rectangular (Cartesian) coordinates (x,y,z) and we write $P(x,y,z)$. The directed distance $OA = x$ is also equal to the directed distance of P from the $yz-$ plane, i.e., $x = DP$. Similarly, $y = EP\,and\,z = FP$. Conversely, given an ordered triple (x,y,z) of real numbers there is a unique point P in space such that x is the directed distance of P from the $yz-plane$. i.e., $x = DP.y$ is the directed distance of P from the $zx-plane$, i.e., $y = EP$. z is the directed distance of P from the $xy-plane$, i.e., $z = FP$. The point P has coordinates (x,y,z). Thus there is a one-to-one correspondence between the set of points in space and the set $\{(x,y,z): x,y,z \in R\}$ of ordered triples of real numbers. The space is divided by the coordinate planes in eight parts, called octants. The set of points (x,y,z) such that $x > 0, y > 0\,and\,z > 0$ is called the first octant.

The coordinate system described above with the orientations of axes as shown in figure is called a right – handed system of coordinates. If $x\,and\,y$ axes are interchanged then a left - handed system of coordinates is determined. We shall use only right-handed system of coordinates.

The coordinates of the vertices of the parallelepiped formed by the coordinate planes and the planes through P parallel to the coordinate planes are shown in figure.

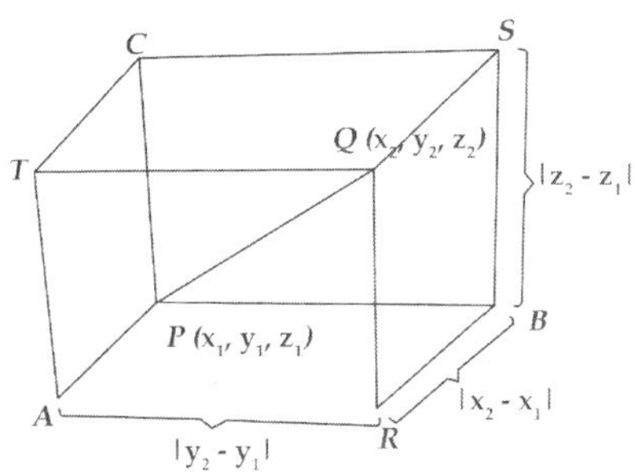

Let $P(x_1, y_1, z_1)$ and $Q(x_2, y_2, z_2)$ be any two points in space. The parallelepiped having its face parallel to the coordinate planes drawn through P and Q and with PQ as a diagonal is shown in figure.

The coordinates of the other vertices of the parallelepiped

$$A = (x_2, y_1, z_1), B = (x_1, y_2, z_1)$$

$$C = (x_1, y_1, z_2), R = (x_2, y_2, z_1)$$

$$S = (x_1, y_2, z_2), T = (x_2, y_1, z_1)$$

1.2. DISTANCE BETWEEN TWO POINTS

Definition: Let $P(x_1, y_1, z_1)$ and $Q(x_2, y_2, z_2)$ be two distinct points in space such that the line PQ is not parallel to any of the coordinate axes. Then the distance between $P \, and \, Q$ is taken to be positive.

Theorem. Let $P(x_1, y_1, z_1) and \, Q(x_2, y_2, z_2)$ be two distinct points in space such that PQ is not parallel to one of the coordinate axes. Then

$$|PQ| = \sqrt{(x_2 - x_1)^2 + (y_2 - y_1)^2 + (z_2 - z_1)^2}$$

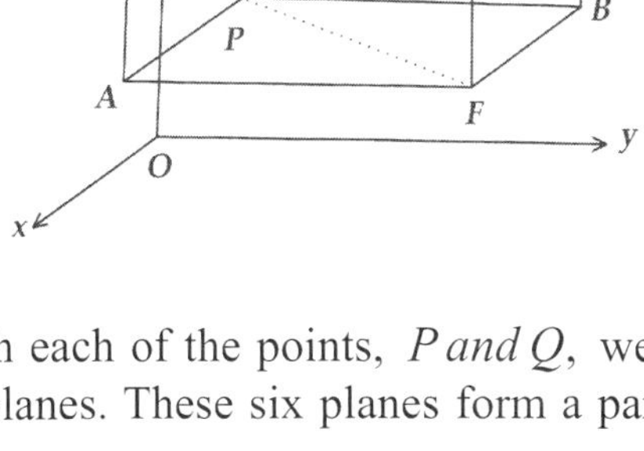

Proof: Through each of the points, $P \, and \, Q$, we draw planes parallel to the coordinate planes. These six planes form a parallelepiped as shown in figure. $|PA| = |x_2 - x_1|, |AF| = |PB| = |y_2 - y_1|$ and $|FQ| = |PC| = |z_2 - z_1|$.

From the right – angled triangle PAF, we have

$$|PF|^2 = |PA|^2 + |AF|^2$$

$$= (x_2 - x_1)^2 + (y_2 - y_1)^2$$

Again, from the right-angled triangle PFQ, we get

$$|PQ|^2 = |PF|^2 + |FQ|^2$$

$$= (x_2 - x_1)^2 + (y_2 - y_1)^2 + (z_2 - z_1)^2$$

Hence $|PQ| = \sqrt{(x_2 - x_1)^2 + (y_2 - y_1)^2 + (z_2 - z_1)^2}$ as required.

If $y_1 = y_2, z_1 = z_2$, then $|PQ| = |x_2 - x_1|$. Similarly, if $z_1 = z_2, x_1 = x_2$ then $|PQ| = |y_2 - y_1|$. Also, if $x_1 = x_2, y_1 = y_2$ then $|PQ| = |z_2 - z_1|$.

1.2.1. Alternate Method (Using Vectors)

We have $\overline{OP} = [x_1, y_1, z_1], \overline{OQ} = [x_2, y_2, z_2]$

Therefore, $\overline{PQ} = \overline{PO} + \overline{OQ}$

$$= -\overline{OP} + \overline{OQ}$$

$$= -[x_1, y_1, z_1] + [x_2, y_2, z_2]$$

$$= [x_2 - x_1, y_2 - y_1, z_2 - z_1]$$

$$|PQ| = magnitude\, of\, \overline{PQ}$$

$$= \sqrt{(x_2 - x_1)^2 + (y_2 - y_1)^2 + (z_2 - z_1)^2}.$$

1.2.2. Point Dividing a Line Segment in a Given Ratio

To find the coordinate of the point dividing the join of $P(x_1, y_1, z_1)$ and $Q(x_2, y_2, z_2)$ in the ratio $m:n$. Suppose $R(x, y, z)$ divides PQ in the given ration $m:n$. Let $O(0,0,0)$ be the origin. We know that if position vector of $P\, and\, Q$ are $a\, and\, b$, respectively, then the position vector r of the point R is

$$\boldsymbol{r} = \frac{n\boldsymbol{a} + m\boldsymbol{b}}{n+m}$$

Now,

$$\boldsymbol{a} = [\boldsymbol{x}_1 - 0, \boldsymbol{y}_1 - 0, \boldsymbol{z}_1 - 0] = [\boldsymbol{x}_1, \boldsymbol{y}_1, \boldsymbol{z}_1]$$

Similarly,

$$\boldsymbol{b} = [\boldsymbol{x}_2, \boldsymbol{y}_2, \boldsymbol{z}_2] \; \boldsymbol{and} \; \boldsymbol{r} = [\boldsymbol{x}, \boldsymbol{y}, \boldsymbol{z}]$$

Thus,

$$[x, y, z] = \frac{n[x_1, y_1, z_1] + m[x_2, y_2, z_2]}{m+n}$$

$$= \frac{[nx_1 + mx_2, my_1 + my_2, nz_1 + mz_2]}{m+n}$$

i.e., $x = \frac{nx_1 + mx_2}{m+n}, y = \frac{ny_1 + my_2}{m+n}, z = \frac{nz_1 + mz_2}{m+n}$

are the coordinates of the required point.

If R is the middle point of the segment PQ, then $m = n$ and the coordinates of the middle point are

$$x = \frac{x_1 + x_2}{2}, y = \frac{y_1 + y_2}{2}, z = x = \frac{z_1 + z_2}{2}.$$

Example. Prove that the points $A(3,2,4), B(6,10,-1), and\, C(9,4,1)$ are the vertices of an isosceles right triangle. Find the coordinates of the midpoints of the sides of the triangle ABC.

Solution. Here,

$$|AB| = \sqrt{(6-3)^{ü} + (10-2) + (-1-4)} = \sqrt{98} = 7\sqrt{2}$$

$$|AC| = \sqrt{(9-3)^2 + (4-2)^2 + (1-4)^2} = \sqrt{49} = 7$$

$$|BC| = \sqrt{(9-6)^{ü} + (4-10) + (1+1)} = \sqrt{49} = 7$$

Since $|AC| = |BC|$, the triangle is isosceles, moreover,

$$|AC|^2 + |BC|^2 = |AB|^2$$

Therefore, the triangle is a right triangle with the right angle at C.

The midpoint D of BC has coordinates $\left(\frac{3+6}{2}, \frac{2+10}{2}, \frac{4-1}{2}\right)$ i.e., $\left(\frac{9}{2}, 6, \frac{3}{2}\right)$. similarly, the midpoints $E\,and\,F\,of\,AC\,and\,BC$ have coordinates $\left(6, 3, \frac{5}{2}\right) and \left(\frac{15}{2}, 7, 0\right)$, respectively.

1.2.3. Straight Lines in

We know from vector algebra that two vectors $\boldsymbol{a}\,and\,\boldsymbol{b}\left(\,\boldsymbol{in}\,\boldsymbol{R}^3\right)$ are parallel if and only if there exists a real number k such that $b = ka$ or $a = kb$. If $a \neq 0, b \neq 0$, and $a = kb$, we say aandb are in the same direction if, and only if, $k > 0$. Similarly, if $k < 0$, a and b are in the opposite directions.

1.3. EQUATIONS OF A STRAIGHT LINE

A straight line is completely determined if it passes through two fixed points or it passes through a fixed point and is parallel to a given vector. We first find equation of a line L through a given point and parallel to a given vector.

Let $A(x_1, y_1, z_1)$ be a fixed point on the line L whose direction is parallel to the vector

$$\boldsymbol{d} \aleph \boldsymbol{ai} \quad \boldsymbol{bj} \quad \boldsymbol{ck}$$

Suppose $P(x,y,z)$ is any point on L. Then $\overrightarrow{AP}=t\boldsymbol{d}$, where t is some real number. It is a vector equation of the line through A and parallel to d.

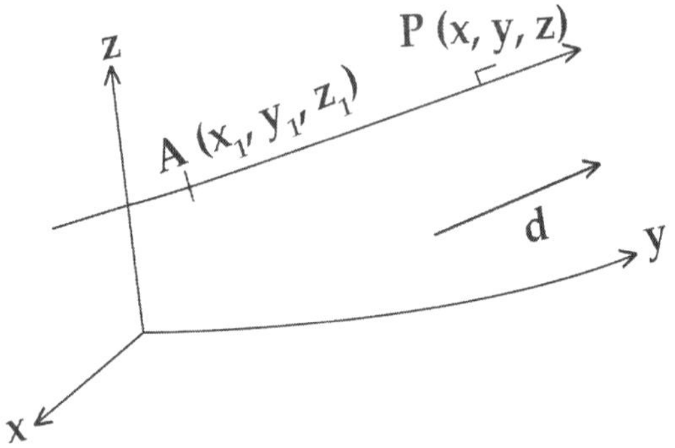

Now, $\overrightarrow{AP}=(x-x_{ü})i+(y-y\ )j+(z-z\ )k$

Thus, vector equation of the line L is

$$(x-x_1)\boldsymbol{i}+(y-y_1)\boldsymbol{j}+(z-z_1)\boldsymbol{k}=\boldsymbol{t}(\ \boldsymbol{ai}+\boldsymbol{bj}+\boldsymbol{ck}).$$

Equating components, we get

$$\left.\begin{aligned} x-x_1&=at\\ y-y_1&=bt\\ z-z_1&=ct \end{aligned}\right\} \qquad (1)$$

These are called parametric equations of the required line L.

From (1), we have

$$\frac{x-x_1}{a}=\frac{y-y_1}{b}=\frac{z-z_1}{c}=t$$

which is called a symmetric form of equations of the line through (x_1,y_1,z_1) and parallel to the vector $[a,b,c]$. The vector $\boldsymbol{d}=[\boldsymbol{a},\boldsymbol{b},\boldsymbol{c}]$ is called a direction vector for L and a,b,c are called direction numbers or direction ratios of the line L.

If a line passes through two given points $A(x_1,y_1,z_1)\, and\, B(x_2,y_2,z_2)$ then since its direction vector is $d=(x_2-x_1)i+(y_2-y_1)j+(z_2-z_1)k,$ a vector equation of the line is

$$(x-x_1)i+(y-y_1)j+(z-z_1)k = td, through\ A(x_1,y_1,z_1)$$

$$(x-x_2)i+(y-y_2)j+(z-z_2)k = td, through\ B(x_2,y_2,z_2).$$

A symmetric form of equations of this line is

$$\frac{x-x_1}{x_2-x_1}=\frac{y-y_1}{y_2-y_1}=\frac{z-z_1}{z_2-z_1}=t$$

or

$$\frac{x-x_2}{x_2-x_1}=\frac{y-y_2}{y_2-y_1}=\frac{z-z_2}{z_2-z_1}=t$$

Angle between two given lines is equal to the angle between their direction vectors. Thus, two lines are perpendicular if their direction vectors are perpendicular.

Example. Find parametric equations for the straight line through the point $A(2,4,3)$ and parallel to the vector $d=[4,0,-7]$.

Solution. Here $x_1=2, y_1=4, z_1=3$ and $a=4, b=0, c=-7$. The required equations are

$x=5-4t, y=-3+t, z=4+3t,$ through the point $A(5,-3,4)$

$x=1-4t, y=-2+t, z=7+3t,$ through the point $B(1,-2,7)$.

Example 4. Find the point of intersection (if it exists) of the pair of lines L and M given by

L: through $P(4,0,-2)\, and\ parallel\ to\ c=[2,3,0]$

M: through $Q(-1,2,-3)\, and\ parallel\ to\ d=[9,4,1]$

Solution. Parametric equations of the lines are

$$L:\begin{cases} x=4+2t \\ y=0+3t \\ z=-2+0t \end{cases} and\ M:\begin{cases} x=-1+9s \\ y=2+4s \\ z=-3+s \end{cases}$$

The conditions for L and M to intersect are that the equations

$$4+2t=-1+9s,$$

$$3t=2+4s$$

$$-2=-3+s$$

must have a simultaneous solution.

Solving the last two equations we obtain $t=2\,and\,s=1$. Since these values of $t\,and\,s$ satisfy the first equation, L and M intersect. Substituting $t=2$ in the equations for L (or $s=1$ in the equations for M), the required point of intersection is $(8,6,-2)$.

Note that if the three equations in t and s have no simultaneous solution, the lines do not intersect.

Example. Determine whether the lines L and M are (i) parallel (ii) non-parallel and intersecting (iii) non-parallel and non-intersecting, where

$$L:r=(-i+2j+3k)+t(5i+4j-3k)$$

$$M:r=(-i+2j-3k)+s(6i+3j+2k).$$

Solution. The line L is parallel to the vector $5i+4j-3k$ and the line M is parallel to the vector $6i+3j+2k$. Since these vectors are not parallel, the lines L and M are not parallel.

Parametric equations of the lines are

$$L:\begin{cases} x=-1+5t \\ y=2+4t \\ z=3-3t \end{cases}$$

$$\pm\ :\begin{cases} x\ \aleph\ddot{\imath} \qquad s \\ \qquad =2+3 \\ z\ \aleph\ 3 \quad 2s \end{cases}$$

The lines will intersect if the system of equations

$-1+5t=-1+6s$ (2)

$2+4t=2+3s$ (3)

$3-3t=-3+2s$ (4)

has a solution.

From Eqs. (3) and (4), we have $t=\frac{18}{17}, s=\frac{24}{17}.$

However, these values of $t\, and\, s$ do not satisfy (2). Hence, the lines are non-intersecting and non-parallel.

1.4. DIRECTION ANGLES OF A VECTOR

Definition. Let $v=[x,y,z]$ be a given vector. The direction angles α,β,γ of v are defined as

$$\alpha = Angle\, between\, v\, and\, the\, positive\, x-axis.$$

$$\beta = Angle\, between\, v\, and\, the\, positive\, y-axis.$$

$$\gamma = Angle\, between\, v\, and\, the\, positive\, z-axis.$$

By definition, each of these angles lies between 0 and π.

From the right triangle OAP, we have

$$\cos\alpha=\frac{x}{OP}=\frac{x}{\sqrt{x^2+y^2+z^2}}$$

Similarly, from the right triangle OBP, we get

$$\cos\beta=\frac{y}{OP}=\frac{y}{\sqrt{x^2+y^2+z^2}}$$

and from the right triangle OCP,

$$\cos\gamma=\frac{z}{OP}=\frac{z}{\sqrt{x^2+y^2+z^2}}$$

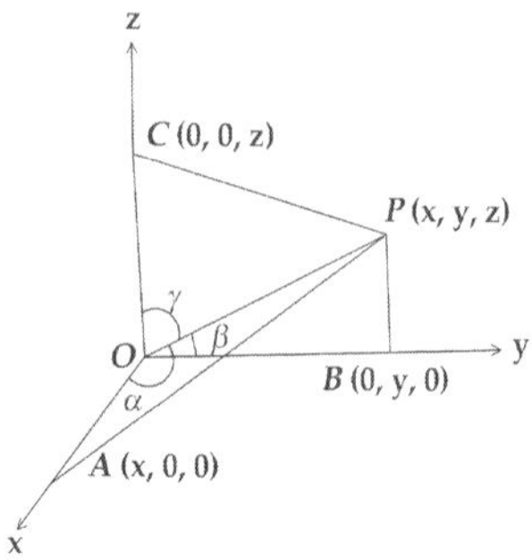

$\cos\alpha, \cos\beta, \cos\gamma$ are called direction cosines of v.

Now,

$$v = [x, y, z] = xi + yj + zk$$
$$= |v|(\cos\alpha)i + |v|(\cos\beta)j + |v|(\cos\gamma)k$$
$$= |v|\,[(\cos\alpha)i + (\cos\beta)j + (\cos\gamma)k]$$

The role of the direction cosines of a line is the same in three-dimensional geometry as that of the slope of a line in plane analytic geometry.

1.5. ANGLE BETWEEN TWO STRAIGHT LINES

Let l_1, m_1, n_1 be the direction cosines of a straight-line L and l_2, m_2, n_2 be the direction cosines of another straight line M. Let $\theta \in [0, \pi]$ be the measure of the angle between L and M. Then θ is also measure of the angle between the direction vectors (or unit vectors) along the two straight lines.

Now, $d_1 = l_1 i + m_1 j + n_1 k$ and $d_2 = l_2 i + m_2 j + n_2 k$ are unit vectors along L and M, respectively and

$$\cos\theta = d_1.d_2 = l_1 l_2 + m_1 m_2 + n_1 n_2.$$

If a_1, a_2, a_3 *and* b_1, b_2, b_3 are direction ratios of the two lines and θ is the angle between them, then

$$\cos\theta = \frac{a_1b_1 + a_2b_2 + a_3b_3}{\left(a_1^2 + a_2^2 + a_3^2\right)^{\frac{1}{2}}\left(b_1^2 + b_2^2 + b_3^2\right)^{\frac{1}{2}}}$$

The two lines are perpendicular to each other if, and only if $\theta = \frac{\pi}{2}$, i.e.,

If and only if, $a_1b_1 + a_2b_2 + a_3b_3 = 0$

Example. Find the direction cosines and measure of the direction angles of the line L passing through $(-3,4,2)\,and\,(4,-5,7)$.

Solution.

Here

$$d = [4+3,-5-4,7-2] = [7,-9,5]$$

and

$$|d| = \sqrt{49+81+25} = \sqrt{155}.$$

Thus $n = \left[\frac{7}{\sqrt{155}}, \frac{-9}{\sqrt{155}}, \frac{5}{\sqrt{155}}\right]$ is the unit vector along L. Therefore, the direction cosines of the lines L are $\frac{7}{\sqrt{155}}, \frac{9}{\sqrt{155}}, \frac{5}{\sqrt{155}}$.

Now, $\cos\alpha = \frac{7}{\sqrt{155}} = 0.5632, \cos\beta = \frac{-9}{\sqrt{155}} = -0.7229\; and\; \cos\gamma = \frac{5}{\sqrt{155}} - 0.4016$

Using tables of natural cosines, we find $\alpha = 55^0 47', \beta = 136^0 18', \gamma = 66^0 19'$, which are the required measures of the direction angles of the line L.

Example. Find measure of the angle between the straight lines

$$L: \frac{x-2}{1} = \frac{y-3}{1} = \frac{z+1}{2}$$

$$M : \frac{x-2}{2} = \frac{y-3}{-1} = \frac{z+1}{3}$$

Solution. We first observe that both L and M pass through $(2,3,-1)$.

Here, $d_1 = [1,1,2]\, and\, d_2 = [2,-1,3]$ are the direction vectors of the given lines.

Since $d_1 . d_2 = |d_1||d_2|\cos\theta,$ we have

$$2-1+6 = \sqrt{1+1+4} \times \sqrt{4+1+9}\cos\theta.$$

or

$$\cos\theta = \frac{7}{\sqrt{84}} = \sqrt{\frac{7}{12}} \text{ i.e., } \theta = \arccos\sqrt{\frac{7}{12}} = 40.2^0.$$

1.6. DISTANCE OF A POINT FROM A LINE

To find the distance from the point $P(x_1, y_1, z_1)$ to the line L with equation $\boldsymbol{r} = \boldsymbol{a} + t\boldsymbol{b}$. Let $\left|\overrightarrow{PM}\right| = p$ be the (perpendicular) distance from P to the given line L.

Then $p = \left|\overrightarrow{AP}\right| |\sin\theta|,$ where A is any point on L and θ is measure of the angle formed between $\overrightarrow{AP}$ and the direction vector d of the line L.

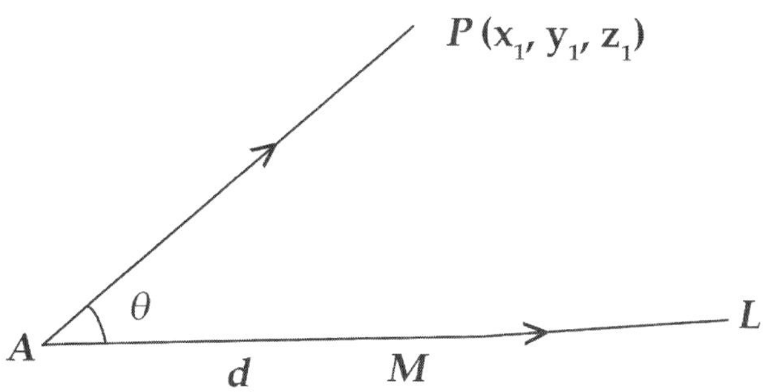

Now,

$$p = |\overrightarrow{AP}||\sin\theta| = \frac{|\overrightarrow{AP}||d||\sin\theta|}{|d|} = \frac{|\overrightarrow{OP} \times d|}{|d|}$$

Since $|d| \neq 0$ for all lines L, we have the formula

$$p = \frac{|\overrightarrow{\pm \quad \times}|}{|d|}.$$

Example. Find the distance of the point $P(2,-1,4)$ to the line with equation $r = a + td$ where $a = [3,-1,-1] \, and \, d = [1,2,1]$.

Solution. The point $A(3,-1,-1)$ lies on the line. Therefore,

$$\overrightarrow{AP} = [2-3,-1+1,4+1] = [-1,0,5].$$

The direction vector of the given line is $d = [1,2,1]$. Hence

$$p = \frac{|\overrightarrow{AP} \times d|}{|d|} = \frac{1}{\sqrt{6}}|10i + 6j - 2k| = \sqrt{\frac{70}{3}}$$

Example. Find equations of the line L through the point $\left(5, \frac{7}{2}, 5\right)$ and intersecting at right angles the line M with equations $x = 4 + 3t, y = 1 + t, z = -3t.$

Solution. Direction ratios of the line M are $3,1,-3.$ If L intersects M at $P(x,y,z)$, then direction ratios of the line L are $x-5, y-\frac{7}{2}, z-5.$ Since L is perpendicular to M, we have

$$3(x-5) + \left(y - \frac{7}{2}\right) - 3(z-5) = 0$$

$$3(4+3t-5)-\left(1+t-\frac{7}{2}\right)-3(-3t-5)=0$$

$$3(3t-1)+\left(t-\frac{5}{2}\right)-3(-3t-5)=0$$

$$t=-\frac{1}{2}.$$

This value of t yields the point P on M. Thus, $P=\left(\frac{5}{2},\frac{1}{2},\frac{3}{2}\right)$. Now L is the line through $\left(5,\frac{7}{2},5\right) and \left(\frac{5}{2},\frac{1}{2},\frac{3}{2}\right).$ The direction vector d **of L is**

$$\boldsymbol{d}=\left[\frac{5}{2}-5,\frac{1}{2}-\frac{7}{2},\frac{3}{2}-5\right]=\left[-\frac{5}{2},-3,-\frac{7}{2}\right]$$

Thus, equations of the line L are

$$\frac{x-5}{-\frac{5}{2}}=\frac{y-\frac{7}{2}}{-3}=\frac{z-5}{-\frac{7}{2}} \text{ through } \left(5,\frac{7}{2},5\right)$$

$$\frac{x-\frac{5}{2}}{-\frac{5}{2}}=\frac{y-\frac{1}{2}}{-3}=\frac{z-\frac{3}{2}}{-\frac{7}{2}} \text{ through } \left(\frac{5}{2},\frac{1}{2},\frac{3}{2}\right).$$

1.7. EQUATIONS FOR PLANES

A plane in space is uniquely determined by specifying a point in the plane and a non-zero vector perpendicular to the plane. A vector perpendicular to a plane is called a normal (vector) to the plane.

Theorem: The plane through a given point $A(x_1,y_1,z_1)$ and with non-zero normal vector $\boldsymbol{n}=[a,b,c]$ has equation

$$a(x-x_1)+b(y-y_1)+c(z-z_1)=0$$

Proof: It is obvious that a point $P(x,y,z)$ lies in the plane if, and only if, the vector $\overrightarrow{AP}$ is perpendicular to n. This is the case if, and only if, (Figure 1.7)

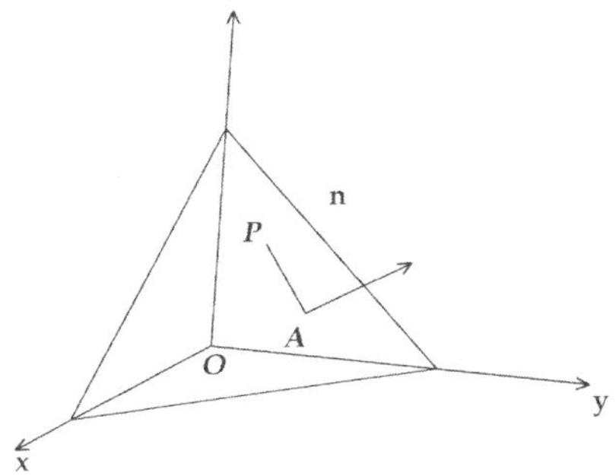

$\boldsymbol{n}.\overrightarrow{AP} = 0$ (5)

Since

$\overrightarrow{AP} = [x - x_1, y - y_1, z - z_1],(1)$

becomes

$a(x - x_1) + b(y - y_1) + c(z - z_1) = 0$ (6)

which is an equation of the required plane. This is known as point normal form of a plane.

Theorem: The linear equation,

$ax + by + cz + d = 0$ (7)

represents a plane having the vector $\boldsymbol{n} = [a,b,c]$ as a normal provided $\boldsymbol{n}$ is non-zero (d is a constant).

Proof. Since $n \neq 0,$ at least one of the numbers a,b,c is non-zero. To be definite, let $a \neq 0$. Equation (1) can be rewritten as

$$a\left(x - \frac{d}{a}\right) + b(y - 0) + c(z - 0) = 0.$$

Comparing this equation with (2) of above theorem, we see that it is a point normal form equation of the plane passing through the point $\left(-\frac{d}{a}, 0, 0\right)$

and having $\boldsymbol{n} = [a, b, c]$ as a normal vector.

If $a = 0$, then either $b \neq 0$ or $c \neq 0$ and these cases may be handled in a similar way.

1.8. PLANE THROUGH THREE GIVEN NON-COLLINEAR POINTS

Let $P(x_1, y_1, z_1), Q(x_2, y_2, z_2)$ and $R(x_3, y_3, z_3)$ be three given non-collinear points lying on the required plane.

Suppose an equation of the plane is

$$ax + by + cz = d$$

or

$$ax + by + cz - d = 0 \qquad (8)$$

Since the points $P, Q \, and \, R$ lie on (1), we have

$$ax_1 + by_1 + cz_1 - d = 0 \quad (9)$$

$$ax_2 + by_2 + cz_2 - d = 0 \quad (10)$$

$$ax_3 + by_3 + cz_3 - d = 0 \quad (11)$$

Eliminating $a, b, c, -d$ from (8) to (11), we get an equation of the required plane as

$$\begin{vmatrix} x & y & z & 1 \\ x_1 & y_1 & z_1 & 1 \\ x_2 & y_2 & z_2 & 1 \\ x_3 & y_3 & z_3 & 1 \end{vmatrix} = 0.$$

Example. Find an equation of the plane through the points $P(4,-1,2), Q(-3,-2,-1), R(7,-1,3)$.

Solution. ***Method I.*** Using above theorem, an equation of the plane is

$$\begin{vmatrix} x & y & z & 1 \\ 4 & -1 & 2 & 1 \\ -3 & -2 & -1 & 1 \\ 7 & -1 & 3 & 1 \end{vmatrix} = 0.$$

or

$$\begin{vmatrix} x-7 & y+1 & z-3 & 0 \\ -3 & 0 & -1 & 0 \\ -10 & -1 & -4 & 0 \\ 7 & -1 & 3 & 1 \end{vmatrix} = 0. \quad \begin{matrix} R_1 - R_4 \\ R_2 - R_4 \\ R_3 - R_4 \end{matrix}$$

Alternatively,

$$\begin{vmatrix} x-7 & y+1 & z-3 \\ -3 & 0 & -1 \\ -10 & -1 & -4 \end{vmatrix} = 0$$

or

$$-(-3)\begin{vmatrix} y+1 & z-3 \\ -1 & -4 \end{vmatrix} - (-1)\begin{vmatrix} x-7 & y+1 \\ -10 & -1 \end{vmatrix} = 0$$

$$3(-4y+z-7)+(-x+10y+17)=0$$

or

$x+2y-3z+4=0$ is equation of the required plane.

Method II: Let a normal vector of the plane be $\boldsymbol{n}=[\,a, b, c\,]$. Equation of a plane through the point $P(4,-1,2)$ with normal vector n is

$a(x-4)+b(y+1)+c(z-2)=0$ (Point-Normal form)

Since the other two points lie on the plane, we have

$-7a-b-3c=0$

$3a+0b+c=0$

Therefore, $\frac{a}{-1}=\frac{b}{-2}=\frac{c}{3}$ or $n=[1,2,-3]$.

Hence an equation of the plane is

$x-4+2(y+1)-3(z-2)=0$

$x+2y-3z+4=0$

Method III. The vectors $\overrightarrow{PQ}\, and\, \overrightarrow{PR}$ are coplanar. The vector $\overrightarrow{PQ}\times\overrightarrow{PR}$ is a normal vector of the plane.

But $\overrightarrow{PQ}=[-7,-1,-3]$ and $\overrightarrow{PR}=[3,0,1]$.

Therefore,

$$\overrightarrow{PQ}\times\overrightarrow{PR}=\begin{vmatrix} i & j & k \\ -7 & -1 & -3 \\ 3 & 0 & 1 \end{vmatrix}$$

$$=-i-2j+3k$$

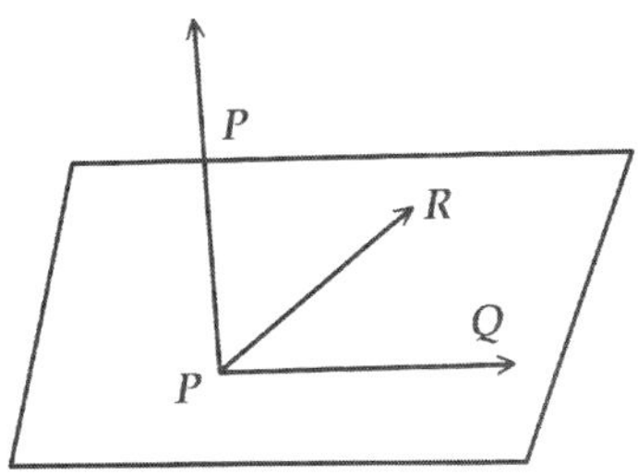

A normal vector of the plane is $n=[-1,-2,3]$.

Equation of the plane through P, having n as a normal vector, is

$-1(x-4)-2(y+1)+3(z-3)=0$ (Point –normal form)

$$x+2y-3z+4=0$$

1.8.1. Intercept Form of Equation of a Plane

Let an equation of a plane be $ax + by + cz = d.$ suppose it makes intercepts A, B, and Con the x-axis, y-axis, and z-axis, respectively. Then the plane meets the axes in the point $(A,0,0),(0,B,0),(0,0,C)$ since these points lie on the plane, we have

$$aA = d, bB = d, cC = d$$

Thus, $a = \frac{d}{A}, b = \frac{d}{B}, c = \frac{d}{C}$

Therefore, an equation of the plane is

$$\frac{d}{A}x + \frac{d}{B}y + \frac{d}{C}z = d$$

This can be written in the form

$$\frac{x}{A} + \frac{y}{B} + \frac{z}{C} = 1$$

This is known as the intercept form of equation of a plane.

1.8.2. Theorem Normal Form of Equation of a Plane

Let l, m, n be the direction cosines of a normal from the origin to a plane and let p be the length of this normal. Thus, an equation of the plane is

$$lx + my + nz = p.$$

Proof. Let A be the foot of the perpendicular from the origin to the plane. Since the direction, cosines of $\overrightarrow{OA}$ are l, m, n and $|\overrightarrow{OA}| = p,$ the coordinates of A are $(lp, mp, np).$ Let $P(x, y, z)$ be any point on the plane. Then $\overrightarrow{AP} = [x - lp, y - mp, z - np].$

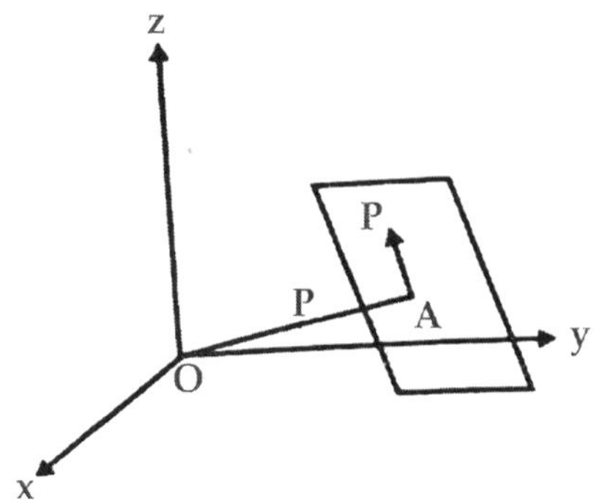

Since $\overrightarrow{OA}$ is perpendicular to the plane, $\overrightarrow{OA}$ is perpendicular to $\overrightarrow{AP}$ and so $\overrightarrow{OA}.\overrightarrow{Ap}=0$ or $l(x-lp)+m(y-mp)+n(z-np)=0$. i.e., $lx+my+nz=\left(l^2+m^2+n^2\right)p=p,$ since $l^2+m^2+n^2=1.$

To transform the equation $ax+by+cz+d=0$ into the normal form, we proceed as follows:

If the plane

$$ax+by+cz+d=0 \qquad (12)$$

is to be written as

$$lx+my+nz=p \qquad (13)$$

where $l^2+m^2+n^2=1,$ then (12) and (13) represent the same plane and we must have

$$\frac{a}{l}=\frac{b}{m}=\frac{c}{n}=-\frac{d}{p}=\pm\frac{\sqrt{a^2+b^2+c^2}}{\sqrt{l^2+m^2+n^2}}=\pm\sqrt{a^2+b^2+c^2}$$

Therefore,

$$-\frac{d}{p}=\pm a\sqrt{a^2+b^2+c^2}=\pm\sqrt{\Sigma a^2}$$

or $p=\dfrac{-d}{\pm\sqrt{\Sigma a^2}}$ and p is to be always positive.

Hence if d is positive, the sign of the radical is negative and therefore,

$$l=-\frac{a}{\sqrt{\Sigma a^2}}, m=-\frac{b}{\sqrt{\Sigma a^2}}, n=-\frac{c}{\sqrt{\Sigma a^2}}, p=\frac{d}{\sqrt{\Sigma a^2}}$$

In case d is negative, the sign of $l, m\, and\, n$ will be positive.

Thus the normal form of the plane $ax+by+cz+d=0$ is

$$-\frac{a}{\sqrt{\Sigma a^2}}x+\frac{b}{\sqrt{\Sigma a^2}}y+\frac{c}{\sqrt{\Sigma a^2}}z=\frac{d}{\sqrt{\Sigma a^2}},$$ if $d<0$.

Theorem. The (perpendicular) distance D of a point $A(x_1.y_1,z_1)$ from the plane $ax+by+cz+d=0$

$$D=\frac{|ax_1+by_1+z_1+d|}{\sqrt{a^2+b^2+c^2}}$$

Proof. Let $P(x_0,y_0,z_0)$ be any point on the plane and position the normal $n=[a,b,c]$ in such a way that its initial point is at P.

The required distance

$D=$Length of the orthogonal projection of $\overrightarrow{PA} on\, n$

$$D=\frac{|\overrightarrow{PA}.n|}{|n|}$$

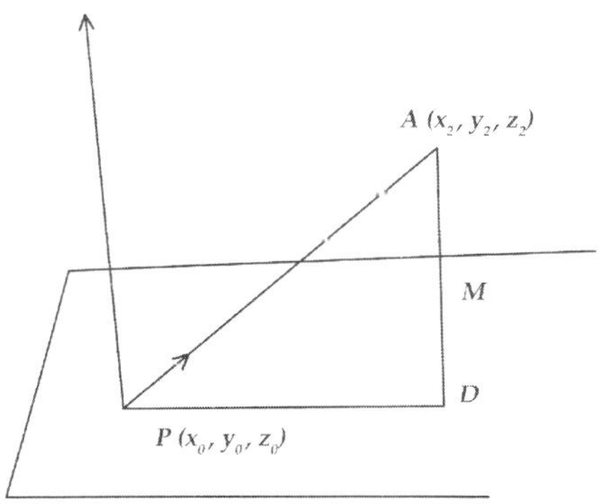

But

$$\overrightarrow{PA}=|x_1-x_0,y_1-y_0,z_1-z_0|$$

Therefore,

$$\overrightarrow{PA} = a(x_1 - x_0) + b(y_1 - y_0) + c(z_1 - z_0)$$

where $|n| = \sqrt{a^2 + b^2 + c^2}$

Hence,

$$D = \frac{|a(x_1 - x_0) + b(y_1 - y_0) + c(z_1 - z_2)|}{\sqrt{a^2 + b^2 + c^2}}$$

$$D = \frac{|ax_1 + by_1 + cz_1 - (ax_0 + by_0 + cz_0)|}{\sqrt{a^2 + b^2 + c^2}}$$

$$D = \frac{|ax_1 + by_1 + cz_1 + d|}{\sqrt{a^2 + b^2 + c^2}}$$

since the point $p(x_0, y_0, z_0)$ lies on the plane and therefore,

$$ax_0 + by_0 + cz_0 + d = 0 i.e, -d = ax_0 + by_0 + cz_0.$$

Example. Find the distance of the point $A(3, -1, 2) to the$ plane $2x + y - z - 4 = 0.$ (14)

Also, reduce the equation of the plane to the normal form and find distance for the origin.

Solution. The required distance D from $A(3, -1, 2)$ to (14) is

$$D = \frac{|6 - 1 - 2 - 4|}{2^2 + 1^2 + (-1)^2} = \frac{1}{\sqrt{6}}$$

Now we convert (14) to the normal form.

Since $d = -4 < 0,$ we divide through (1) by $\sqrt{2^2 + 1^2 + (-1)^2} = \sqrt{6}$.

Thus, the normal from of the given plane is

$$\frac{2}{\sqrt{6}}x + \frac{1}{\sqrt{6}}y - \frac{1}{\sqrt{6}}z = \frac{4}{\sqrt{6}}$$

Distance from the origin to the plane is $\frac{4}{\sqrt{6}}$.

Example. Find the distance between the parallel planes.

$2x+2y-4z+3=0$ (15)

and

$3x+3y-6z+1=0$ (16)

Solution. The distance between two parallel planes is the distance form an arbitrary point on one plane to the other plane. A point on (15) is $A\left(-\frac{3}{2},0,0\right)$. the distance $D of A$ from (16)

$$D=\frac{\left|\frac{-9}{2}+1\right|}{\sqrt{9+9+36}}=\frac{7}{2\sqrt{54}}=\frac{7}{6\sqrt{6}}$$

which is the distance between the given parallel planes.

Example. Find an equation of the plane that passes through the points $(3,2,-1) and (1,-3,4)$ and the contains a line parallel tp $2i-4j+3k$.

Solution. A plane through (3, 2, –1) and having $a_i+b_j+c_k$ as a normal vector has an equation

$a(x-3)+b(y-2)+c(z+1)=0$ (17)

Since $(1,\ 3,4)$ lies on (17), we have

$$a(1-3)+b(-3-2)+c(4+1)=0$$

i.e., $-2a-5b+5c=0$ (18)

The required plane is to contain a line parallel to $2i-4j+3k$. Thus normal vector of the plane is perpendicular to $2i-4j+3k$. therefore,

$2a-4b+3c=0$ (19)

Eliminating a,b,c from (17), (18) and (19), we have

$$\begin{vmatrix} \pm\aleph 2 & 2 & 1 \\ & \aleph 2 \; 53 & \\ & 2-43 & \end{vmatrix} \quad 0$$

or

$5x+16y+18z-29=0$ is an equation of the required plane.

1.9. ANGLE BETWEEN TWO PLANES

Definition. The angle between two planes is defined as the angle between their normal vectors. If equation of the two plane is

$$a_1x_1 + b_1y + c_1z + d_1 = 0$$

and

$$a_2x + b_2y + c_2z + d_2 = 0$$

Then the angle between then is the angle between their normal vectors.
If equations of the two planes are

$$a_1x + b_1y + c_1z + d_1 = 0 \qquad (20)$$

and

$$a_2x + b_2y + c_2z + d_2 = 0 \qquad (21)$$

Then the angle between them is the angle between their normal vectors

$$n_1 = [a_1b_1, c_1] \, and \, n_2 = [a_2, b_2, c_2]$$

Since $n_1.n_2 = |n_1||n_2|\cos\theta, where\, \theta\, is\, the\, angle$ between $n_1 \, and \, n_2$, therefore,

$$\cos\theta = \frac{n_1.n_2}{|n_1||n_2|}$$

$$= \frac{a_1a_2 + b_1b_2 + c_1c_2}{\sqrt{a_1^2 + b_1^2 + c_1^2} . \sqrt{a_2^2 + b_2^2 + c_2^2}}$$

The two planes intersect at right angles if $\theta = \frac{\pi}{2}$. Hence the condition for the plane to be perpendicular to each other is

$$a_1a_2 + b_2b_2 + c_1c_2 = 0$$

The two planes are parallel if their normal vectors are parallel. Hence for the planes (20) and (21) to be parallel, we have

$$\frac{a_1}{a_2} = \frac{b_1}{b_2} = \frac{c_1}{c_2}.$$

Example

Find the measure of the acute angle between the planes ü±+ − − = and $x - y - 2z + 5 = 0$

Solution

Let θ be measure of the angle between the normal vectors of the planes. Then

$$\text{ü} \quad \theta \aleph \frac{2-1+2}{\sqrt{4+1+1}.\sqrt{1+1+4}} \quad \frac{3}{6} \quad \frac{1}{2}$$

Therefore,

$$\theta = \frac{\pi}{3}.$$

1.10. STRAIGHT LINES AS INTERSECTION OF TWO PLANES

We have already found in that the condition for the planes and

$$\left.\begin{aligned} a_1x + b_1y + c_1z + d_1 = 0 \\ a_2x + b_2y\,c_2y + d_2 = 0 \end{aligned}\right\} \qquad (22)$$

to be parallel is

$$\frac{a_1}{a_2} = \frac{b_1}{b_2} = \frac{c_1}{c_2}$$

If the two parallel, then they intersect and their intersection is a straight line. Thus, two non-parallel planes represent a straight line given by two simultaneous linear equations 1 x, y and z and are known as non-symmetric form of equations of a straight line. In order to transform the equations into a symmetric vectors of the line.

A line in R^3 is often characterized as the intersection of two non-parallel planes, that is, as the set if simultaneous solutions of two linear equations in $x, y\, and\, z$.

Note. if $a_1x+b_1y+c_1z+d)1=0\, and\, a_2x+b_2\,y+c_2z+d_2=0$ are two non-parallel planes, then $(a_1x+b_1y+c_1z+d_1)+k(a_2x+b_2y+c_2z+d_2)=0$ is a plane through the line of intersection of these planes. For, $a_1\,x+b_1y+c_1z+d_1+k(a_2x+b_2y+c_2z+d_2)=0$ is a plane and contain all points common to the two given planes.

Example

Show that the planes. $3x-2y+5z=-2\, and\, x-y-z=2$ intersect and find symmetric form of their line intersection.

Solution

The given planes intersect since they are not parallel. Let us choose $z=0$ in the two equations. Then the equations become

$$3x-2y=-2$$

and $x-y=2$

These equations have the unique solution $x=-6, y=-8$.

Hence the point $(x_1, y_1, z_1)=(-6,-8,0)$ lies on both the planes. Similarly, (by choosing $y=0$), we find that the point $(x_2, y_2, z_2)=(1,0,-1)$ also lies on both the planes.

The direction vector d for the line of intersection of the two planes is

$$d=[x_2-x_1, y_2-y_1, x_2-z_1]=[7,8,-1]$$

Thus asymmetric form d for the line of intersection is

$$\frac{x+6}{7}=\frac{y+8}{8}=\frac{z}{-1}$$

Example

Find an equation of the plane passing through the line of intersection of the plane $2x-y+3z=0\, and\, x+2y-2z-3=0$ and perpendicular to the $yz-plane$.

Solution

$2x-y+3z+k(x+2y-2z-3)=0$ is any plane through the line of intersection of the given planes. This equation can be written as

$(2+k)x(-1+2k)y+(3-2k)z-3k=0$ (23)

Equation of the $yz-$ plane is $x=0$ (24)

Since (23) and (24) are perpendicular, we have

$2+k=0\,i.e.,k=-2$.

Substituting this value k into (23), equation of the required plane is

$$-5y+7z+6=0$$

i.e., $-5y+7z+6.$

1.11. A STRAIGHT LINE AND A PLANE

Suppose a line $L\,is$ given by $\frac{x-x_1}{c_1}=\frac{y-y_1}{c_2}=\frac{z-z_1}{c_3}=t$ (25)

And a plane S is defined by '

$ax+by+cz=d$ (26)

Any point on (25) has coordinates $(x_1+c_1t, y_1+c_2t, z_1+c_3t)$.

If this point lies on the plane S, we have

$$a(x_1++c_1t)+b(y_1+c_2t)+c(z_1+c_3t)=d$$

or

$t(ac_1+bc_2+cc_3)+ax_1+by_1+cz_1=d$ (27)

if $ac_1+bc_2+cc_3\neq 0,$ then (27) has a unique solution for t and so the line and the plane have a unique point in common. We say that straight line intersects the plane in this point. We note that in this case the direction vector of their line L and the normal vector of line S are not perpendicular. If $ac_1+bc_2+cc_3=0,$ then there are two possibilities.

Either $ax_1+by_1+cz_1=d\;or\;ax_1+by_1+cz_1\neq d$

If $ax_1+by_1+cz_1=d,$ then (27) is satisfied by the all values of t.

This means that the entire straight line (25) is contained in the plane S.
If $ax_1 + by_1 + cz_1 \neq d$, then (27) is satisfied by no value t. This means that the straight line L and the plane S have no point in common. In this case we say that the straight line and the plane are **parallel** and the direction vector of straight line L and the normal of the plane S are perpendicular.
We summarize these results as under:

I. The straight line L and the plane S intersected in one point if, and only if,

$$ac_1 + bc_2 + cc_3 \neq 0$$

II. The straight line L is parallel to the plane S if, and only if,

$$ac_1 + bc_2 + cc_3 = 0 \, and \; ax_1 + by_1 + cz_1 = d:$$

III. The straight line L is parallel to the plane S if, and only if,

$$ac_1 + bc_2 + cc_3 = 0 \, and \; ax_1 + by_1 + cz_1 \neq d.$$

Note. It follow that any plane containing the straight line L is

$$A(x - x_1)B(y - y_1) + C(z - z_1) = 0,$$

Example

Show that the straight line

$$\frac{x}{-1} = \frac{y+1}{2} = \frac{z-2}{-5} \qquad (28)$$

and the plane

$$3x + 4y - 2z = 22 \qquad (29)$$

Have a unique point of intersection. Find the point of intersection.

Solution

A plane through (28) is

$$A(x - a) + B(y - \beta) + C(z - y) = 0 \qquad (30)$$

where

$$Al + Bm + Cn' = 0 \qquad (31)$$

Since the plane (30) is parallel to the straight line (29), we have

$$Al' + Bm' + Cn' = 0 \qquad (32)$$

Eliminating A, B, C from (30), (31), and (32), equation of the required plane is

$$\begin{vmatrix} x-\alpha\, y-\beta\, z-y \\ l\,m\,n \\ l'm'n' \end{vmatrix} = 0$$

or
$$\Sigma(x-\alpha)(mn'-m'n)=0.$$

Example

Find the condition that the two straight lines

$$\frac{x-\alpha}{c_1}=\frac{x-\beta}{c_2}=\frac{z-\gamma}{c_3} \qquad (33)$$

and
$$\frac{x-\alpha'}{d_1}=\frac{x-\beta'}{d_2}=\frac{z-\gamma'}{d_3} \qquad (34)$$

may be coplanar. Also find an equation of the plane containing them.

Solution

Equation of a plane containing the first straight line is

$$A(x-\alpha)+B(y-\beta)+C(z-y)=0 \qquad (35)$$

$$\text{Where } Ac_1+Bc_2+Cc_3=0. \qquad (36)$$

If this plane contains the straight line (2), we have

$$A(\alpha-\alpha')+B(\beta-\beta')+C(\gamma-\gamma')=0 \qquad (37)$$

where
$$Ad_1+Bd_2+Cd_3=0 \qquad (38)$$

Eliminating *A, B, C* from (36), (37), and (38), the required conditions is

$$\begin{vmatrix} \alpha-\alpha'\,\beta-\beta\,\gamma-\gamma \\ c_1c_2c_3 \\ d_1d_2d_3 \end{vmatrix} = 0$$

or $\Sigma(a-a')(c_2d_2-c_3d_3)=0.$

If this conditions is satisfied then an equation of the plane containing the two straight line is obtained by eliminating *A, B, C* from (36), (37), and (38), and it is

$$\begin{vmatrix} x-a\, y-\beta\, z-\gamma \\ c_1 c_2 c_3 \\ d_1 d_2 d_3 \end{vmatrix} = 0$$

or $$\Sigma(x-a')(c_2d_2 - c_3d_3) = 0.$$

1.12. SHORTEST DISTANCE BETWEEN TWO STRAIGHT LINES

Definition. Let L and M be two given straight lines. If the straight lines are coplanar, then they either intersect, the shortest distance between them is zero. If they are parallel, then the shortest distance between the, can be obtained by finding the perpendicular distance of any point on one straight line to the other straight line.

1.12.1. Two Non-Coplanar Straight Line Neither Intersect Nor They Are Parallel

Such straight lines are called skew lines. Any tow skew lines lie in parallel planes. In this case there is one and only one straight line which s perpendicular to both the lines and intersects them, Measure of the intercept of this common perpendicular is called the shortest distance between the given skew lines.

Find the length of the shortest distance between the skew lines

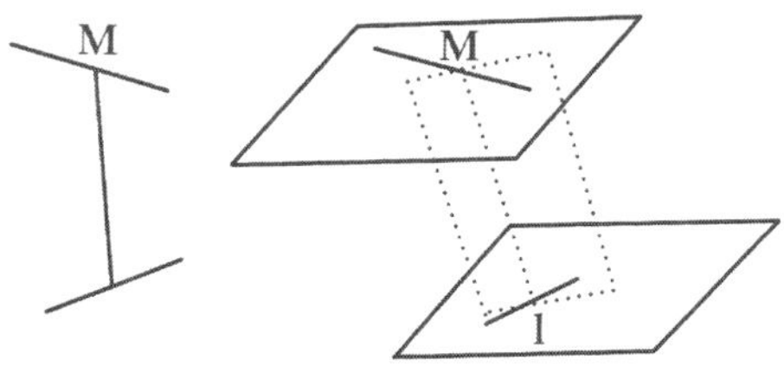

$$\frac{x-x_1}{l_1} = \frac{y-y_1}{m_1} = \frac{z-z_1}{m_1} \qquad (39)$$

and $$\frac{x-x_2}{l_2} = \frac{y-y_2}{m_2} = \frac{z-z_2}{m_2} \qquad (40)$$

Let PQ be perpendicular to both the straight lines (39) and (40). Suppose the directions cosines of PQ are l, m, and n. Since PQ is perpendicular to both straight lines (39) and (40), we have

$$ll_1 + mm_1 + nm_1 = 0 \qquad (41)$$

and $$ll_2 + mm_2 + mm_2 = 0. \qquad (42)$$

Thus $$\frac{l}{m_1n_2 - m_2n_1} = \frac{m}{n_1l_2 - n_2l_2} = \frac{n}{l_1m_2 - l_2m_1}$$

$$= \frac{1}{\sqrt{\Sigma(m_1n_2 - m_2n_2)^2}}$$

The points $A(x_1, y_2, z_1) and B(x_2.y_2, z_2)$ lie on (39) and (40), respectively.

$$\overrightarrow{AB} = [x_2 - x_1, y_2 - y_1, z_2 - z_1]$$

A unit vector $\boldsymbol{u}$ along PQ is $\boldsymbol{u} = [l, m, n]$

The required shortest distance between the lines is

$$\left|\overrightarrow{PQ}\right| = projection\,of\;\overrightarrow{AB}\;on\;\boldsymbol{u}$$

$$= \overrightarrow{AB}.\boldsymbol{u}$$

$$l(x_2 - x_1) + m(y_2 - y_1) + n(z_2 - z_1)$$

$$= \frac{(m_1n_2 - m_2n_1)(x_2 - x_1) + (n_1l_2 - n_2l_1)(y_2 - y_1)(l_1m_2 - l_2m_1)(z_2 - z_1)}{\sqrt{\Sigma(m_1n_1 - m_2n_2)^2}}$$

$$= \begin{vmatrix} x_2 - x_1 y_2 - y_1 z_2 - z_1 \\ l_1 m_1 n_1 \\ l\,m\,n \end{vmatrix} - 0 \qquad (43)$$

and the plane through (3) and PQ is

$$= \begin{vmatrix} x - x_2\; y - y_2\; z - z_2 \\ l_2 m_2 n_2 \\ l\,m\,n \end{vmatrix} = 0 \qquad (44)$$

where, l,m,n are given by (41).

Thus (43) and (44) together are equations of PQ.

Note. If it is required to find only the shortest distance between two skew lines *L* and *Mm* we make use of the following fact:

Since the two skew lines LandM lie in parallel planes, the shortest distance between them is the distance between the parallel planes containing them. Suppose *p* and *q* are the parallel planes containing *L* and *M*, respectively. Then P is a plane containing to *L*. Now the distance between P and q is the distance of any point of the plane containing the line. Thus, the distance of any point on one line to the plane containing the order line and parallel to the plane containing first line is the required shortest distance between the two lines.

Example

Show that

$$L:\frac{x-3}{3}=\frac{y-8}{-1}=\frac{z-3}{1} \text{ and } M:\frac{x+3}{-3}=\frac{y+7}{2}=\frac{z-6}{4}$$

are skew lines. Find the coordinates of the feet of the common perpendicular to the two lines and equation and length of the common perpendicular.

Solution

It is obvious that the lines are not parallel. We shall prove that the lines also do not intersect in a point.

A point on $LisP(3+3t,8-t,3+t)$ and a point on M is

$$Q(-3-3s,-7+2s,6=4s).$$

The condition for the lines to intersect in a point are that the equations

$$\left.\begin{aligned} 3+3t &= -3-3s \\ 8-t &= -7+2s \\ 3+t &= 6+4s \end{aligned}\right\} \qquad (45)$$

must have a simultaneous solution.

From the last two equations we obtain $x=2,t=11.$ These values of *s* and *t* do not satisfy the first equation of (45). Hence the given lines do not meet in a point. Thus, *L* and *M* are skew lines.

Suppose the points P and Q are the feet of the common perpendicular to the line L and M. The direction vector

$$\overrightarrow{PQ} = [-6-3t-3s, -15+t+2s, 3-t+4s]$$

Since PQ is perpendicular to both the lines, we have

$$3(-6-3t-3s)-(-15+t+2s)+3-t+4s)$$

and

$$-3(-6-3t-3s)+2(-15+t+2s)+4(3-t+4s)=0$$

i.e., $\quad -11t-7s=0$

and $\quad 7t-22s=0$

Hence $\quad t=0, s=0$

Therefore, the coordinates of the feet of the common perpendicular are

$$P(3,8,3)\, and\, Q(-3,-76)$$

Length of the common perpendicular is $\left|\overrightarrow{PQ}\right| = 3\sqrt{30}.$ Equation of the common perpendicular PQ are

$$\frac{x-3}{2}=\frac{y-8}{5}=\frac{z-3}{-1}.$$

Example

Final the length and equations of the common perpendicular to the line

$$L: 5x-y-z=0, x-2y+z+3=0$$

$$M: 7x-4y-2z=0, x-y+z-3=0$$

Solution

We first transform the equation of L and M in to symmetric forms putting $z=0$ in the equations of $L,$ we have

$$5x-y=0$$

$$x-2y+3=0$$

And so $x=\frac{1}{3}, y=\frac{5}{3}, z=0$ is a point of L. Let $\{a,b,c\}$ be a direction vector of L. Since L is the intersection of the two given planes, it is perpendicular to the normal vectors of each plane. Therefore,

$$5a-b-c=0$$

and $a-2b+c=0$

Hence $[a,b,c]=[1,2,3]$

Thus, L L in a symmetric form is

$$\frac{x-\frac{1}{3}}{1}=\frac{y-\frac{5}{3}}{2}=\frac{z}{3} \qquad (46)$$

Similarly, MM can be written as

$$\frac{x+4}{2}=\frac{y+7}{3}=\frac{z}{1} \qquad (47)$$

Let $n=[a,b,c]$ be direction vector of the common perpendicular to both (46) and (47). Then

$$A+2b+3C=0$$

and $$2A+3B+C=0$$

Therefore, $$\frac{A}{-7}=\frac{B}{5}=\frac{C}{-1}$$

Hence $$n=[7,-5,1]$$

The point $P\left(\frac{1}{3},\frac{5}{3},0\right) and Q(-4,-7,0)$ lie, respectively on (46) and (47).

The direction vector $\overrightarrow{PQ}=\left[-\frac{13}{3},\frac{-26}{3},0\right]$. Length of the shortest distance between the lines is the orthogonal projection of $\overrightarrow{PQ} on n$.

Shortest distance $=\frac{|\overrightarrow{PQ}.n|}{|n|}=\frac{\left|\frac{-91}{3}+\frac{130}{3}\right|}{\sqrt{75}}=\frac{13}{\sqrt{75}}.$ To find equation of the common perpendicular, we proceed as follow A plane through M is

$$7x-4y-2z+m(x-y+z-3)=0$$

i.e.,

$$x(7+m)+y(-4-m)+z(-2+m)-3m=0$$

$$or\, m=-67/13$$

Thus, the plane containing M and the common perpendicular is

$8x+5y-31z+67=0$ (48)

A plane containing L is

$$5x-y-z+k(x-2y+z+3)=0$$

i.e.,

$$x(5+k)+y(-1-2k)+z(-1+k)=0$$

or

$$k=-13/6$$

The plane containing L and the common perpendicular is

$17x+20y-19z-39=0$ (49)

Eqs. (48) and (49) constitute equations of the common perpendicular.

1.13. CYLINDRICAL COORDINATES

Let $P(x,y,z)$ b a point in rectangular coordinate system and M be the projcction of P in the $xy-plane$. Let the xy $plane$ be coordinated by polar coordinates so that the polar coordinates of $M are(r,\theta)$. We say that the point P has cylindrical* (polar) coordinates (r,θ,z) and the write $p(r,\theta,z)$. We have therefore, the equations

$$x=r\cos\theta, r\geq 0$$

$$y=r\,sin\,\theta\, and\, z=z$$

Giving the relationship between the rectangular and cylindrical coordinates of a point in $\boldsymbol{R}^3$. From the above equations we obtain

$$r = \sqrt{x^2 + y^2}$$

$$\tan\theta = \frac{y}{x}$$

$$z = z$$

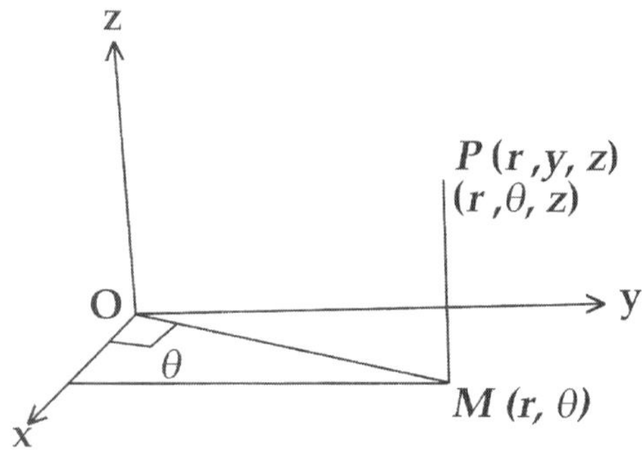

1.13.1. Expressing Cylindrical Coordinates in Terms of Rectangular Coordinates

Example. An equation in cylindrical coordinates is $r^2 \cos 2\theta = z$. Transform the equation into rectangular coordinates.

Solution. we have

$$r^2 \cos 2\theta = z$$

i.e., $$r^2(\cos^2\theta - \sin^2\theta) = z$$

or $x^2 + y^2 = z$ is the equation in rectangular coordinates.

Example

Express the equation $x^2 + y^2 + z^2 = 16$ in cylindrical coordinates.

Solution

Since $x^2 + y^2 = r^2$, the required equation in cylindrical coordinates is $r^2 + z^2 = 16$.

1.14. SPHERICAL COORDINATES

Suppose (x,y,z) are the rectangular coordinates of a point P (different from the origin) in R^3. We define

$p = |OP| =$ define of p from the origin.

θ = polar coordinate of the projection $M(x,y,0) of P$ in the $xy - plane$.
∅ = angle between positive z-axis and OP.

p is called the radial distance of P, the angle θ the azimuth of P, and the angle θthe colatitude of P. These three qualities are called spherical*(or geographical) coordinates of P and we write $P(p,\varnothing,\theta) i.e.,$ the azimuth is denoted by ∅ and the colatitudes by θ. But it is preferable to let θ denote the same angle in both cylindrical and spherical coordinates and to place it in the same position in both the systems.

It is customary to take $p \geq 0, 0 \leq \theta < 2\pi, 0 \leq \phi \leq \pi. \theta and \phi$ are always measured in radians. The origin 0 is the given the coordinates $(0,\theta,\phi)$ for any $\theta and \phi$.

It is clear from figure that

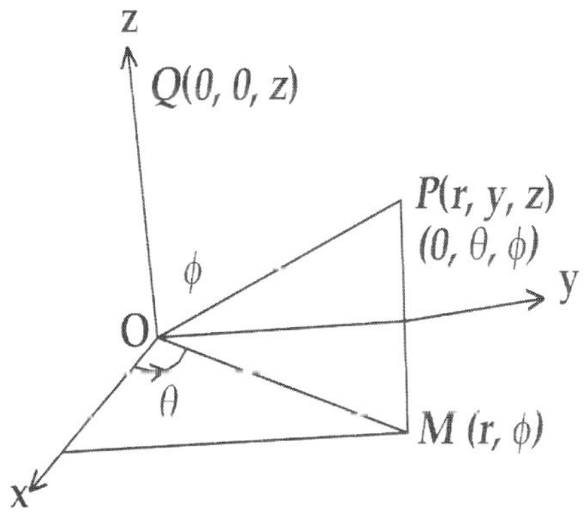

$$r = |OM| = |QP|$$

$$= p \sin\phi, (\sin\phi is nonnegative)$$

$$\theta = \theta,$$

which give relationship between spherical and cylindrical coordinates of the point p.

Since $x = r\cos\theta \, and \, y = r\sin\theta$, we have the following formulas to compute the rectangular coordinates of a point from its spherical coordinates:

$$x = p\sin\phi\cos\theta$$

$$y = p\sin\phi \sin\theta$$

$$z = p\cos\phi$$

Moreover, since p is the distance between the origin and P, we have

$$p = \sqrt{x^2 + y^2 + z^2}$$

1.14.1. Translation of Axes

The process of choosing new axes parallel to a given set of axes is called translation of axes.

Suppose Ox, Oy, Oz and $O'x', O'y', Oz'$ are two sets of parallel axes and let the rectangular coordinates of $P be(x, y, z)$ relative to the first set of axes and the rectangular coordinates of P be (x', y', z') relative to the second set of axes.

Let O' have coordinates (a, b, c) relative to the first set of axes. Then

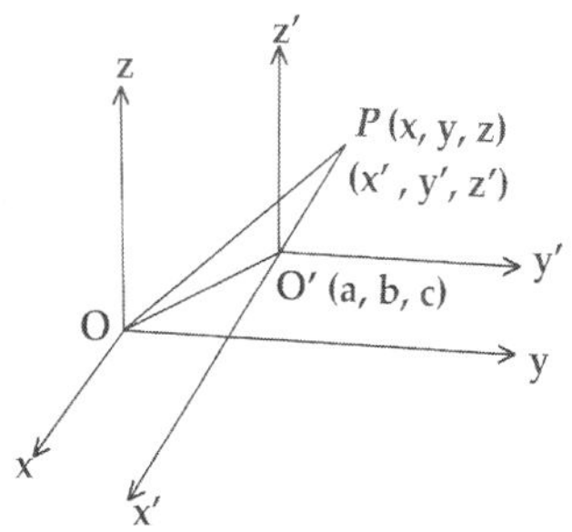

$$\overrightarrow{OP} = \overrightarrow{OO'} + \overrightarrow{OP}$$

i.e., $[x, y, z] = [a, b, c] + [x', y', z']$

or $x = a + x', y = b + y', z = c + z'$

or $x' = x - a, y' = y - b, z' = z - c$

These equation define a translation of axes or shifting of the origin.

Surfaces

Definition. The set of all points (x, y, z) which satisfy an equation of the from $f(x, y, z) = 0$ is called a surface. Thus, plane is a surface.

The set of all point (x, y, z) satisfying the second degree equation

$$ax^2 + by^2 + cz^2 = fyz + gxz + hxy + ux + vy + wz + k = 0 \tag{50}$$

where a, b, c, f, g and h are not all zero, determines a surface, called a quadric surface.

If (50) factorized into two shape of a curve can be obtained by plotting points. However, for a surface in R^3 plotting of points is not generally useful to get its sketch. To sketch a surface, we have to observe its special properties as stated below:

I. Symmetry: If $(x, y, z) = 0\, implies\, f(-x, y, z) = 0$; the surface is symmetric with respect to (or about) the $yz - plane.$ Symmetry with respect to the other coordinates planes is checked similarly. If $f(x, y, z) = 0$ implies $f(x, -y, z) = 0,$ the surface is symmetric with respect to (or about) the $x - axis.$ Symmetric about the other axes is seen similarly. The surface is symmetric about the origin if

$$f(x, -y, -z) = 0 = f(x, y, z).$$

II. Traces: The curve of intersection of a surface and a plane is called the trace of the surface in the plane. The trace of a surface are usually found in the coordinates planes or in planes parallel to the coordinates planes.

III. Intercepts: The x-coordinates of the points of intersection of the and a surface are called the x-intercepts of the surface. The intercepts on the $x - axis$ (or $x - intercepts$) $y = 0, z = 0$ in the equation of a surface and solving it for x. The $y\, and\, z -$ intercepts are found similarly.

Example. Discuss the surface

$$S : f(x,y,z) = \frac{x^2}{4} + \frac{y^2}{4} + \frac{z^2}{16} - 1 = 0$$

=

Solution. Symmetric. $f(-x,y,z) = \frac{(-x)^2}{4} + \frac{y^2}{4} + \frac{z^2}{16} - 1 = f(x,y,z) = 0.$

Therefore, S is symmetric with respect to the yz-plane. It is easy to see that S is also symmetric about the other axes and the origin is not difficult to check.

Traces. The trace in the coordinate planes have equations

$$\frac{y^2}{4} + \frac{z^2}{16} = 1, (\textit{ellipse in the } yz - \textit{plane})$$

$$\frac{x^2}{4} + \frac{z^2}{16} = 1, (\textit{ellipse in the } zx - \textit{plane})$$

$$x^2 + y^2 = 4, (\textit{ellipse in the } xy - \textit{plane}).$$

The trace in $z = c(\textit{a plane parallel to the xyplane})$ is

$$\frac{x^2}{4} + \frac{y^2}{4} + \frac{c^2}{16} - 1 = 0$$

i.e., $x^2 + y^2 = \frac{16 - c^2}{16}$, which is a circle for $-4 < c < 4$.

Intercepts. The points of intersection of S and the x-axis are given by $\frac{x^2}{4} - 1 = 0.i.e., x = \pm 2$. Thus, ± 2 are intercepts on the $x - \text{axis}$. The intercepts on the $y \, and \, z$ axes can be easily found as $\pm$ 2 and ± 4, respectively.

The surface can be visualized from the above information.

1.15. SURFACES OF REVOLUTION

Definition. A surface S is called a surface of revolution if there is a straight line L such that every section if S perpendicular to L is a circle with center on L such a surface can be thought of as being generated by a plane curve, called a generate (or generator) revolved about L as axis of revolution. Usually, one of the coordinate axis is taken of revolution.

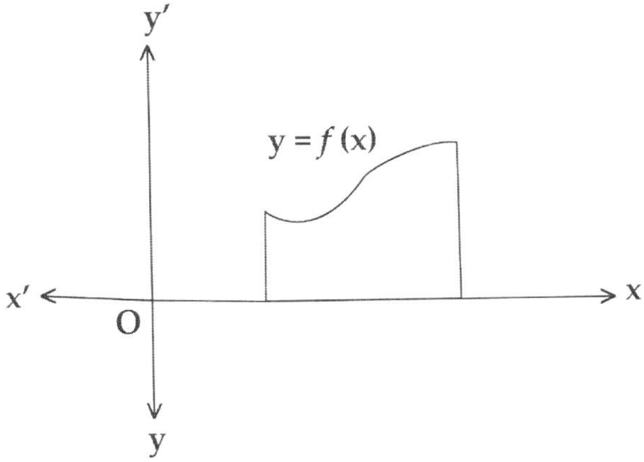

Suppose we want to find the surface of revolution obtained by revolving the curve $y = f(x)$ about the x − axis. The surface of revolution is the set of all points (x, y, z) such that square of distance of each point (x, y, z) *from* $x - axis$ is equal to the square of $f(x)$, where x is now the directed distance from the $yz - plane$. Since square of the distance of (x, y, z) from $x - axis$ is $y^2 + z^2 = [f(x)]^2$. The surface of revolution is the set of all points (x, y, z) such that square of distance of each point (x, y, z) *from* $x - axis$ is equal to the square of $f(x)$, where x is now the directed distance from the *yz*-plane yz − plane . Since square of the distance of (x, y, z) from $x - axis$ is $y^2 + z^2 = [f(x)]^2$

Thus, in order to obtain the equation of the surface of revaluation by revolving the curve $y = f(x)$ about the equation $y = f(x)$ and replace y^2 *by* $y^2 + z^2$. Convulsively, any equations in which y *and* z appear only in the form $y^2 + z^2$ represents surface of revaluation about the $x - axis$. A similar procedure is followed in finding an equation of a surface obtained by revolving a curve in any coordinate plane about a coordinate axis lying in it. The result are:

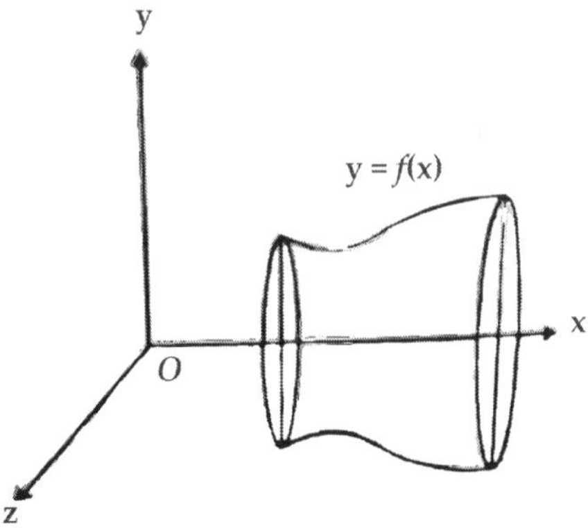

Curve in the *xy*-plane	Revolved about	Replace	By
	$x-axis$ $y-axis$	y^2 x^2	y^2+z^2 x^2+z^2
xz-plane	$x-axis$ $z-axis$	z^2 x^2	z^2+y^2 x^2+y^2
yz-plane	$y-axis$ $z-axis$	z^2 y^2	z^2+x^2 y^2+x^2

Example

Write an equation of the surface of revolution obtained by revolving the curve $x=z^2, y=0$

(a) about x-axis (b) about the $z-\text{axis}$.

Solution

(a) Replace $z^2\ by\ y^2+z^2$. Required equation is $x=y^2+z^2$

(b) Squaring both sides of the equation $x=z^2$ and replacing x^2 by x^2+y^2, we get the required equation as $x^2+y^2=z^4$

Example

Show that $S:x\left(y^2+z^2\right)=1$ is as surface of revolution. Find a generatix and axis of revolution.

Solution. The equation maybe written as

$$y^2 + z^2 = \frac{1}{x}, (since\, x \neq 0)$$

This is of the form $y^2 + z^2 = [f(x)]^2$ $where\, f(x) = \frac{1}{\sqrt{x}}, x > 0.$ Hence it is surface of revolution, and $x - axis$ is the axis of revolution. A generate is the curve $y = \frac{1}{\sqrt{x}}$ $in\, the\, xy - plane, or\, z = \frac{1}{\sqrt{x}}$ in the $xz - plane.$

1.15.1. The Cylinder

Let $(x, y) = 0$ be an equation of a curve C in the $xy - plane$ and consider a vector

$$n = [a, b, c]$$

Not parallel to the $xy -$ plane so that $c \neq 0.$ The parametric equations of line $L.$ through a point $(x_1, y_1, 0)$ on C parallel to n are

$$x = x_1 + at, y = y_1 + bt, z = ct$$

If $P(x, y, z)$ is any point on $L,$ then $t = \frac{z}{c}$ and so

$$x_1 = x - \frac{a}{c} z; y_1 = y - \frac{b}{c} z.$$

Since $f(x_1, y_1) = 0,$ the point $P(x, y, z)$ must satisfy the equation

$$f\left(x - \frac{a}{c} z, y - \frac{b}{c} z\right) = 0$$

Conversely, if $p(x, y, z)$ satisfies this equation, the point $(x_1, y_1, 0), where x_1 = x - \frac{a}{c} z, y_1 = y - \frac{b}{c} z,$ lies on C since $f(x_1 . y_1) = 0.$

Letting $t = \frac{z}{c}, we$ have

$x_1 = x - at, y_1 = y - bt, z = ct.$

Accordingly, $x = x_1 + at, y = y_1 + bt, z = ct$ and so that the point $p(x, y, z)$; lies on the line L through $(x_1, y_1, 0)$ and parallel to n. The union of all lines is called a cylinder.

Definition. Let C be a curve in a plane and let L be a line not in the plane. The union of all lines that intersects C and are parallel to L is called a cylinder.

Each line through C and parallel to L, is called an element (or rolling) of the cylinder and the curve C is called n curve C is called a directrix of the cylinder.

It is clear from, that an equation of the cylinder, with directrix C and element parallel to $\boldsymbol{n}$, is

$$F(x, y, z) = f\left(x - \frac{a}{c}z, y - \frac{b}{c}z\right) = 0.$$

The trace of the cylinder in the plane $z = d$ (a plane parallel to the $xy - plane$) is the curve

$$C' ; f\left(x - \frac{a}{c}d, y - \frac{b}{c}d\right) = 0$$

which is just a translate of C. Thus, all cross-section (traces) of the cylinder by planes parallel to the $xy - planes$ are $'carbon\, copies\, of\, C$.

Example

Find an equation of the cylinder with directrix $C : y - z^2 - 0$ and having elements parallel to $n = [2,3,4]$.

Solution

Since $\boldsymbol{n}$ is not parallel to the $yz - plane,$ the cylinder exist. We shall derive an equation of the cylinder by definition.

Equation of the line L through $(0, y_1, z_1)$ and parallel to n are

$$L:\begin{cases} x = 2t \\ y = y_1 + 3t \\ z = z_1 + 4t \end{cases}$$

Thus, $t = \dfrac{x}{2}, y_1 = y - \dfrac{3x}{2}, z_1 = z - \dfrac{4x}{2}$

Equation of the directrix is $f(y,z) = y - z^2 = 0$ and the point (y_1, z_1) lies on it. Therefore, $y_1 = z_1^2$.

Hence, an equation of the cylinder is

$$y - \frac{3x}{2} = (z - 2x)^2$$

The trace of the cylinder in the plane $x = d$ is

$$y - \frac{3}{2}d = (z - 2d)^2$$

which is translate of the directrix $y = z^2$

1.15.2. Right Cylinder

Definition. If the elements of a cylinder are normal to a plane, then it is called a right cylinder with respect to that plane.

Equation of a Right Cylinder: Let the directrix $C: f(x, y) = 0$ lie in the $xy - plane$. The elements of the right cylinder will be parallel to z $axis$ i.e., parallel to $\boldsymbol{n} = [0, 0, 1]$. Thus, equation of the cylinder is

$$f\left(x - \frac{0}{c}z, y - \frac{0}{c}z\right) = 0$$

i.e., $\pm(\ , \ ) = 0$

Hence aright cylinder with respect to a coordinate plane has precisely the same equation as the equation of the directrix in that plane.

The coordinate axis normal to the coordinate plane with respect row which a cylinder is right, is called the axis of the cylinder.

If in an equation two variable are missing, the surface is a cylinder whose elements are parallel to the axes of both missing variable and consist of a collection of planes.

Example

Find an equation of the right cylinder whose directrix is the circle in the $xy-plane$ with center at $(4,6,0)$ and radius 5.

Solution

Equation of the directrix is $(x-4)^2+(y-6)^2=5^2$

i.e., $x^2+y^2-8x-12y+27=0$. This is also an equation of the right cylinder.

Example

Discuss the surface: $9z^2-6x^2=54$.

Solution

The equation $\frac{z^2}{6}-\frac{x^2}{9}=1$

Represent a hyperbola in the $zx \quad plane. The\, surface$ is therefore, a right hyperbolic cylinder with elements parallel to the y-axis.

Example

Discuss the surface $x+y=2$

Solution

The equation $x+y=2$ represent a straight line in the $xy-plane$. It is a right cylinder with elements parallel to $z-axis.$ The cylinder is made up of all lines parallel to $x+y=2,$ which is in fact the plane $x+y=2.$

1.15.3. The Cone

Definition. A surface generated by a straight line passing through a fixed point and intersecting a fixed plane curve not containing the fixed point is called a cone. The given curve is called the directrix and the fixed point is known as the vertex of the cone. The straight line is called the generator and any particular position of the generator is called an element(or rolling) of the cone.

Example. Find an equation of the come whose directrix is

$y^2 = x, z = 4$ (51)

and whose vertex is at $A(0,2,0)$.

Solution. Let $P(x,y,z)$ be a point on the cone and let the element AP of the cone meet the directrix at the point $Q(x_1,x_2,z_1)$. Since $Q(x_1,x_2,z_1)$ lies on (51), we have

$y_1^2 = x_1, z_1 = 4$ (52)

Equation of the line AQ are $\frac{x}{x_1} = \frac{y-2}{y_1-2} = \frac{z}{z_1}$

Using $z_1 = 4$ *from* the second equation of (52), we have

$$x_1 = \frac{4x}{z}$$

and $$y_1 - 2 = \frac{4(y-2)}{z}$$

or $$y_1 = \frac{4y-8+2z}{z}$$

Substituting these in the of equation (52), we find

$$\left(\frac{4y+2z-8}{z}\right)^2 = \frac{4x}{z}$$

or $(2y+z-4)^2 = xz$

i.e., $4y^2 + z^2 - 4yz - zx - 6y - 8z + 16 = 0$ is required equation of the cone.

1.15.4. The Ellipsoid

A surface $(in R^3)$ *defined* by an equation of the form $\frac{x^2}{a^2} + \frac{y^2}{b^2} + \frac{z^2}{c^2} = 1,$ (53)

or by an equation reducible to form (53), is called an ellipsoid.

It is symmetric with respect to each of the coordinates planes, to each of the coordinates axes and to the origin. A sphere is a special case of the ellipsoid. When $a = b = c$, the ellipsoid $\frac{x^2}{a^2} + \frac{y^2}{b^2} + \frac{z^2}{c^2} = 1$ becomes the sphere $x^2 + y^2 + z^2 = a^2$.

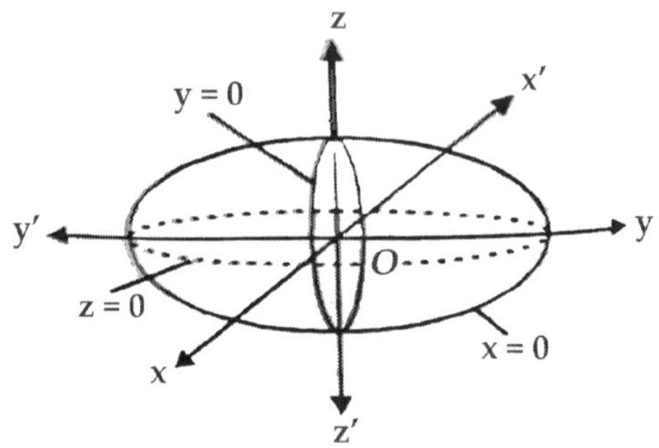

If two of the number a, b, c are equal and different from the third, the ellipsoid is called a spheroid, oblate if the third is shorter than the two which are equal (like the earth), prolate if the third is longer than the two that are equal like a football). The trace of the ellipsoid (53) in the $yz - plane$ has equation

$$\frac{y^2}{b^2} + \frac{z^2}{c^2} = 1 \qquad (54)$$

which is an ellipse. Similarly, the traces in the xz and $xy - plane$ are ellipse with equation, respectively

$$\frac{x^2}{a^2} + \frac{z^2}{c^2} = 1 \quad \text{and} \quad \frac{x^2}{a^2} + \frac{y^2}{b^2} = 1 \qquad (55)$$

It is easy to verify that the trace of (53) in a plane parallel to a coordinate plane to the $yz -$ plane has equation

$$\frac{y^2}{b^2} + \frac{z^2}{c^2} = 1 - \frac{k^2}{a^2}$$

which is an ellipse if $k^2 < a^2$.

1.15.5. The Elliptic Paraboloid

A surface defined by an equation of the form

$$\frac{x^2}{a^2}+\frac{y^2}{b^2}=cz \qquad (56)$$

is called an elliptic Paraboloid.

The name elliptic Paraboloid is due to the fact that the traces of (56) in planes parallel to two coordinate planes are parabolas, where the traces in planes parallel to the third coordinate plane are ellipse. Thus, in the plane $z=k$ the trace has equation

$$\frac{x^2}{a^2}+\frac{y^2}{b^2}=ck$$

which is an ellipse provided $ck>0$. In the planes $x=k\, and\, y=k,$ the traces are parabolas with equations.

$$\frac{y^2}{b^2}=cz-\frac{k^2}{a^2}\, and\, \frac{x^2}{a^2}=cz-\frac{k^2}{b^2}$$, respectively.

The elliptic Paraboloid (55) symmetric with respect to the yz and $xz-planes,$ and hence also with respect to the $z-axis.$ The elliptic Paraboloid is the above or below the xy - plane according as $C>0\, or\, c<0.$ The case in which $c>0$ is indicated in the figure below.

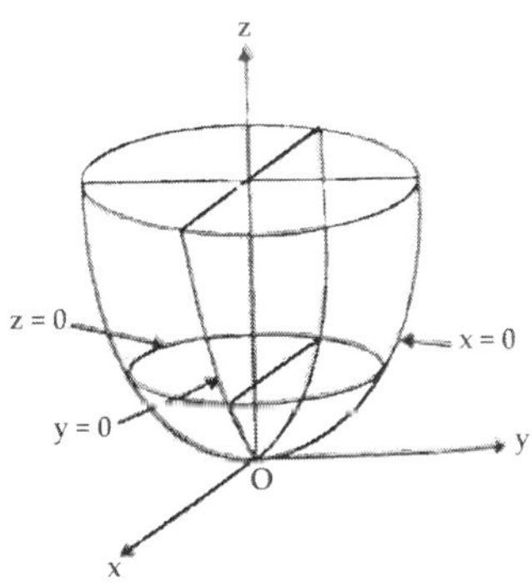

If $a=b$ in (55), the elliptic Paraboloid is called a **Paraboloid of revolution.** The trace of (55) in a plane parallel to the xy - plan is a circle, and the surface may be thought of as being generated by a rotation of the parabola

$$\frac{x^2}{a^2}=cz$$

in the xz-plane about the z-axis

1.15.6. The Hyperbolic Paraboloid

A surface defined by an equation of the form

$$z = \frac{y^2}{b^2} - \frac{x^2}{a^2} \qquad (57)$$

is called a Hyperbolic Paraboloid.

Setting $z = 0$ in (57) we note that the trace in the $xy-plane$ is the pair of intersecting lines $y = \pm\frac{b}{a}x.$ The traces in the $xz-$plane and the $yz-$plane are the parabolas $z = \frac{x^2}{a^2}$ and $z = \frac{y^2}{b^2}$, respectively. However, the traces in planes parallel are the parabolas $z = -\frac{x^2}{a^2}$ and $z = -\frac{y^2}{b^2}$ respectively. However, the traces in planes parallel to $xy-plane(z = k)$ are hyperbolas $\frac{y^2}{b^2} - \frac{x^2}{a^2} = k.$ The hyperbolic Paraboloid is symmetric with respect to $xz-$plane, $yz-plane$ and the $z-axis.$ The surface is **saddle-shaped** as shown in figure.

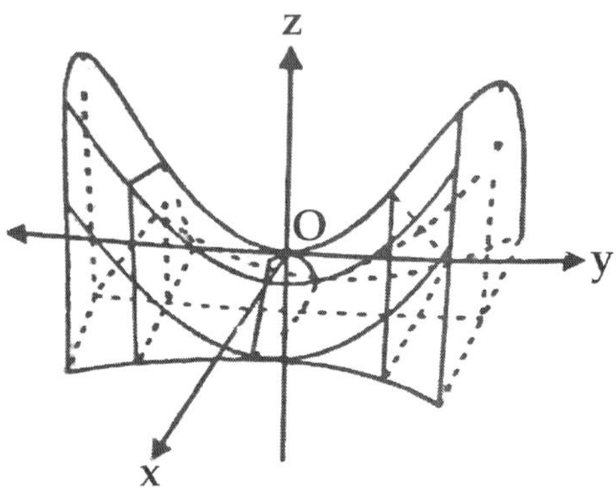

1.15.7. The Hyperboloid of One Sheet

A surface defined by an equation of the form $\frac{x^2}{a^2} + \frac{y^2}{b^2} - \frac{z^2}{c^2} = 1$, is called a hyperboloid of one sheet. The trace in the $xy-plane$ is the ellipse

$\frac{x^2}{a^2}+\frac{y^2}{b^2}=1$ while traces in the $yz\,and\,zx$ –planes are, respectively the hyperbolas

$$\frac{y^2}{b^2}-\frac{z^2}{c^2}=1\,and\,\frac{x^2}{a^2}-\frac{z^2}{c^2}=1$$

Similarly, setting $z=d\,in(1),$ the trace in planes parallel to the $xy-$ plane are the ellipses $\frac{\pm^{ü}}{\pm^{ü}}$ ℵ— — ü

The traces in planes parallel to the other two coordinates planes hyperbolas. The surface is symmetric with respect to each of the coordinate planes and axes. It has no *z*-intercepts.

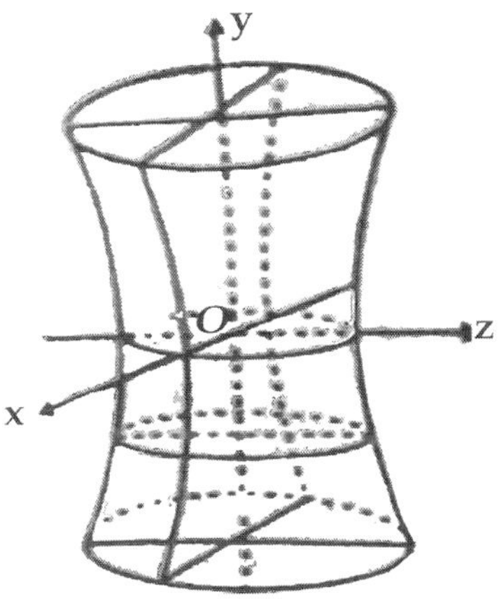

The x and $y-$ intercepts are $\pm a\,and\,\pm b$, respectively. If $a=b,the$ trace of the surface in the $xy-$ plane is the circles

$$x^2+y^2=a^2$$

The trace in planes parallel to xy -plane $(z=k)$ are circles

$$x^2=y^2=a^2+\frac{k^2}{c^2}.$$

In this case the surface may be thought of as obtained by rotating the hyperbola

$$\frac{x^2}{a^2}-\frac{z^2}{c^2}=1$$

In the xz-plane about the z -axis and is called a hyperboloid of revolution.

1.15.8. The Hyperboloid of Two Sheets

A surface represented by an equation of the form

$$\frac{x^2}{a^2}+\frac{y^2}{b^2}-\frac{z^2}{c^2}=-1 \qquad (58)$$

is called hyperboloid of two sheets.

Putting $z = 0$ in (58) we have

$$\frac{x^2}{a^2}+\frac{y^2}{b^2}=-1$$

which shows that the surface has no trace in the xy-plane. The traces in yz-plane and zx-plane are, respectively the hyperbolas

$$\frac{y^2}{b^2}-\frac{z^2}{c^2}=-1\,and\,\frac{x^2}{a^2}-\frac{z^2}{c^2}=-1$$

The trace in planes parallel to the $yz-plane(x=d)$ and in planes parallel to the $zx-plane(y=k)$ are also hyperbolas. The traces in planes parallel to $xy-planes(z=m)$ are ellipse

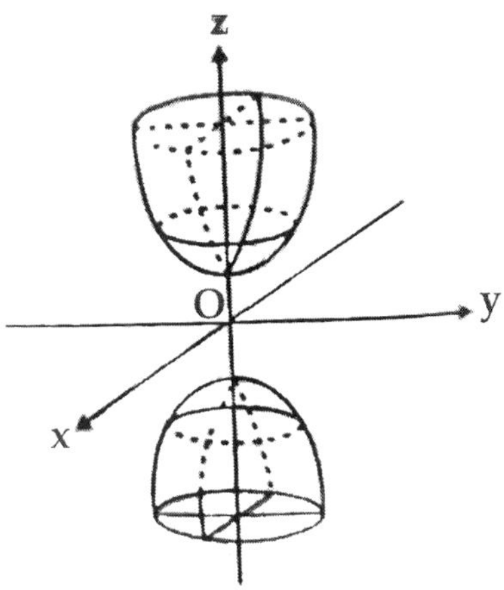

$$\frac{x^2}{a^2}+\frac{y^2}{b^2}=\frac{m^2}{c^2}-1$$

Provided $|m|>|c|$. The surface is symmetric with respect to each of the coordinates planes and axes. It has no intercepts on $x-axis\ and\ y-axis$. The $z-$ intercepts are $\pm c$.

1.15.9. The Elliptic Cone

A surface defined by an equation of the form

$$\frac{x^2}{a^2}+\frac{y^2}{b^2}=cz^2,(c>0) \qquad (59)$$

is called an elliptic cone.

We can show that (59) defines a cone with vertex V at the origin. $Let\ P(x_1, y_1, z_1)$ be any point on this surface other than the origin. Direction ratios of the straight line VP are x_1, y_1, z_1. Thus equation of the straight line $VP\ are\ x = x_1t, y = y_1t, z = z_1t$. Since P is on the surface, we have

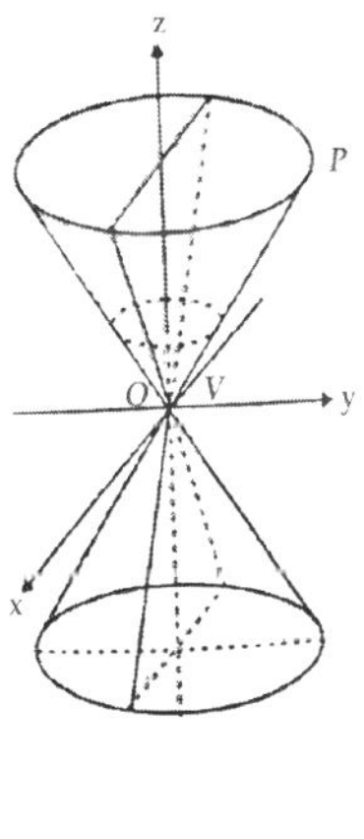

$$\frac{x^2}{a^2}+\frac{y^2}{b^2}=cz_1^2 \qquad (60)$$

or $$\frac{(x_1t)^2}{a^2}+\frac{(y_1t)^2}{b^2}=c(z_1t)^2$$

Thus every point of the straight line VP lies on the surface defined by (59) i.e., VP lies on the surface. Thus, (59) defines a cone.

The trace of the cone defined by (59) in a plane parallel to the $xy-plane$ is an ellipse, and hence the name elliptic cone. If $a = bin(1)$, the trace in planes parallel to the $xy-$ plane are circles. The surface is then the familiar right circular cone.

1.15.10. General Quadric Surfaces

The general equation of the second degree $ax^2 + by^2 + cz^2 + fyz + gzy + hxy + ux + vy + wz + k = 0, where\, a.b.c.f.g.h$ are not all zero, is called a general quadric surface. A general quadric equation can be transformed into one of the form given in to provided the surface or semi-degenerate surface (such as point spheres and imaginary surfaces).

Example. Identity the surface defined by $z^2 - 4y^2 - 16x - 16y - 2z + 49 = 0$

Solution. The given equation is equivalent to

$$(z-1)^2 - 4(y+2)^2 = 16x - 64$$

or

$$\frac{(z-1)^2}{4} - \frac{(y+2)^2}{1} = 4(x-4)$$

Let $x' = x - 4, y' = y+2, z' = z - 1,$ i.e., we translate the axes to new origin $(4,-2,1),$ then the equation of the surfaces becomes

$$\frac{z'^2}{4} - \frac{y'^2}{1} = 4x'$$

which represented a hyperbolic Paraboloid.

1.15.11. Ruled Surfaces

Definition. A surface generated by a line moving in accordance with given conditions is called a ruled surface. The moving lines is called the generator and a particular position of the generator is called an element or a ruling of the surface.

Planes, cylinder and cones are example of ruled surfaces. The rulings of a plane are all lines parallel to a given line. The ruling of a cylinder meet a curve (the directrix) and are parallel to a given line (the axis). The rulings of a cone meet a curve (the directrix) and pass through a fixed point (the vertex).

Theorem. The hyperboloids of one sheet and hyperbolic Paraboloid are ruled surface.

Proof. The equation

$$\frac{x^2}{a^2}+\frac{y^2}{b^2}-\frac{z^2}{c^2}=1 \tag{61}$$

represent a hyperboloid of one sheet. If we write (61) in the form

$$\frac{x^2}{a^2}-\frac{z^2}{c^2}=1-\frac{y^2}{b^2}$$

We see that each member of this equation may be resolved into linear factors with real coefficients:

$$\left(\frac{x}{a}+\frac{z}{c}\right)\left(\frac{x}{a}-\frac{z}{c}\right)=\left(1-\frac{y}{b}\right)\left(1+\frac{y}{b}\right) \tag{62}$$

It is easy to see that (62) may be obtained by eliminating the parameter t from the family of lines

$$\left.\begin{aligned} \frac{x}{a}+\frac{z}{c}&=t\left(1+\frac{y}{b}\right) \\ t\left(\frac{x}{a}-\frac{z}{c}\right)&=1-\frac{y}{b} \end{aligned}\right\} \tag{63}$$

The surface (62) can also be obtained by eliminating the parameter p from the family of lines

$$\left.\begin{aligned} &\frac{\pm}{\pm}\aleph- \quad p\left(1 \quad -\right) \\ &t\left(\frac{\pm}{\pm}\aleph-\right) \quad 1 \quad - \end{aligned}\right\} \tag{64}$$

Hence the hyperboloids of one sheet is a ruled surface which may be generated by either of the two family of lines (63) or (64). The families of line (63) and (64) are two distinct systems and a member of one system cannot becomes identical with any member of one system cannot become identical with any member of the other system whatever real values be assigned to the families (63) and (64), each line being a ruling of the surface.

The equation

$$\frac{y^2}{b^2}-\frac{x^2}{a^2}=z \qquad (65)$$

Represent a hyperbolic Paraboloid. When the left members of (65) is factorized, we get

$$\left(\frac{y}{b}+\frac{x}{a}\right)\left(\frac{y}{b}-\frac{x}{a}\right)=z$$

Hence the hyperbolic Paraboloid may be generated by either one of two families of lines

$$\frac{y}{b}+\frac{x}{a}=t,t\left(\frac{y}{b}-\frac{x}{a}\right)=z$$

or

$$\frac{y}{b}-\frac{x}{a}=p,p\left(\frac{y}{b}+\frac{x}{a}\right)=z$$

where $t\, and\, p\, are$ parameters. Thus, the surface (65) is a ruled surface.

1.16. SPHERICAL TRIGONOMETRY

Let S be a sphere and P a plane that passes through the center of $S.$ Then $S \cap P$ i.e., intersection of a sphere and a plane through its center is called a great circle on $S.if\ q$ is a plane not passing through the center of S, then $S \cap q$ is called A spherical angle between any two intersecting great circles. Two plane $\alpha \pm \ \beta$ with a common line of intersection AB are said to form a **dihedral angle,** AB is called the edge of dihedral angle, the planes forming the dihedral angle are called its faces

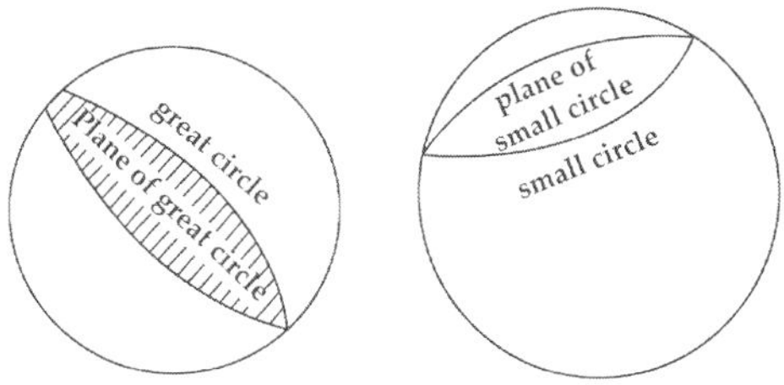

The plane angle of a dihedral angle is the angle formed by two rays one on each face, perpendicular to the edge at the same point. Measure of a dihedral angle is equal to measure of its plane angle.

A trihedral angle is formed by three planes which intersect so as t=to have only one point is common. The point of intersection of these planes forming a trihedral angle is known as the vertex of the trihedral angle.

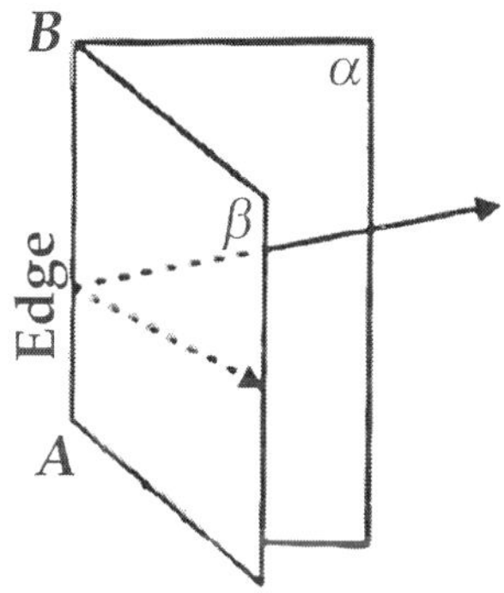

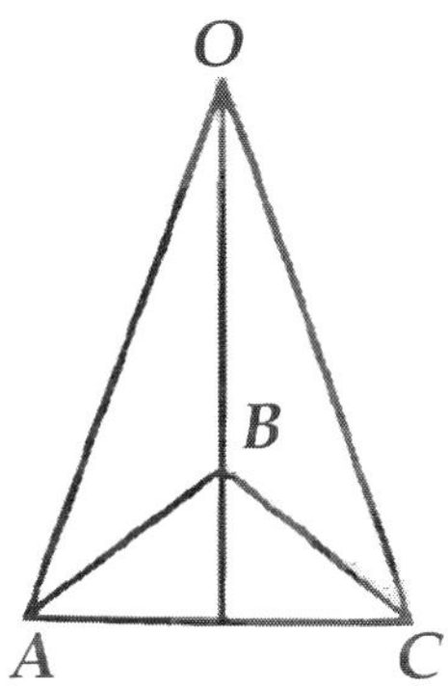

A spherical triangle is that part of the surface of a sphere which is bounded by the arcs of three great circles. Note that a spherical triangle is the intersection of the surface of the sphere with the trihedral angle whose vertex is O, the center of the sphere and whose faces are the planes of the great circles.

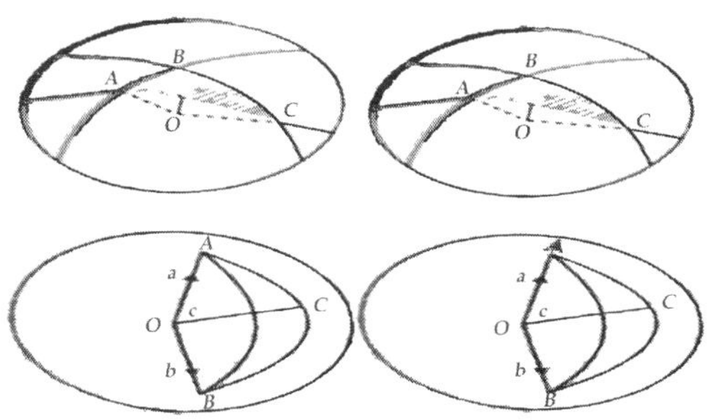

Measure of a side of spherical triangle is equal to measure of the angle subtended by that side at the circle of the sphere. By definition, a spherical angle is the angle between any two intersecting great circles. It can be measured by measuring angle between the tangents at the point of intersection of the great circles, or by measuring the dihedral angle of the trihedral angle whose vertex is at the center of the sphere.

1.16.1. The Cosine Formula

Suppose we have a spherical triangle ABC on a sphere of unit radius. Let measure of the sides $BC, CA\ and\ AB$ be $a, b\ and\ c$, respectively. Let measure of the dihedral angle defined by the planes $OAB\ and\ OAC$ be denoted by α. similarly, we define $\beta\ and\ \gamma. Let\ \overrightarrow{OA} = \mathrm{I}, \overrightarrow{OB} = \mathrm{m\,and}\,\overrightarrow{OC} = \mathrm{n}.$ As the sphere is of unit radius. I, m, n are unit vectors.

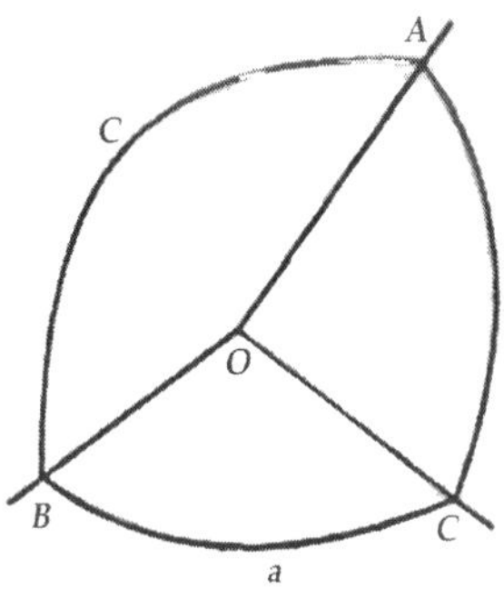

Now by a result of a vector analysis

$$(I\,x\,m).(I\,x\,n) = \begin{vmatrix} I.I & I.n \\ m.I & m.n \end{vmatrix} = \begin{vmatrix} 1 & \cos b \\ cos c & cos a \end{vmatrix}$$

$= \cos a - \cos b \cos c$ (66)

Also $\mathrm{I} \times \mathrm{m}$ is a vector perpendicular to the plane AOB (drawn inward) and its magnitude is $\sin c. Similalry$ Ixn is a vector perpendicular to the plane $\pm$ (drawn outward) and its magnitude s $\sin b$.

Therefore,

$(\mathrm{I}\,x\,\mathrm{m}).(\mathrm{I}\,x\,\mathrm{n}) = \sin b\ \sin c\ \cos A$ (67)

From (66) and (67), we obtain

$$\cos a - cosb \cos c = sin b \sin c \cos A$$

Thus

$$cosa = \cos b \cos c + \sin c \sin a \cos B \qquad (68)$$

In a similar manner we can prove that

$$cosb = \cos c \cos a + \sin c \sin a \cos B \qquad (69)$$

$$cosc = \cos a \cos b + \sin a \sin b \cos C \qquad (70)$$

Result (68), (69) and (70) are known as the cosine formulas.

1.16.2. The sine Formula

Consider once again

Now,

$$(\text{I x m})(\text{I x n}) = [\text{I m n}]\text{I} - [\text{i m I}]\text{n}$$

$$= [\text{I m n}]1$$

Left hand side is a vector quantity perpendicular to the normal to the planes $AOB \, and \, AOC$ and is, therefore in the direction of I, while the magnitude of the vectors is sin c and b sin A.

Thus,

$$\sin c \sin b \sin A \boldsymbol{I} \quad [\mathbf{l} \boldsymbol{m} \boldsymbol{n}]\mathbf{I}$$

$$\text{or } \sin b \sin c \sin A = [\mathbf{l} \boldsymbol{m} \boldsymbol{n}]$$

Similarly,

$$\sin c \sin a \sin B = [\mathbf{l} \boldsymbol{m} \boldsymbol{n}]$$

From (66) and (67), $\sin b \sin c \, sinA = \sin c \sin a \sin B$

Therefore,

$$\frac{sinA}{sina} = \frac{sinB}{sinb}.$$

Similarly,

$$\frac{\sin B}{\sin b} = \frac{\sin C}{\sin c}$$

Hence

$$\frac{sinA}{\sin a}=\frac{sinB}{sinb}=\frac{sinC}{sinc}$$

which is known as the sine formula.

1.16.3. The Four-parts (or the Cotangent) Formula

From (70) and (68) of, we have

$$\cos a = cosb\ \cos c + sin b \sin c\ \cos C$$

$$\cos a = cosb\ \cos c + sin b \sin c\ \cos A$$

Eliminating $\cos c$ from the above equations, we get

$$\cos a = \cos b\ (\cos a \cos b + \sin a + \sin b \cos C) + \sin b \sin c\, cosA$$

Thus,

$$\cos a\ (1-\cos^2 b) = \sin a \sin b\ \cos b \cos C + \sin b \sin c \cos A$$

Dividing both sides by $sina\ \sin b,$ we get

$$\cot a \sin b = \cos b\ \cos C + \frac{sinc}{\sin a}\cos A$$

or $$cot\ a\, sin b = cos\, b\, cos\ C + \frac{SinC}{sinA} cosA$$ (using since formula)

or $\cos b\ \cos C = \sin b\ \cot a - \sin C\ \cot A$

This is known as the four-parts (or the cotangent) formula.

1.17. LATITUDE AND LONGITUDE

For many practical problems the earth is regarded as a sphere spinning about a diameter $\overrightarrow{PQ}\, p$ is called north pole $and\, Q$ the south pole. The great circle whose plane is perpendicular to $\overrightarrow{PQ}$ is called the equator. Any semi-great circle with diameter $\overrightarrow{PQ}$ is called a meridian in particular the meridian which through the fundamental instrument (the transect circle) of Greenwich observatory is, by universal agreement, regarded as the principal or standard meridian. Let it be $PGKQ$ in intersecting the equator in $K.$ Let

$PHLQ$ be any other meridian cutting the equator in L.

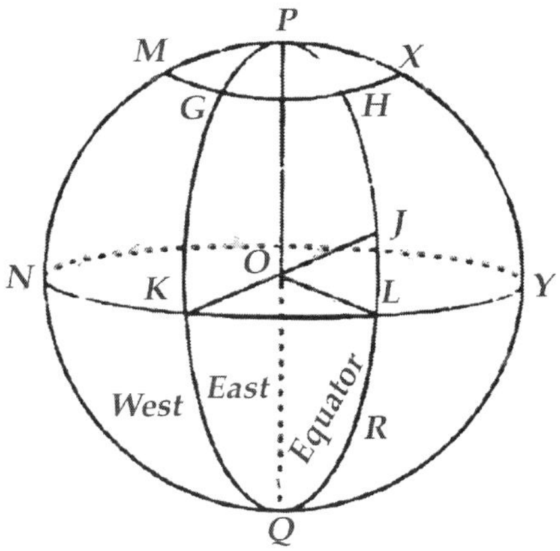

Measure of the angle KOL is defined to be longitude of the meridian PHQ and it can be described equally well as the length of the equatorial are KL or the magnitude of the spherical angle $KPL.$ Longitude are measured from $0^o\ to 180^o$ west(W), following the direction of the arrows near K in. Thus, from the longitude of the meridian PHQ is about 100° E and that of the meridian PMQ is about 60° W. All planes on the same meridian have the same longitude and the meridian on which a particular place is situated is specified with reference to the principal meridian $PHQ.$ The meridian through j cuts the equator in L and measure of the angle $LOJ,$ or the length if the great circle arc $lj,$ is called the latitude (lat) of $j.$ if j is between the equator and the north pole $p,$ the latitude is said to be north (N): a place such as $R,$ between the equator and the south pole Q is said to be south (S) latitude In this way the position of any point on the surface of the earth is referred to the fundamental great circles, the equator and the meridian of Greenwich. Longitude are measured from $0^o\ to 180^o$ west(W), following the direction of the arrows near K in. Thus, from the longitude of the meridian PHQ is about 100° E and that of the meridian PMQ is about 60° W. All planes on the same meridian have the same longitude and the meridian on which a particular place is situated is specified with reference to the principal meridian $PHQ.$ The meridian through j cuts the equator in L and measure of the angle $LOJ,$ or the length if the great circle arc $lj,$ is called the latitude (lat) of $j.$ if j is between the equator and the north pole $p,$ the latitude is said to be north (N): a place such as $R,$ between the equator and the south pole Q is. Thus, from the longitude of the meridian PHQ is about 100° E and that of the meridian PMQ is about 60° W. All planes on the same meridian have the same longitude and the meridian on which a particular place is situated is specified with reference to the principal meridian $PHQ.$ The meridian through j cuts the equator in L and measure of the angle $LOJ,$ or the length if the great circle arc $lj,$ is called the latitude (lat) of $j.$ if j is between the equator and the north pole $p,$ the latitude is said to be north (N): a place such as $R,$ between the equator and the south pole Q is said to be south (S) latitude In this way the position of any point on the surface of the earth is referred to the fundamental great circles, the equator and the meridian of Greenwich.

Let ϕ denote the latitude of J, then measure of angle LOJ or $LJ = \phi.$ Since OP is perpendicular to the plane of the equator, $m\angle POL = 90^0$ and therefore, $m\angle POJ = 90^0 - \phi.$ Measure of the angle POJ or the spherical arc PJ is the co-latitude (colat) of J. We have thus

$$colat = 90^0 - lat.$$

All places which have the same latitude lie on small circle parallel to the equator, called a parallel of latitude. Thus, all places with same latitude as Greenwich lie on the small $MGHX$. If θ denotes the latitude of Greenwich, then the length of the small circle arc $HX,$ for example, is given in terms of the length of the corresponding equatorial arc LY by

$$HX = of\ the LY \cos\theta \qquad (71)$$

On the great circle, distance between two points subtending an angle of one minute of arc at the center of the earth is taken as unit distance on the surface of the earth and is known as the **nautical mile.** The formula (71) can be used to calculate the distance between $H\ and\ X$ expressed in natural miles (or minutes of arc) and measured along the parallel of latitude.

1.17.1. Summary and Discussions

Analytic Geometry play a vital role in different branches of mathematics, especially in Calculus, Geometry, Algebra and in Applied mathematics as well. In this chapter we have discussed in detail about three dimensions, its rules, formulas and applications as well.

1.17.2. Distance between Two Points

Let $P(x_1, y_1, z_1)$ and $Q(x_2, y_2, z_2)$ be two distinct points in space such that the line PQ is not parallel to any of the coordinate axes. Then the distance between $P and Q$ is taken to be positive.

Let $P(x_1, y_1, z_1) and Q(x_2, y_2, z_2)$ be two distinct points in space such that PQ is not parallel to one of the coordinate axes. Then

$$|PQ| = \sqrt{(x_2 - x_1)^2 + (y_2 - y_1)^2 + (z_2 - z_1)^2}$$

1.17.3. Point Dividing a Line Segment in a Given Ratio

To find the coordinate of the point dividing the join of $P(x_1, y_1, z_1)$ and $Q(x_2, y_2, z_2)$ in the ratio $m:n.$ Suppose $R(x, y, z)$ divides PQ in the given ration $m:n.$ Let $O(0,0,0)$ be the origin. We know that if position vector of $P and Q$ are $a and b$, respectively, then the position vector r of the point R is

$$\boldsymbol{r} = \frac{n\boldsymbol{a} + m\boldsymbol{b}}{n + m}$$

1.17.4. Equations of a Straight Line

A straight line is completely determined if it passes through two fixed points or it passes through a fixed point and is parallel to a given vector.

Let $A(x_1, y_1, z_1)$ be a fixed point on the line L whose direction is parallel to the vector

$$\boldsymbol{d} = a\boldsymbol{i} + b\boldsymbol{j} + c\boldsymbol{k}$$

Suppose $P(x, y, z)$ is any point on L. Then $\overline{\pm} = \boldsymbol{d},$ where t is some real number. It is a vector equation of the line through A and parallel to d.

Now,

$$\overrightarrow{AP} = (x - x_1)i + (y - y_1)j + (z - z_1)k$$

Thus, vector equation of the line L is

$$(x - x_1)\boldsymbol{i} + (y - y_1)\boldsymbol{j} + (z - z_1)\boldsymbol{k} = \boldsymbol{t}(\ \boldsymbol{ai} + \boldsymbol{bj} + \boldsymbol{ck}).$$

Equating components, we get

$$\left.\begin{aligned} x - x_1 &= at \\ y - y_1 &= bt \\ z - z_1 &= ct \end{aligned}\right\}$$

These are called parametric equations of the required line L.

From above, we have

$$\frac{x - x_1}{a} = \frac{y - y_1}{b} = \frac{z - z_1}{c} = t$$

which is called a symmetric form of equations of the line through (x_1, y_1, z_1) and parallel to the vector $[a, b, c].$ The vector $\boldsymbol{d} = [\boldsymbol{a}, \boldsymbol{b}, \boldsymbol{c}]$ is called a direction vector for L and a, b, c are called direction numbers or direction ratios of the line L.

If a line passes through two given points $A(x_1, y_1, z_1)\, and\, B(x_2, y_2, z_2)$ then since its direction vector is $d = (x_2 - x_1)i + (y_2 - y_1)j + (z_2 - z_1)k,$ a vector equation of the line is

$$(x - x_1)i + (y - y_1)j + (z - z_1)k = td, through\, A(x_1, y_1, z_1)$$

$$(x - x_2)i + (y - y_2)j + (z - z_2)k = td, through\, B(x_2, y_2, z_2).$$

A symmetric form of equations of this line is

$$\frac{x - x_1}{x_2 - x_1} = \frac{y - y_1}{y_2 - y_1} = \frac{z - z_1}{z_2 - z_1} = t$$

or
$$\frac{x-x_2}{x_2-x_1}=\frac{y-y_2}{y_2-y_1}=\frac{z-z_2}{z_2-z_1}=t$$

Angle between two given lines is equal to the angle between their direction vectors. Thus, two lines are perpendicular if their direction vectors are perpendicular.

1.17.5. Direction Angles of a Vector

Definition. Let $v=[x,y,z]$ be a given vector. The direction angles α,β,γ of v are defined as

$$\alpha = \mathit{Angle\,between\,v\,and\,the\,positive\,x-axis}.\beta = \mathit{Angle\,between\,v\,and\,the\,positive}\ y-\mathit{axis}.\gamma = \mathit{Angle\,between\,v\,and\,the\,positive\,z-axis}.$$

By definition, each of these angles lies between 0 and π.

$$\cos\alpha=\frac{x}{OP}=\frac{x}{\sqrt{x^2+y^2+z^2}}$$

$$\cos\beta=\frac{y}{OP}=\frac{y}{\sqrt{x^2+y^2+z^2}}$$

$$\cos\gamma=\frac{z}{OP}=\frac{z}{\sqrt{x^2+y^2+z^2}}$$

$\cos\alpha,\cos\beta,\cos\gamma$ are called direction cosines of v.

1.17.6. Angle between Two Straight Lines

Let l_1,m_1,n_1 be thc direction cosines of a straight-line L and l_2,m_2,n_2 be the direction cosines of another straight line M. Let $\theta\in[0,\pi]$ be the measure of the angle between L and M. Then θ is also measure of the angle between the direction vectors (or unit vectors) along the two straight lines.

Now, $d_1=l_1i+m_1j+n_1k$ and $d_2=l_2i+m_2j+n_2k$ are unit vectors along L and M, respectively and

$$\cos\theta=d_1.d_2=l_1l_2+m_1m_2+n_1n_2.$$

If $a_1, a_2, a_3 \; and \; b_1, b_2, b_3$ are direction ratios of the two lines and θ is the angle between them, then

$$\cos\theta = \frac{a_1b_1 + a_2b_2 + a_3b_3}{\left(a_1^2 + a_2^2 + a_3^2\right)^{\frac{1}{2}} \left(b_1^2 + b_2^2 + b_3^2\right)^{\frac{1}{2}}}$$

The two lines are perpendicular to each other if, and only if $\theta = \frac{\pi}{2}$ i.e.,

If and only if, $a_1b_1 + a_2b_2 + a_3b_3 = 0$

1.17.7. Distance of a Point from a Line

To find the distance from the point $P(x_1, y_1, z_1)$ to the line L with equation $\boldsymbol{r} = \boldsymbol{a} + t\boldsymbol{b}$. Let $\left|\overrightarrow{PM}\right| = p$ be the (perpendicular) distance from P to the given line L.

Then $p = \left|\overrightarrow{AP}\right| \left|\sin\theta\right|,$ where A is any point on L and θ is measure of the angle formed between $\overrightarrow{AP}$ and the direction vector d of the line L.

Now,

$$p = \left|\overrightarrow{AP}\right| \left|\sin\theta\right| = \frac{\left|\overrightarrow{AP}\right| \left|d\right| \left|\sin\theta\right|}{\left|d\right|} = \frac{\left|\overrightarrow{OP} \times d\right|}{\left|d\right|}$$

Since $\left|d\right| \neq 0$ for all lines L, we have the formula

$$p = \frac{\left|\overrightarrow{OP} \times d\right|}{\left|d\right|}.$$

1.17.8. Equations for Planes

A plane in space is uniquely determined by specifying a point in the plane and a non-zero vector perpendicular to the plane. A vector perpendicular to a plane is called a normal (vector) to the plane.

The plane through a given point $A(x_1, y_1, z_1)$ and with non-zero normal vector $\boldsymbol{n}=[a,b,c]$ has equation $a(x-x_1)+b(y-y_1)+c(z-z_1)=0$.

1.17.9. Plane through Three Given Non-Collinear Points

Let $P(x_1,y_1,z_1),Q(x_2,y_2,z_2)$ and $R(x_3,y_3,z_3)$ be three given non-collinear points lying on the required plane.

$$\begin{vmatrix} x & y & z & 1 \\ x_1 & y_1 & z_1 & 1 \\ x_2 & y_2 & z_2 & 1 \\ x_3 & y_3 & z_3 & 1 \end{vmatrix}=0.$$

1.17.9.1. Intercept Form of Equation of a Plane

Let an equation of a plane be $ax+by+cz=d$. Suppose it makes intercepts $A, B\ and\ C$ on the $x-axis, y-axis\ and\ z-axis$, respectively. Then the plane meets the axes in the point $(A,0,0),(0,B,0),(0,0,C)$. Since these points lie on the plane, we have

$$aA=d, bB=d, cC=d$$

Thus,

$$a=\frac{d}{A}, b=\frac{d}{B}, c=\frac{d}{C}$$

Therefore, an equation of the plane is

$$\frac{d}{A}x+\frac{d}{B}y+\frac{d}{C}z=d$$

This can be written in the form

$$\frac{x}{A}+\frac{y}{B}+\frac{z}{C}=1$$

which is known as the intercept form of equation of a plane.

Normal Form of Equation of a Plane

Let l, m, n be the direction cosines of a normal from the origin to a plane and let p be the length of this normal. Thus, an equation of the plane is

$$lx + my + nz = p.$$

$$l = -\frac{a}{\sqrt{\Sigma a^2}}, m = -\frac{b}{\sqrt{\Sigma a^2}}, n = -\frac{c}{\sqrt{\Sigma a^2}}, p = \frac{d}{\sqrt{\Sigma a^2}}$$

In case d is negative, the sign of $l, m\, and\, n$ will be positive.

Thus the normal form of the plane $ax + by + cz + d = 0$ is

$$-\frac{a}{\sqrt{\Sigma a^2}}x + \frac{b}{\sqrt{\Sigma a^2}}y + \frac{c}{\sqrt{\Sigma a^2}}z = \frac{d}{\sqrt{\Sigma a^2}},$$ if $d < 0$.

1.17.10. Angle Between Two Planes

The angle between two planes is defined as the angle between their normal vectors.

If equation of the two plane are

$$a_1x_1 + b_1y + c_1z + d_1 = 0$$

and

$$a_2x + b_2y + c_2z + d_2 = 0$$

Then the angle between then is the angle between their normal vectors.

Then the angle between them is the angle between their normal vectors

$$n_1 = [a_1b_1, c_1]\, and\, n_2 = [a_2, b_2, c_2]$$

Since

$n_1.n_2 = |n_1||n_2|\cos\theta, where\, \theta\, is\, the\, angle$ between $n_1\, and\, n_2$, therefore,

$$\cos\theta = \frac{n_1.n_2}{|n_1||n_2|}$$

$$= \frac{a_1a_2 + b_1b_2 + c_1c_2}{\sqrt{a_1^2 + b_1^2 + c_1^2}.\sqrt{a_2^2 + b_2^2 + c_2^2}}$$

The two planes intersect at right angles if $\theta = \frac{\pi}{2}.$ Hence the condition for the plane to be perpendicular to each other is

$$a_1a_2 + b_2b_2 + c_1c_2 = 0$$

The two planes are parallel if their normal vectors are parallel. Hence for the planes (1) and (2) to be parallel, we have

$$\frac{a_1}{a_2} = \frac{b_1}{b_2} = \frac{c_1}{c_2}.$$

Note. if $a_1x + b_1y + c_1z + d)1 = 0\, and\, a_2x + b_2\, y + c_2z + d_2 = 0$ are two non-parallel planes, then $(a_1x + b_1y + c_1z + d_1) + k(a_2x + b_2y + c_2z + d_2) = 0$ is a plane through the line of intersection of these planes. For, $a_1\, x + b_1y + c_1z + d_1 + k(a_2x + b_2y + c_2z + d_2) = 0$ is a plane and contain all points common to the two given planes.

1.17.11. A Straight Line and a Plane

Suppose a line $L is$ given by $\frac{x - x_1}{c_1} = \frac{y - y_1}{c_2} = \frac{z - z_1}{c_3} = t$. And a plane S is defined by $ax \aleph by \quad cz \quad d$

I. The straight lined the plane S intersected in one point if, and only if,

$$ac_1 + bc_2 + cc_3 \neq 0$$

II. The straight line L is parallel to the plane S if, and only if,

$$ac_1 + bc_2 + cc_3 - 0\, and\, ax_1 + by_1 + cz_1 = d:$$

III. The straight line L is parallel to the plane S if, and only if,

$$ac_1 + bc_2 + cc_3 = 0\, and\, ax_1 + by_1 + cz_1 \neq d.$$

1.17.12. Shortest Distance between Two Straight Lines

Let L and M be two given straight lines. If the straight lines are coplanar, then they either intersect, the shortest distance between them is zero. If they are parallel, then the shortest distance between the, can be obtained by finding the perpendicular distance of any point on one straight line to the other straight line. Skew Lines. Two non-coplanar straight line neither intersect nor they are parallel. Such straight lines are called skew lines. Any tow skew lines lie in parallel planes.

1.17.13. Cylindrical Coordinates

Let $P(x,y,z)$ be a point in rectangular coordinate system and M be the projection of P in the $xy-plane$. Let the $xy-plane$ be coordinated by polar coordinates so that the polar coordinates of $Mare(r,\theta)$. We say that the point P has cylindrical* (polar) coordinates (r,θ,z) and the write $p(r,\theta,z)$. We have therefore, the equations

$$x = r\cos\theta, r \geq 0$$

$$y = r\sin\theta \, and \, z = z$$

Given the relationship between the rectangular and cylindrical coordinates of a point in $\boldsymbol{R}^3$. From the above equations we obtain

$$r = \sqrt{x^2 + y^2}$$

$$\tan\theta = \frac{y}{x}$$

$$z = z$$

1.17.14. Spherical Coordinates

Suppose (x,y,z) are the rectangular coordinates of a point P (different from the origin) in R^3. We define

$p = |OP| =$ define of p from the origin.

θ = polar coordinate of the projection $M(x,y,0)ofP$ in the $xy-plane$.

$\phi =$ angle between positive z-axis and OP.

p is called the radial distance of P, the angle θ the azimuth of P, and the angle θ the colatitutde of P. These three qualities are called spherical*(or geographical) coordinates of P and we write $P(p,\phi,\theta)$ *i.e.*, the azimuth is denoted by ϕ and the colatitudes by θ. But it is preferable to let θ denote the same angle in both cylindrical and spherical coordinates and to place it in the same position in both the systems.

1.17.14.1. Translation of Axes

The process of choosing new axes parallel to a given set of axes is called translation of axes.

a. Surfaces

Definition. The set of all points (x, y, z) which satisfy an equation of $f(x, y, z) = 0$ is called a surfaces. Thus, plane is a surface.

The set of all point (x, y, z) satisfying the second-degree equation

$$ax^2 + by^2 + cz^2 = fyz + gxz + hxy + ux + vy + wz + k = 0$$

where a, b, c, f, g and h are not all zero, determines a surface, called a quadric surface.

1. **Symmetry.** If $(x, y, z) = 0$ *implies* $f(-x, y, z) = 0$; the surface is symmetric with respect to (or about) the $yz-plane.$ Symmetry with respect to the other coordinates planes is checked similarly.

if $f(x, y, z) = 0$ implies $f(x, -y, z) = 0,$ the surface is symmetric with respect to (or about) the $x-axis.$ Symmetric about the other axes is seen similarly. The surface is symmetric about the origin if

$$f(x, -y, -z) = 0 = f(x, y, z).$$

II. **Traces.** The curve of intersection of a surface and a plane is called the trace of the surface in the plane.

The trace of a surface are usually found in the coordinates planes or in planes parallel to the coordinates planes.

III. Intercepts. The x-coordinates of the points of intersection of the and a surface are called the x-intercepts of the surface. The intercepts on the

$x-axis$ (or x-intercepts) $y=0$, $z=0$ in the equation of a surface and solving it for x. The y and z-intercepts are found similarly.

1.17.15. Surfaces of Revolution

A surface S is called a surface of revolution if there is a straight line L such that every section if S perpendicular to L is a circle with center on L such a surface can be thought of as being generated by a plane curve, called a generate (or generator) revolved about L as axis of revolution. Usually one of the coordinate axis is taken of revolution.

1.17.15.1. Cylinder

Let C be a curve in a plane and let L be a line not in the plane. The union of all lines that intersects C and are parallel to L is called a cylinder. Each line through C and parallel to L, is called an element (or rolling) of the cylinder and the curve C is called n curve C is called a directrix of the cylinder.

Right Cylinder. If the elements of a cylinder are normal to a plane, then it is called a right cylinder with respect to that plane.

1.17.15.2. Equation of a Right Cylinder

Let the directrix $C: f(x, y)=0$ lie in the $xy-plane$. The elements of the right cylinder will be parallel to $z-axis$ i.e., parallel to $\boldsymbol{n}=[0,0,1]$. Thus, equation of the cylinder is

$$f\left(x-\frac{0}{c}z, y-\frac{0}{c}z\right)=0$$

i.e., $f(x, y)=0$. Hence a right cylinder with respect to a coordinate plane has precisely the same equation as the equation of the directrix in that plane. The coordinate axis normal to the coordinate plane with respect row which a cylinder is right, is called the **axis** of the cylinder. If in an equation two variable are missing, the surface is a cylinder whose elements are parallel to the axes of both missing variable and consist of a collection of planes.

1.17.15.3. The Cone

A surface generated by a straight line passing through a fixed point and intersecting a fixed plane curve not containing the fixed point is called a cone. The given curve is called the **directrix** and the fixed point is known as the **vertex** of the cone. The straight line is called the **generator** and any particular position of the generator is called an **element**(or rolling) of the cone.

1.17.15.4. The Ellipsoid

A surface $(inR^3)\,defined$ by an equation of the form $\frac{x^2}{a^2}+\frac{y^2}{b^2}+\frac{z^2}{c^2}=1,$ is called an **ellipsoid.**

1.17.15.5. The Elliptic Paraboloid

A surface defined by an equation of the form $\frac{x^2}{a^2}+\frac{y^2}{b^2}=cz$, is called an **elliptic Paraboloid.**

1.17.15.6. The Hyperbolic Paraboloid

A Surface defined by an equation of the form $z=\frac{y^2}{b^2}-\frac{x^2}{a^2}$, is called a hyperbolic Paraboloid.

1.17.15.7. The Hyperboloid of One Sheet

A surface defined by an equation of the form $\frac{x^2}{a^2}+\frac{y^2}{b^2}-\frac{z^2}{c^2}=1$, is called a hyperboloid of one sheet.

1.17.15.8. The Hyperboloid of Two Sheets

A surface represented by an equation of the form $\frac{x^2}{a^2}+\frac{y^2}{b^2}-\frac{z^2}{c^2}=-1$, is called hyperboloid of two sheets.

1.17.15.9. The Elliptic Cone

A surface defined by an equation of the form $\frac{x^2}{a^2}+\frac{y^2}{b^2}=cz^2,(c>0)$, is called an elliptic cone.

1.17.15.10. Ruled Surfaces

A surface generated by a line moving in accordance with given conditions is called a ruled surface. The moving lines are called the generator and a particular position of the generator is called an element or a ruling of the surface.

Planes, cylinder and cones are example of ruled surfaces. The rulings of a plane are all lines parallel to a given line. The ruling of a cylinder meet a curve (the directrix) and are parallel to a given line (the axis). The rulings of a cone meet a curve (the directrix) and pass through a fixed point (the vertex).

1.17.16. Spherical Trigonometry

Let S be a sphere and P a plane that passes through the center of S. Then $S\cap P$, i.e., intersection of a sphere and a plane through its center is called a great circle on $S.if\ q$ is a plane not passing through the center of S, then $S\cap q$ is called A spherical angle between any two intersecting great circles. Two plane $\alpha\ and\ \beta$ with a common line of intersection AB are said to form a dihedral angle, AB is called the edge of dihedral angle, the planes forming the dihedral angle are called its faces.

The plane angle of a dihedral angle is the angle formed by two rays one on each face, perpendicular to the edge at the same point. Measure of a dihedral angle is equal to measure of its plane angle.

A trihedral angle is formed by three planes which intersect so as t = to have only one point is common. The point of intersection of these planes forming a trihedral angle is known as the vertex of the trihedral angle.

A spherical triangle is that part of the surface of a sphere which is bounded by the arcs of three great circles. Note that a spherical triangle is the intersection of the surface of the sphere with the trihedral angle whose vertex is O, the center of the sphere and whose faces are the planes of the great circles.

CHAPTER 2

FUNCTIONS OF SEVERAL VARIABLES

CONTENTS

2.1. INTRODUCTION

The idea of a function of several variable was introduced in the concepts of limit and continuity of functions of two variables. Definition of partial derivatives of first and higher orders were also discussed before this. In this Chapter, we shall study further properties and some application of partial derivatives. Also Volumes and surfaces are other topics of this chapter.

2.2. HOMOGENEOUS FUNCTIONS

Definition. A function g defined by $u = f(x, y, z, \ldots)$ of any number of variables is said to be homogenous of degree n in these variables if multiplication of these variables by any number $t(\neq 0)$ result in the multiplication of the function by t^n, i.e.,

$$f(tx, ty, tz, \ldots.) = t^n f(x, y, z, \ldots.)$$

Provided that $(tx, ty, tz \ldots)$ is in the domain of f.

Taking $t = \frac{1}{x}, (x \neq 0)$ the equation (1) becomes

$$f\left(1, \frac{y}{x}, \frac{z}{x} \ldots.\right) = \frac{1}{x^n} f(x, y, z, \ldots.)$$

or

$$f(x, y, z \ldots.,) = x^n f\left(1, \frac{x}{y}, \frac{z}{x} \ldots\right)$$

$$x^n g\left(\frac{y}{x}, \frac{z}{x} \ldots.\right)$$

which is another criterion for a function to be homogenous.

Example

Consider the function f defined by

$$f(x, y) = \frac{x}{y} + \frac{3}{4}\frac{y}{x} + \cos\sqrt{\frac{y}{x}} + in\,x - in\,y.$$

Here, $$f(tx,ty)=\frac{tx}{ty}+\frac{3}{4}\frac{ty}{tx}+\cos\sqrt{\frac{ty}{tx}}+In\,t+In\,x-In\,t-In\,y$$

$$=f(x,y)=t^o f(x,y)$$

Thus f is a homogenous function of degree zero.

Example

Let $$f(x,y)=\frac{\sqrt{y}+\sqrt{x}}{y+x}$$

Here, $$\frac{\sqrt{y}+\sqrt{x}}{y+x}=\frac{\sqrt{x}\left[1+\sqrt{\frac{y}{x}}\right]}{x\left[1+\frac{y}{x}\right]}=x^{-1/2}\frac{1+\sqrt{\frac{y}{x}}}{+\frac{y}{x}}$$

Thus $f(x.y)$ is a homogenous function of degree $-\frac{1}{2}$

Theorem (Euler's). if $u=f(x,y)$ is a homogenous function of x,y of degree $n,$ then

$$x\frac{\partial u}{\partial x}+y\frac{\partial u}{\partial y}=nu.$$

Proof. We have $$u=f(x,y)=x^n g\left(\frac{y}{x}\right).$$ Therefore,

$$\frac{\partial u}{\partial x}=n\,x^{n-1}\,g\left(\frac{y}{x}\right)+x^n g'\left(\frac{y}{z}\right)\left(\frac{-y}{x^2}\right).$$

$$=n\,x^{n-1}\,g\left(\frac{y}{x}\right)-yx^{n-2}g'\left(\frac{y}{x}\right). \qquad (1)$$

and $$\frac{\partial u}{\partial x}=x^n g'\left(\frac{y}{x}\right)\frac{1}{x}=x^{n-1}g'\left(\frac{y}{x}\right)+yx^{n-1}g'\left(\frac{y}{x}\right) \qquad (2)$$

Thus from (1) and (2) we have

$$x\frac{\partial u}{\partial y}+y\frac{\partial u}{\partial y}=nx^n\, g\left(\frac{y}{x}\right)-yx^{n-1}g'\left(\frac{y}{x}\right)+yx^{n-1}g'\left(\frac{y}{x}\right)$$

$$nx^n\left(\frac{y}{x}\right)=nu=nf(x,y)$$

Example

if $u=\arctan\left(\dfrac{x^3+y^3}{x-y}\right), show that$

$$x\frac{\partial u}{\partial x}=y\frac{\partial u}{\partial y}=\sin 2u.$$

Solution

We have $z=\tan u=\dfrac{x^3+y^3}{x-y}=x^2\dfrac{1+\left(\dfrac{y}{x}\right)^3}{1-\dfrac{y}{x}}$

Thus z is a homogenous function of x, y of degree 2.

Therefore,

$$x\frac{\partial z}{\partial x}+y\frac{\partial z}{\partial y}=2z, \qquad (1)$$

But $\dfrac{\partial z}{\partial x}=\sec^2 u\dfrac{\partial u}{\partial x} and \dfrac{\partial z}{\partial y}=\sec^2 u\dfrac{\partial u}{\partial y}$

Substituting into (1), we get

$$\sec^2 u\ (x\frac{\partial u}{\partial x}+y\frac{\partial u}{\partial y})=2z=2\tan u$$

or $x\dfrac{\partial u}{\partial x}+y\dfrac{\partial u}{\partial y}=\dfrac{2\sin u}{\cos u}\cos^2 u=2\sin u\cos u=\sin 2u.$

2.3. DIFFERENTIALS

Suppose $u=f(x,y)$ processes first partial derivatives $f_x(x,y)$ and

$f_y(x,y)$ in a region $D,$ and that f_x *and* f_y are continuous at some point (x_0,y_0) of D. Furthermore, take a square neighborhood of (x_0,y_0) that lies in D and let Δx *and* Δy be numerically less than half the length of a side of this neighborhood, so that the point $x_0+\Delta x, y_0+\Delta y)$ lies in D. Then the increment Δu of the dependent variable can be written as follows:

$$\Delta u = f(x_0+\Delta x, y_0+\Delta y) - f(x_0,y_0)$$

$$= f(x_0+\Delta x, y_0+\Delta y) - f(x_0,y_0+\Delta y) + f(x_0,y_0+\theta_2\,\Delta y)\Delta y$$

where $0<\theta_1<1, 0<\theta_2<1\,and\,\theta_1\,and\,\theta_2$ depend upon $\Delta x\,and\,\Delta y$, respectively, As $f_x\,and\,f_y$ are continuous at (x_0,y_0) and when x $(\Delta x,\Delta y)\to(0,0)$,

$$(x_0+\theta_1\,\Delta x, y_0+\Delta y)\to(x_0,y_0),(x_0,y_0+\theta_2\,\Delta y)\to(x_0,y_0),$$

we have

$$f_x(x_0+\theta_1\,\Delta x, y_0+\Delta y) = fx(x_0,y_0)\varepsilon_1(\Delta x,\Delta y)$$

$$f_y(x_0,y_0+\theta_2\,\Delta y) = f_y(x_0+y_0)\varepsilon_2(\Delta x,\Delta y)$$

where $\varepsilon_1(\Delta x,\Delta y)\to 0\,and\,\ \varepsilon_2(\Delta x,\Delta y)\to(0,0).$

Thus $\Delta u = f_x(x_0,y_0)\Delta x + f_y(x_0,y_0)\Delta y + \ \varepsilon_1(\Delta x,\Delta y)\Delta x\ \varepsilon_2(\Delta x,\Delta y)\Delta y.$

For the sake of convenience, we shall write $\varepsilon_1\ for\ \varepsilon_1(\Delta x,\Delta y)$ and $\varepsilon_2\ for\ \varepsilon_2(\Delta x,\Delta y)$.

We have thus proved the following:

Theorem. If $u=f(x,y)$ processes first order partial derivatives in a region $D,$ and if these partial derivatives are continuous at some point (x_0,y_0) of $D,$ then (for increments $\Delta x\,and\,\Delta y$ sufficiently small numerically,) the increment Δu of the dependent variable can be written as

$$\Delta u = \frac{\partial u}{\partial x}\Delta x + \frac{\partial u}{\partial y}\Delta y\, \varepsilon_1\, \Delta x + \varepsilon_2\, \Delta y \qquad (1)$$

where the partial derivatives are evaluated at the point (x_0, y_0) and each of $\varepsilon_1, \varepsilon_2$ tends to zero when $(\Delta x, \Delta y) \to (0,0)$.

Definition. When the increment of a function u can be expressed in the form (1) of theorem the function is said to be differentiable at the particular point at which the first partial derivatives are evaluated. function is said to be differentiable at the particular point at which the first partial derivatives are evaluated

Example

$Let\, u = f(x, y) = x^2\, y.$ Then at the point (x_0, y_0), we have

$$\Delta u = (x_0 + \Delta x)^2 (y_0 + \Delta y) - x_0^2 y_0$$

$$= 2x_0 y_0\, \Delta x + y_0\, \Delta x^2 + x_0^2\, \Delta y + 2x_0 + \Delta x\, \Delta y + \Delta x^2\, \Delta y$$

$$= 2x_0 y_0\, \Delta x + x_0^2\, \Delta y + y_0\, \Delta x^2 + 2x_0 + \Delta x\, \Delta y + \Delta x^2\, \Delta y$$

$$= \frac{\partial u}{\partial x}\Delta x + \frac{\partial u}{\partial y}\Delta y + (y_0\, \Delta x\, 2x_0\, \Delta y)\Delta x + \Delta x^2\, \Delta y$$

$$= \frac{\partial u}{\partial x}\Delta x + \frac{\partial u}{\partial y}\Delta y + \varepsilon_1\, \Delta x + \varepsilon_2\, \Delta y$$

where $\varepsilon_1 = y_0\, \Delta x + 2x_0\, \Delta y\, and\, \varepsilon_2 = \Delta x_2$

Thus u is differentiable at (x_0, y_0).

Definition. The differential (or total differential) du of a differentiable function $u = f(x, y)$ of two variables is defined as

$$du = \frac{\partial u}{\partial x} dx + \frac{\partial u}{\partial y} dy$$

where $dx = \Delta x, dy = \Delta y$ are differentiable of $x\, and\, y.$

The equation (1) of theorem can now be rewritten as

$$\Delta u = du + \varepsilon_1 \Delta x = \varepsilon_2 \Delta y$$

It follows that du is a fairly good approximation of Δu.

The concepts of differentiability and total differential can be extended analogously for a function of more than two variables.

Example

Use the total differential at $(2,8)$ to approximate 2.037×7.98.

Solution

Consider $f(x, y) = xy$

$$df = \frac{\partial f}{\partial x} dx + \frac{\partial f}{\partial y} dy = y\,dx + x\,dy.$$

Here $dx = 0.37, dy = -.02, x = 2, y = 8$

$$df = 8(0.37) - 2(.02) = .296 - .04 = .256.$$

$$f(2.037, 7.98) = f(2,8) + \Delta f \approx f(2,8) + df$$

$$= 16 + .256 = 16.256$$

Using calculator, $2.037 \times 7.98 = 16.25526.$

Example

find the percentage error in the area of an ellipse when as error of 1% is made in measuring each of the major and mirror axes of the ellipse.

Solution. $Let\, a, b\, and\, A$ denote semi-major axis, semi-minor axis and area of an ellipse, respectively. Then

$$A = \pi ab$$

$$\frac{\partial A}{\partial a} = \pi b\, abd \frac{\partial A}{\partial b} = \pi a$$

$$dA = \frac{\partial A}{\partial a} da + \frac{\partial A}{\partial b} db$$

$$\pi\, da + \pi a\, db$$

Now, $da = \frac{a}{100}, db = \frac{b}{100}$

Therefore $$dA = \frac{\pi ab}{100} + \frac{\pi ab}{100} + \frac{2A}{100}$$

Thus there is an error of $2\ \%$ in the area of the ellipse.

2.4. CHANGE OF VARIABLE: THE CHAIN RULE

Let $u = f(x, y)$ be a differential function, Suppose each of x and y is a differentiable function of a single variable y. Suppose t changes to $t + \Delta t(\Delta t \neq 0)$. Then using (1) we can write

$$\frac{\Delta u}{\Delta t} = \frac{\Delta u}{\Delta x}\frac{\Delta x}{\Delta t} + \frac{\Delta u}{\Delta y}\frac{\Delta y}{\Delta t} + \varepsilon_1 \frac{\Delta x}{\Delta t} + \varepsilon_2 \frac{\Delta y}{\Delta t}$$

Considering u as a function of the single variable t, we let $\Delta t \to 0$ and take limits of both sides of the above equation. Since *x* and *y* are differentiable functions of t, and $\varepsilon_1\ and\ \varepsilon_2$ tend to zero), we conclude that

$$\frac{du}{dt} = \frac{\partial u}{\partial x}\frac{dx}{dt} + \frac{\partial u}{\partial y}\frac{dy}{dt}$$

If each of $x\ and\ y$ is a differentiable function of variable r and s, we would similarly get,

$$\frac{\partial u}{\partial r} = \frac{\partial u}{\partial x}\frac{\partial x}{\partial r} + \frac{\partial u}{\partial y}\frac{\partial y}{\partial r}$$

$$\frac{\partial u}{\partial s} = \frac{\partial u}{\partial x}\frac{\partial x}{\partial s} + \frac{\partial u}{\partial y}\frac{\partial y}{\partial r}$$

These results can be generalized for a function of several variables. Thus, if $u = f(x, y, z \ldots)$ is a differentiable function and each of $x, y, z, \ldots$ is a differentiable function of several variables $r, s, t, \ldots$ then

$$\frac{\partial u}{\partial r} = \frac{\partial u}{\partial x}\frac{\partial x}{\partial r} + \frac{\partial u}{\partial y}\frac{\partial y}{\partial r} + \frac{\partial u}{\partial z}\frac{\partial z}{\partial r} + \ldots$$

$$\frac{\partial u}{\partial s} = \frac{\partial u}{\partial x}\frac{\partial x}{\partial s} + \frac{\partial u}{\partial y}\frac{\partial y}{\partial s} + \frac{\partial u}{\partial z}\frac{\partial z}{\partial s} + \ldots$$

$$\frac{\partial u}{\partial t}=\frac{\partial u}{\partial t}\frac{\partial x}{\partial t}+\frac{\partial u}{\partial y}\frac{\partial y}{\partial t}+\frac{\partial u}{\partial z}\frac{\partial z}{\partial t}+\ldots$$

These equations are known as the Chain Rule.

Example

A closed box of inner dimension 2, 3 and 4 decimeters is to be made of metal sheet 1 cm thick. Approximate the volume of the metal using differentials.

Solution

If $x, y\, and\, z$ are dimension of the box, then its volume is

$$V(x,y,z)=xyz$$

For $x=2, y=3, z=4, \Delta x=\Delta y=\Delta z=\left(\frac{1}{10}\right)\times 2=\frac{1}{2}$ decimeter,

The exact volume of the metal is

$$V(x+\Delta x, y+\Delta y, z+\Delta z)-V(x,y,z)=\Delta V.$$

Since $\Delta V \approx dV\; and\; \Delta x=dx, \Delta y=dy, \Delta z=dz$ we shall use differentials.

$$\Delta V \approx dV=\frac{\partial V}{\partial x}dx+\frac{\partial V}{\partial y}dy+\frac{\partial V}{dz}dz$$

$$=yz\,dx+zxdy+xydz$$

$$=3(4)\frac{1}{5}+4(2)\frac{1}{5}+2(3)\frac{1}{5}=\frac{26}{5}cu\,dm.$$

Example

find $\frac{dz}{dt}$ when $z=xy^2+x^2y, x=at^2, y=2at$

Solution

$$\frac{\partial z}{\partial x}=y^2+2xy, \frac{\partial z}{\partial y}=2xy+x^2, \frac{dx}{dt}=2at, \frac{dy}{dt}=2a.$$

Therefore, $\frac{dz}{dt}=\frac{\partial z}{\partial x}\frac{dx}{dt}+\frac{\partial z}{\partial y}\frac{dy}{dt}=\left(y^2+2xy\right)2at+\left(2xy+x^2\right)2a$

$$=\left(4a^2t^2+4a^2t^3\right)2at+\left(4a^2t^3-a^2t^4\right)2a=a^3\left(16t^3+10t^2\right).$$

2.5. IMPLICIT FUNCTIONS

Theorem. Let $y=F(x)$ be a differentiable function defined by the new independent variable $t\,;x=t,y=F(t).Then\,u,$ as a function of $t,$ is differentiable and by the chain rule

$$\frac{du}{dt}=\frac{\partial f}{\partial x}\frac{dx}{dt}+\frac{\partial f}{\partial y}\frac{dy}{dt}$$

$$0=\frac{\partial f}{\partial x}1+\frac{\partial f}{\partial y}\frac{dy}{dt}$$

Thus, $\dfrac{dy}{dt}=-\dfrac{\partial f}{\partial x}\Big/\dfrac{\partial f}{\partial y}=-\dfrac{f_x}{f_y}=\dfrac{dy}{dx},(since\,x=t)$

2.6. DIRECTION DERIVATIVES

Definition. Let $u=f(x,y,z)$ be defined in a region D of R^3. Let p be a point of D. Let Δs denote the corresponding change in u.

Then $\lim\limits_{\Delta s\to 0}\dfrac{\Delta u}{\Delta s},if\;it$ exerts, is called the derivative of $u\;at\;p$ in the specified direction and is denoted by $\dfrac{du}{ds}$.

Let u be as in the above definition. Suppose the displacement $\Delta s=[\Delta x,\Delta y,\Delta z]$ is in the direction of the vector $v=[a,b,c]$. Then, for some real $k,$

$$[\Delta x,\Delta y,\Delta z]=k[a,b,c]\,and\;so\,\Delta x=ka,\Delta y=kb,\Delta z=kc$$

Now, magnitude of the displacement

$$\Delta s=|\Delta s|=\sqrt{\Delta x^2+\Delta y^2+\Delta z^2}$$

$$=k\sqrt{a^2+b^2+c^2}$$

Further,

$$\Delta u = \frac{\partial u}{\partial x}\Delta x + \frac{\partial u}{\partial y}\Delta y + \frac{\partial u}{\partial z}\Delta z + \varepsilon_1 \Delta x + \varepsilon_2 \Delta y + \varepsilon_3 \Delta z,$$

where $\varepsilon_1, \varepsilon_2, \varepsilon_3 \to 0$ as $(\Delta x, \Delta y, \Delta z) \to (0,0,0)$.

Therefore,

$$\frac{\Delta u}{\Delta s} = \frac{\partial u}{\partial x}\frac{\Delta x}{\Delta s} + \frac{\partial u}{\partial y}\frac{\Delta y}{\Delta s} + \frac{\partial u}{\partial s}\frac{\Delta z}{\Delta s} + \varepsilon_1 \frac{\Delta x}{\Delta s} + \varepsilon_2 \frac{\Delta y}{\Delta s} + \varepsilon_3 \frac{\Delta z}{\Delta s}$$

$$= \frac{1}{\sqrt{a^2+b^2+c^2}}\left(a\frac{\partial u}{\partial x} + b\frac{\partial u}{\partial y} + c\frac{\partial u}{\partial z} + a\varepsilon_1 + \text{b}\varepsilon_2 + \text{c}\varepsilon_3 \right)$$

and

$$\frac{du}{ds} = \lim_{\Delta s \to 0} \frac{\Delta u}{\Delta s},$$

$$= \frac{1}{\sqrt{a^2+b^2+c^2}}\left(a\frac{\partial u}{\partial x} + b\frac{\partial u}{\partial y} + c\frac{\partial u}{\partial z} \right)$$

From vectors Analysis, we know that grad $u = \frac{\partial u}{\partial x} i + \frac{\partial u}{\partial y} j + \frac{\partial u}{uz} K$

Thus,

$$a\frac{\partial u}{\partial x} + b\frac{\partial u}{\partial y} + c\frac{\partial u}{\partial z} = [a,b,c]\left[\frac{\partial u}{\partial x}, \frac{\partial u}{\partial y}, \frac{\partial u}{\partial z} \right]$$

$$= v.\textbf{grad}\, u$$

Therefore,

$$\frac{du}{ds} = \frac{v.grad\, u}{|v|}$$

Note. Compression of the various directional derivatives of $u = f(x, y, z)$ at a particular point P leads to the question, "in what direction or directions does u change most rapidly?" and "in what directions, if any u ".

We have shown above that

$$\frac{du}{ds} = \frac{v.grad\, u}{|v|} grad\, u.\frac{v}{|v|} = |grad\, u|_p \cos\theta$$

where $|grad\, u|_p$ is the magnitude of a vector that is normal to the surface at

P and \tehta measure of the angle between grad u and v. Maximum value of $cos\theta = 1$ and so $\theta = 0$. Hence the directional derivatives is greatest in the direction of the gradient itself, and the magnitude of that gradient vector. On the other hand, $\cos\theta = 0\ for\ \theta = \pi / 2$. Therefore, the directions of no change for $u\ are$ the directions perpendicular to grad u.

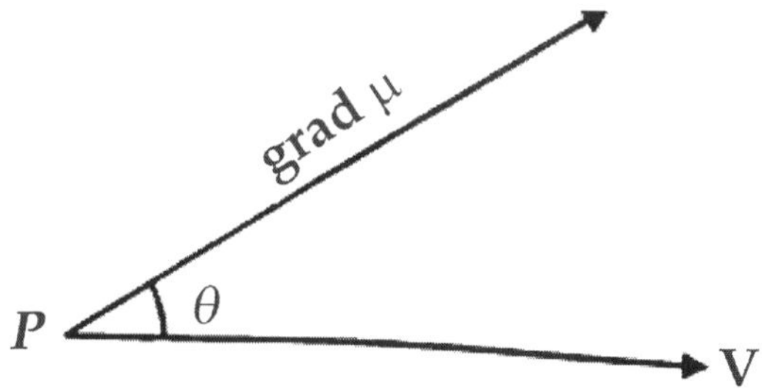

2.7. TANGENT PLANES AND NORMAL LINES

Consider a surface S defined by $f(x, y, z) = 0$ Let P be a point of S. Consider a curve through P and lying on S. Then for all points on the curve we have $f(x, y, z) = 0$ since each point of the curve is a point of S. Thus, at P the various curves of S give the directions for which the rate of change of f is zero. Perpendicular to all these directions is the direction of grad f. Thus we define normal line to a surface as follows:

Definition. Consider all curves through a point P on a surface S. The straight line perpendicular to all these curves is called the normal line to surface at P and if $\frac{\partial f}{\partial x}, \frac{\partial f}{\partial y}, \frac{\partial f}{\partial z}$ are all continuous at P and not all of them are zero, then grad f at P is the normal vector to the surface P.

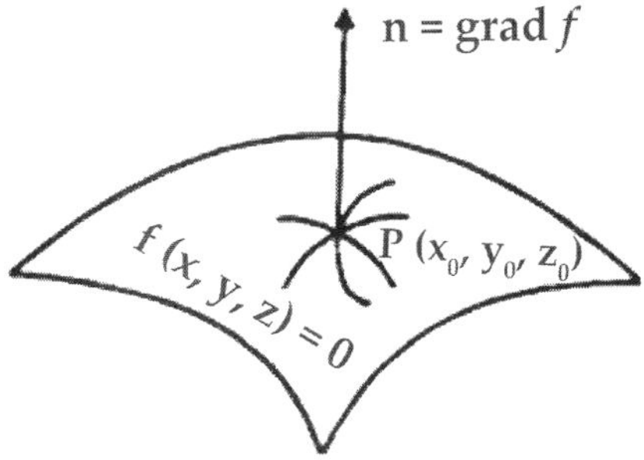

Definition. The tangent plane to the surface S at the point P is the planes whose normal vector is grad f *at p.*

Example. Find an equation of three tangent planes at any point $p(x_1.y_1,z_1)$ of the elliptic Paraboloid $z = x^2 + 4y^2$.

Solution. Hence $f(x,y,z) = x^2 + 4y^2 - z = 0$, Grad $f = 2xi + 8y\,j - k$

At $p(x_1,y_1,z_1)$, Grad $f = 2x_1 i + 8y_1\,j - k$

Equation of the tangent planes at $(x_1,y_1,z_1) = 0$ or $2xx_1 + 8yy_1 - z = 2x_1^2 + 8y_1^2 - z_1 = 2\left(x_1^2 + 4y_1^2 - z_1\right) + z_1 = z_1$.

Example. Two surface are said to be orthogonal at a point of their intersect if there normal at the point are orthogonal. They are said to intersect orthogonally if they are orthogonal at every point common to them.

Prove that the surface $f(x,y,z) = 0$ *and* $g(x,y,z) = 0$ intersect orthogonally if, and only if, $f_x g_x + f_y g_y + f_z g_z = 0$.

Solution. Let (x,y,z) be any point of intersection of

$f(x,y,z) = 0$ (1)

and $g(x,y,z) = 0$ (2)

Direction rations of a line normal to (1) are f_x, f_y, f_z

Similarly, direction rations of a line normal to (2) are $g_x.g_y, g_z$

The two normal lines are orthogonal if, and only if, $f_x g_x + f_y g_y + f_z g_z = 0$ are required.

2.7.1. Parametric Equation of a Surface

The sphere $x^2 + y^2 + z^2 = a^2$ can be described by the parametric equations.

$$x = a \sin \varnothing \cos \theta,$$

$$y = a \sin \varnothing \cos \theta,$$

$$z = a ü \quad \varnothing$$

where $-\pi < \theta < \pi, 0 \le \varnothing \le \pi$. Here $\theta \, and \, \varnothing$ are parameters.

We assume that a surface S is defined by the parametric equations

$$x = f(u,v), y = g(u,v), z = h(u,v) \quad (1)$$

or in, equivalent vector form

$$r = \overrightarrow{OP} = xi + yj + zk \quad (2)$$

Suppose it is required to find the normal vector n at the point p_0 determined by the parametric values $u = u_0 \, and \, v = v_0$.

curve on S through p_0, the parameter being u The tangent vector to this curve is collinear with $\frac{\partial r}{\partial u}$. Similarly if, $u = u_0$ and v varies, then equations (1) or (2) are parametric equation of a curve on S through p_0 with parametric v. The tangent vector to this curve is collinear with $\frac{\partial r}{\partial v}$

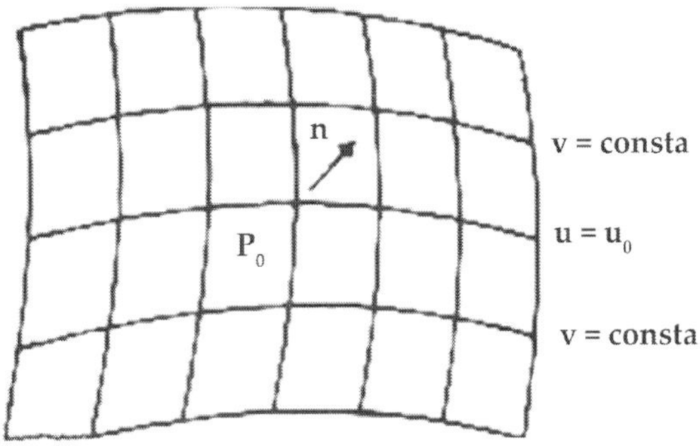

Since the normal vector n at p_0 must be perpendicular to the tangent vectors to all curve on S through p_0, and to these tangent vector in particular, we can say that n is collinear with

$\frac{\partial r}{\partial u}\times\frac{\partial r}{\partial v}$.

For the sphere $x^2+y^2+z^2=a^2$

$$r = xi + yj + zk$$

$$= a\left(\sin\varnothing\cos\theta i+\sin\varnothing\sin\theta j+\cos\varnothing k\right)$$

$$\frac{\partial r}{\partial\theta}=a\left(-\sin\varnothing\sin\theta i+\sin\varnothing\cos\theta j\right)$$

$$=a\sin\varnothing\left(-\sin\theta i+\cos\theta j\right)$$

and $\frac{\partial r}{\partial\varnothing}=a\ (\cos\varnothing\cos\theta j+\cos\varnothing\sin\theta j-\sin\varnothing k$

$$\frac{\partial r}{\partial\varnothing}\times\frac{\partial r}{\partial\theta}=\begin{vmatrix} i\,j\,k \\ a\cos\varnothing\cos\theta\, a\cos\varnothing\cos\theta - a\sin\varnothing \\ -a\sin\varnothing\sin\theta\, a\sin\varnothing\sin\theta\ 0 \end{vmatrix}$$

$$a^2\sin^2\varnothing\cos\theta i+a^2\sin^2\varnothing\sin\theta j+$$

$$(a^2\cos\varnothing\cos^2\theta\sin\varnothing+a^2\cos\varnothing\sin^2\theta\sin\varnothing)k$$

$$=a\sin\varnothing\ (a\sin\varnothing\cos\theta i+a\sin\varnothing\sin\theta j+a\cos\varnothing k$$

$= a \sin \varnothing \varnothing r$

Thus, the normal vector is collinear with r.

2.8. EXTREMA OF FUNCTION OF TWO VARIABLES

The extreme values of a function of one variable were studied in chapter. First and second derivatives tests were proved there and consequently they were used to find the maximum and minimum values of a function. The theory of extrema will be extended to function of two variable and it will be found to be similar to that of one variable.

Definition. Let f be a real –valued function of two variables defined and continuous on the closed disc $D=\{(x,y)\in R^2 : x^2+y^2\le r^2\} of\, \boldsymbol{R}^2 . f$ is said to have an absolute (global) maximum value D of f on D.

The function f is said to have a relative (local) maximum at some point $(x_0, y_0) of\, D$ if there exist an open disc k centered at (x_0, y_0) and of radius a i.e.

$$K=\{(x,y)\in R^2 : (x-x_0)^2+(y-y_0)^2<a^2\}$$

with $K\subset D$ such that $f(x_0,y_0)\ge f(x,y)\, fro\, all (x,y)\in K$ i.e., f has an absolute maximum on K.

Definition. A function f of the variables as defined in is said to have an absolute (global) minimum value on D if there is some point (x_0, y_0) of D such that $f(x_0,y_0)\le f(x,y)$ for all $(x,y)\in D.$ in such a case $f(x_0,y_0)$ is the absolute minimum value of $f\, on\, D.$

The function f is said to have a relative (local) minimum at some point $(x_0, y_0) of\, D$ if there exists an open disc K centered at (x_0, y_0) and of radius r with $K\subset D$ such that $f(x_0,y_0)\le f(x,y)\, for\, all (x,y)\in K.$ i.e., f has an absolute minimum value on K.

The maximum and minimum values of f are reoffered to as the extreme values or extrema of f.

Theorem. Let a function f of two variables be a defined and continuous on the disc $D=\{(x,y)\in R^2:(x-x_0)^2+(y-y_0)^2\leq r^2\}$. Suppose $f_x(x_0,y_0)$ and $f_y(x_0,y_0)$ both exist on K. if f has a relative extrema at (x_0,y_0), then $f_x(x_0,y_0)=0=f_y(x_0,y_0)$

Proof. Suppose f has a relative minimum value at $(x_0.y_0),f(x_0,y_0)\leq(x_0+h,y_0)$ for definition of partial derivative, we have $f_x(x_0,y_0)=\lim\limits_{h\to 0}\frac{f(x_0+h,y_0)-f(x_0,y_0)}{h}$, since f has a relative minimum at $(x_0,y_0),f(x_0,y_0)\leq f(x_0+h,y_0)$ for sufficiently small h. Therefore, for $h>0$,

$\frac{f(x_0+h,y_0)-f(x_0,y_0)}{h}\geq 0$,And for $h<0$, $\frac{f(x_0+h,y_0)-f(x_0,y_0)}{h}\geq 0$

.

Thus $Rf_x(x_0,y_0)\geq 0$ and $Lf_x(x_0,y_0)<0$, where Rf_x *and* Lf_x denote the one-sided right-hand and left-hand derivatives. Since $f_x(x_0,y_0)$ exist, we have

$Rf_x(x_0,y_0)\geq 0$ *and* $f_x(x_0,y_0)\leq$ imply

$$f_x(x_0,y_0)=0.$$

The proof that $f_y(x_0,y_0)=0$ is analogous and is left as an exercise.

Note. The proof of the case of relative maximum is similar and is omitted. The condition stated in the theorem in necessary and not sufficient as will be seen later.

Definition. A point (x_0,y_0) in the domain of a function f of two variables is called a critical point of f if either $f_x(x_0,y_0)=f_y(x_0,y_0)=0$ or one (or both) of the first order partial dericatives of f at (x_0, y_0) does (d) no exits. From theorem it follows that f can have relative extrema only at its critical points.

Example. Find the critical points of $f(x,y)=x^3+y^3-3axy, a>0.$

Solution. f_x, f_y exist at all points of the domain of f.

$$f_x = 3x^2 - 3ay, f_y = 3y^2 - 3ax.$$

For critical points, $f_x = f_y = 0$. Therefore,

$x^2 - ay = 0$ (1)

and $ax - y^2 = 0$ (2)

Substituting the value of x from (2) into (1), we have

$$\frac{y^4}{a^2} - ay = 0 \; or \; y\left(y^3 - a^3\right) = 0$$

i.e., $y = 0, a \; and \; so \; x = 0, a.$

The critical points are $(0,0) \; and \; (a,a).$

Example. Let $f(x,y)=\sqrt{x^2+y^2}$. find the critical points and the relative extrema of f.

Solution. We have

$$f_x(x,y) = \frac{x}{\sqrt{x^2+y^2}}, f_y(x,y) = \frac{y}{\sqrt{x^2+y^2}}$$

The partial derivatives exist at all points of the domain of f except at the origin which is in the domain of f. Thus (0,0) is a critical point of f. Now $f_x(x,y)=0$ only if $x=0$ and $f_y(x,y)=0$ only if $y=0$.

Thus $f_x(x,y)=0 \; and \; f_y(x,y)=0$ has no simultaneous solution. The only critical point is $(0,0)$ and $f(0,0)$. Since $f(x,y) \geq 0$ for all (x,y), 0 is the absolute minimum value of f.

Example. Let $f(x,y)-y^2-x^2$. Then

$$f_x=-2x \, and \, f_y=2y$$

$f_x=0$ and $f_y=0$ give $(x,y)=(0,0)$.

Thus origin is a critical point of f and $f(0,0)=0$.

Consider a disc K centered at (0,0) and of radius r. i.e.,

$$K=\{(x,y):x^2+y^2\le r^2\}$$

Then in $K, f(x,y)>0$ at points on the $y-axis \, and \, f(x,y)<0$ at points on the $x-axis$. Thus $f(0,0)=0$ is neither the largest nor the smallest value of $f \, in \, K$.

Therefore, f, has no extrema at the critical point $(0,0)$.

A critical point at which f does not have a relative extrema is called saddle opoint. The example shows that the condition of theorem for extrema is necessary but not sufficient. There is a second derivatives test which gives sufficient condition for extrema. We state this in.

Theorem. Let f be a function of two variables x,y defined in a neighborhood of z (x_0,y_0) and have continuous second order partial derivatives in this neighborhood. Suppose (x_0,y_0) is a critical point of f. Set $A=f_{xx}(x_0,y_0), B=f_{xx}(x_0,y_0)=f_{yx}(x_0,y_0), C=f_{yy}(x_0,y_0)$ and

$$D=\begin{vmatrix} f_{xx} f_{xy} \\ f_{yx} f_{yy} \end{vmatrix}=\begin{vmatrix} A B \\ B C \end{vmatrix}=AC-B^2.$$

Then

(i) f has a local minimum at $(x_0,y_0) \, if \, D>0 \, and \, A>0$

(ii) f has a local maximum at $(x_0,y_0) \, if \, D>0 \, and \, A<0$.

(iii) f has a saddle point if $D<0$

(iv) The test fails if $D = 0$.

Note. $AC - B^2 > 0$ only if both $A\, and\, C$ have the same sig. In the case of extrema of one variable, a function $y = f(x)$ is minimum (or maximum) at a critical point $x = x_0$ if the second derivatives is positive (or negative) at x_0. Similar is the situation for a function of two variables. When $D > 0$ the function is minimum (maximum) if f_{xx} is positive (or negative) at the critical point.

Example. Find three positive numbers whose sum is 48 and whose product is as large as possible.

Solution. Let the required numbers be x, y and z. Then the product xyz is to be maximized where $x + y + z = 48.$ Thus

$$f(x, y) = xy48 - x - y)$$

and we have to find its maximum value.

$$f_x(x, y) = 48y - 2xy - y^2$$

$$f_{xx}(x, y) = -2y$$

$$f_y(x, y) = 48x - x^2 - 2xy$$

$$f_y(x, y) = -2x$$

$$f_{xy}(x, y) = 48 - 2x - 2y$$

For circle points, we have the equations

$48y - 2xy - y^2 = 0$ (1)

$48c - 2xy - x^2 = 0$ (2)

From (1), we have $y = 0\, or\, 48 - 2x - y = 0$ (3)

If $y = 0,$ then from (2), we get

$48 - x^2 = 0$ i.e., $x = 0, 48$

The solution are (0,0) and (48,0)

From (3), we have $y = 48 - 2x.$ Setting this value in (2), we obtain

$$x\left[48 - 2(48 - 2x) - x\right] = 0$$

$x[-48 + 3x] = 0$ i.e., $x = 0\, and\, 16$

For $x = 0, y = 48.$

When $x = 16, then\, y - 16.$ Therefore (16,16) is a critical point. The points (0,0), (0,480, (48,0) are not admissible since one or both numbers is zero and we seek positive numbers.

$$f_{xx}(16,16) = -32, f_{yy}(16,16) = -32m\, f_{xy}(16,16) = -16$$

$$f_{xx}(16,16).f_{yy}(16,16) - \left[f_{xy}(16,16)\right]^2$$

$$= -32 \times -32 - 256 > 0\, and\, f_{xx}(16,16) < 0.$$

Therefore, at (16,16), $f(x, y)$ is maximum.

Since $x + y + z = 48,$ we find $z = 16$. Thus, the required three positive numbers are 16,16,16.

Example. A topless rectangular box has a volume of 256 cu.cm. Find the dimensions of the box so as to use the least amount of material.

Solution. Let the base of the box be x cm by y cm and its height be z cm. Then xyz = 256 cu.cm.

The surface area S of the box is given by

$$S = xy + 2xz + 2yz = xy + \frac{512}{y} + \frac{512}{x}$$

We need to find the minimum values of S.

$$\frac{\partial S}{\partial x} = y - \frac{512}{x^2}; \frac{\partial S}{\partial x} = x - \frac{512}{y^2}$$

$$\frac{\partial^2 S}{\partial x^2} = \frac{1024}{x^3}; \frac{\partial^2 S}{\partial y^2} = x - \frac{1024}{y^3}$$

$$\frac{\partial^2 S}{\partial x \partial y} = \frac{\partial^2 S}{\partial y \, \partial x} = 1$$

For critical points, we set

$$y - \frac{512}{x^2} = 0, x - \frac{512}{y^2} - 0.$$

From these equations, we have

$$xy^2 = x^2 y = 512 \, and \, so \, x = y \, and \, x^3 = y^3 = 512$$

Therefore, $x = y = 8$.

$$\left.\frac{\partial^2 S}{\partial x^2}\right|_{x=8} = 2, \left|\frac{\partial^2 S}{\partial y^2}\right|_{x=8} = 2, \frac{\partial^2 S}{\partial x \partial y} = 1.$$

Now $$\frac{\partial^2 S}{\partial x^2} \cdot \frac{\partial^2 S}{\partial y^2} - \left(\frac{\partial^2 S}{\partial x \partial y}\right)^2 = 4 - 1 > 0 \, and \left.\frac{\partial^2 S}{\partial x^2}\right|_{x=8} = 2 > 0.$$

Thus S has a minimum value at $x = y = 8$. we have $z = \frac{256}{xy} = \frac{256}{64} = 4.$
The required dimensions of the box are 8 cm, 8 cm, 4 cm.

Example. Find the point of the plane $x - y + 2z = 6$ nearest to the origin.

Solution. The distance of any point (x, y, z) from the origin is $\sqrt{x^2 + y^2 + z^2}$. if (x, y, z) is a point on the plane, the distance d from the origin to this point of the plane is

$$d = \sqrt{x^2 + y^2 + \left(\frac{6 - x + y}{2}\right)^2}$$

Since the minimum of d is also minimum of d^2, we let $f(x, y) = x^2 + y^2 + \left(\frac{6 - x + y}{2}\right)^2$ and find x, y that minimize f.

$$f_x(x, y) = 2x + \frac{1}{2}(6 - x + y)(-1) = \frac{5}{2}x - \frac{y}{2} - 3$$

$$f_{xx}(x,y)=\frac{5}{2}, f_{xy}=-\frac{1}{2}$$

$$f_{y}(x,y)=2y=\frac{1}{2}(6-x+y)=-\frac{x}{2}+\frac{5y}{2}+3$$

$$f_{yy}(x,y)=\frac{5}{2}$$

For critical points set the first partials equal to zero. Then

$5x-y-6=0$ (1)

and $-x+5y+6=0$ (2)

Multiply (1) by 5 and add the result to (2) so that

$$24x-24=0 \; or \; x=1$$

Substituting this value of x into (1), we find $y=-1$. A critical point is $(1,-1)$

Now $f_{xx}f_{yy}-(f_{xy})^2=\frac{25}{4}-\frac{1}{4}>0 \; and \; f_{xx}(1,-1)=\frac{5}{2}>0. So (1,-1)$ is a point of relative minimum. Using the equation of the plane, we have

$$1+1+2z=6 \; or \; z=2.$$

Thus $(1,-1,2)$ is the required point and the minimum distance is

$$\sqrt{1+1+4}=\sqrt{6}.$$

2.9. EXTREMA WITH CONSTRAINTS

If often is required to investigate the extrema of a function of three (or more) variables subject to some given side conditions. Problems of this type have already been presented in There we eliminated one of the variables and then found the extrema by the method of. Such problems are known as extrema with constraints (or side conditions). If often is required to investigate the extrema of a function of three (or more) variables subject to some given side conditions. Problems of this type have already been presented in There we eliminated one of the variables and then found the extrema by the method of. Such problems are known as extrema with constraints(or side conditions).

The procedure adopted in the above-quoted example is often tedious and sometimes impossible. In such cases, another method can be used with convenience. Outline of the method are stated in what follows:

The Method of Lagrange's Multipliers. In order to find the extrema of the function $w = f(x, y, z)$ subject to the constraints $g(x, y, z) = 0,$ we introduced a new variables $\lambda -$ called Lagrange's Multiplier- and form the function

$$F(x, y, z, \lambda) = f(x, y, z) + \lambda g(x, y, z)$$

The function $F(x, y, z\lambda)$ is called Lagrange Function. It has been shown in advanced books on calculus that extreme of $F \, and \, f$ correspond. Thus, for critical points, we have

$$F_x = \frac{\partial f}{\partial x} + \lambda \frac{\partial g}{\partial x} = 0$$

$$F_y = \frac{\partial f}{\partial x} + \lambda \frac{\partial g}{\partial y} = 0$$

$$F_z = \frac{\partial f}{\partial z} + \lambda \frac{\partial g}{\partial z} = 0$$

$$f_\lambda = g(x, y, z) = 0.$$

The solution of these four equations give the critical points of $f \cdot$ The extrema of f may be found by listing the values of f at each critical point and considering the largest and the smallest values.

In vectors notation, the above four equations may be written as

Grad $f + \lambda \, grad \, g = 0, g9(x, y, z) = 0$

The Lagrange multiplier technique will be illustrated by examples.

Example. Find the point of

$$-x^2 - z^2 + y = 2 \qquad (1)$$

Nearest to the origin.

Solution. The distance from a point (x, y, z) lying on (1) to the origin is

$$d = \sqrt{x^2 + y^2 + z^2}.$$

Since d is not negative, it is minimum whenever d^2 is minimum.

Thus we have to minimize $d^2 = x^2 + y^2 + z^2$ subject to the constant (1)

Using Lagrange multiplier, we get

$$F(x, y, z, \lambda) = x^2 + y^2 + z^2 + \lambda(2 + x^2 + z^2 - y)$$

$F_x = 2x + 2\lambda x = 0$ giving $x = 0$

$$F_y = 2y - \lambda = 0$$

$$F_z = 2z + 2\lambda z = 0$$

$F_\lambda = 2 + x^2 + z^2 - y = 0$ (1)

Setting $x = 0 = z\, in(1), we\ find\ y = 2.$

Thus the required point is $(0, 2, 0)$.

Example A topless box having a volume of 12 cubic meters is to be made of material costing Rs. 100 per square meter. Find the dimension of the box that minimize the cost.

Solution. Let the dimension of the box be x, y, z in meters. Volume box is

$xyz = 12$ (1)

The surface of the box is

$$xy + 2xz + 2yz\ sq.m.$$

Total cost C of the material is $C = 100(xy + 2xz + 2yz)$ Rupees. C is to be minimized with the constant (1). Using language multipliers we have

$$F(x, y, \lambda) = 100(xy + 2xz + 2yz) + \lambda(xyz - 12)$$

$F_x = 100(y + 2z) + \lambda yz = 0$ (2)

$F_y = 100(x + 2z) + \lambda xz = z$ (3)

$F_z = 100(2x + 2y) + \lambda xy = 0$ (4)

$$F_{\lambda} = xyz - 12 = 0$$

Form (2) and (3), we get

$\frac{x+2z}{yz} = \frac{x+2z}{xz}$, i.e., $x = y$

From (3) and (4), we have

$$\frac{x+2z}{xz} = \frac{x+2z}{xz} \qquad (5)$$

Subtracting $y = x$ into (5), we obtain

$$\frac{x+2z}{xz} = 4 \text{ or x = 2x}$$

Setting $x = y - 2z\,in(1)$, we get

$$x^2\left(\frac{x}{2}\right) = 12\,or\,x^3 = 24\,i.e, x = 2(3)^{1/3} = y$$

and $z = \frac{x}{z} = 3^{ü}$

Required dimension of the box are $2(3)^{1/3}, 2(3)^{1/3}\,and\,3^{1/3}$ in meters.

The method of Lagrange multipliers can be used when more than one constraints are given. The following example illustrates the case when two constraints are imposed.

Example. Find the extrema of

$f(x, y, z) = x^2 + y^2 + z^2$ subject to $z^2 = x^2 + y^2$ $\qquad (A)$

and $\qquad x + y - z + 1 = 0 \qquad (B)$

This is an approximation of the volume of the solid of revolution. The required volume is obtained by taking limit of the above expression (which is a Riemann sum) as the norm of the partition $P \to 0\,and\,n \to \infty$. Since f is continuous on $[a,b]$, the limit is a define integral. Hence, the required volume

$$V = \lim_{P \to 0_{n \to \infty}} \sum_{r=1}^{n} \pi \left[f(\omega_r) \right]^2 \Delta x_r$$

$$= \int_a^b \pi \left[f(x) \right]^2 dx.$$

Example. The area in the first quadrant by the parabola x $y^2 = 4ax$ and its latus-rectum is revolved about the $x-axis$. Find the volume if the solid generated.

Solution.

$$V = \int_0^b \pi\, y^2\, dx$$

$$= 4a\pi \int_0^a x\, dx$$

$$= 2\pi a^3.$$

Example. Find the volume of the solid obtained by revolving the area enclosed by one arch of the cycloid $x = a(\theta + \sin\theta), y = a(1 + \cos\theta)$ about the $x-axis$.

Solution.

$$V = 2\int_0^\pi \pi y^2\, dx = 2\int_0^\pi \pi a^2 (1+\cos\theta)^2\, a(1+\cos\theta)\, d\theta$$

$$= 2\pi a^3 \int_0^\pi 8\cos^6\left(\frac{\theta}{2}\right) d\theta$$

Let $\frac{\theta}{2} = \phi.$ Then

$$V = 16\pi a^3 \int_0^{\pi/3} 2\cos^6 \phi\, d\theta$$

$$= 32\pi a^3 . \frac{5.3.1}{6.4.2} \frac{\pi}{2} = 5\pi^2 a^3.$$ by Wallis formula.

2.9.1. Washer Method

Consider the plane area enclosed by the graph of $x = a, x = b$ and of two continuous function f *and* g where $f(x) \geq 0$ for all $x \epsilon [a,b]$

Let $p = \{a = x_0, x_1, \ldots, x_n = b\}[a,b]$. Let ω_r be a point of $[x_{n-1}, x_r]$. The area obtained by cutting the solid at $x = \omega_r$ is the difference of the area of the outer circle and the area of the inner circles, namely

$$\pi[f(\omega_r)]^2 - \pi[g(\omega_r]^2.$$

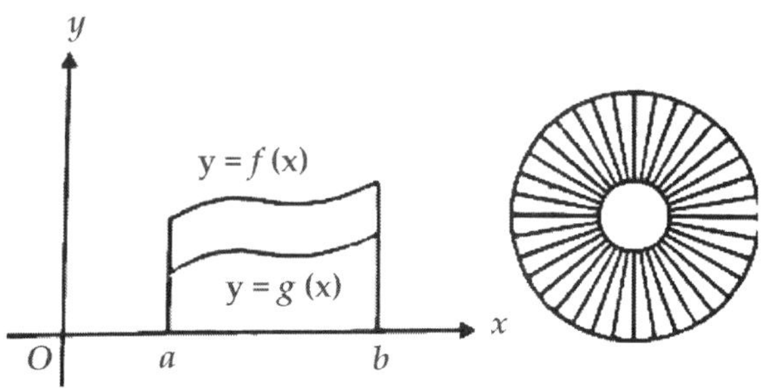

The solid of revolution obtained from the rth subinterval is called a washer and its volume is $\left\{\pi[f(\omega_r)]^2 - \pi[g(x_r)]^2\right\}\Delta x_r$ Adding up the volumes of all washers, we obtain a Riemann sum

$$\sum_{r=1}^{\pi}\left\{\pi[f(\omega_r)]^2 - \pi[g(\omega_r)]^2\right\}\Delta x_r$$

If we take the limit of this sum as the norm of the partition $p \to 0$ and $n \to \infty$, we get the volume V of the solid of revolution as

$$V = \int_a^b \left(x[f(x)]^2 - \pi[g(x)]^2\right)dx.$$

If $g(x) > 0$ for all x in $[a,b]$, there is a hole through the solid.

Example. Find the volume of the solid generated by revolving the area bounded by the parabola $y^2 = 8x$ and its latus-rectum $(x = z)$ about the $y - axis$.

Solution. Here we divide the area by horizontal strips. When an approximating rectangular is revolved about the $y-axis,$ it generated a washer whose volume = volume generated by revolving $ECDF-$ volume generated by rectangle $EPBF\ \pi(2)^2\,\Delta y=\pi\left(4-x^2\right)\Delta y$

Thus the required volume

$$=\pi\int_{-4}^{4}\left(4-x^2\right)$$

$$=2\pi\int_{0}^{4}\left(4-x^2\right)dy$$

$$=2\pi\int_{0}^{4}\left(4-\frac{y^4}{64}\right)dy=\frac{128}{5}\pi$$

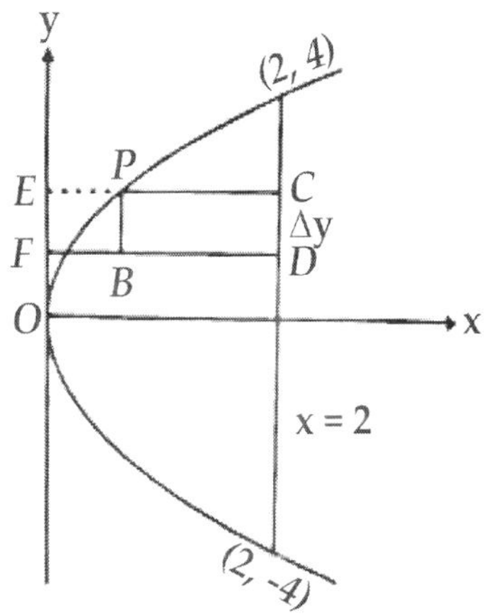

Example. The area bounded by $x^2=y-4,3y-x-3=0,x-0\,and\,x=2$ is revolved about the $x-axis.$ Find the volume of the resulting solid of revolution.

Solution. Volume of a typical washer is

$$=\left[\pi\left(\omega_{r^2}+4\right)^2-\pi\left(\frac{1}{3}\omega_r+1\right)^2\right]\Delta x_r$$

Required volume

$$\int_0^2 \pi\left[\left(x^2+4\right)^2-\pi\left(\frac{1}{3}x+1\right)^2\right]dx$$

$$=\pi\int_0^2 \pi\left(x^2+8x^2+16-\frac{x^2}{9}-\frac{2}{3}x-1\right)dx$$

$$=\pi\left[\frac{x^2}{5}+\frac{71}{27}x^3-\frac{x^2}{3}+15x\right]_0^2$$

$$=\pi\left[\frac{32}{5}+\frac{568}{27}-\frac{4}{3}+30\right]=\frac{7718}{135}\pi.$$

2.9.2. Shell Method

In this case, the volume of a solid of revolution will be found by using cylindrical shell. A cylindrical shell is the solid region between two concentric cylinders if the same height. If the same height. If the inner radius is r_1 and the outer radius is r_2; then the volume of a cylindrical shell of altitude h is

$$\pi r_2^2 h-\pi r_1^{\,2} h$$

$$\pi h\left(r_2+r_1\right)\left(r_2-r_1\right)=2\pi h\left(\frac{r2-r_1}{2}\right)\left(r_2-r_1\right)$$

$=2\pi$ altitude $\times$ average radius $\times$ thickness.

This formula will be used in evaluating the volume of a solid of` revolution.

Let f be continuous and $(x)\geq 0$ for all $x\in[a,b]$. Let the area A bounded by $y=f(x), x=a, x=b$ and the $x-$ axis be revolved about the $y-axis$. To find the volume of the solid so generated we proceed as usual by partitioning $[a,b]$ into n subintervals $\left[x_{r-1,}x_r\right], r=1,2,\ldots,n$. The length of the rth subinterval is $x_r-x_{r-1}=\Delta xr$. Let $\omega_r=\dfrac{x_{r-1}+x_r}{2}$ be the midpoint of the rth subinterval. The rectangle with base $\left[x_r-x_{r-1}\right]$ and height $f(\omega_r)$ is revolved about the $y-$axis. The solid generated by this rectangle is a cylindrical shell with inner radius $x_{r-1,}$ $outer\ radius\ x_r$ and altitude $f(\omega_r)$. Volume of this shell

$$2\pi\, f(\omega_r)\frac{x_{r-1}+x_r}{2},(x_r - x_{r-1})$$

$$2\pi f(\omega_r)\,\omega_r\,\Delta x_r$$

An approximation of the volume of the solid generated by $A =$ sum of volumes of all the shell due to the partition of $[a,b]$

$$=\sum_{r=1}^{n} 2\pi\, f(\omega_r)\,\omega_r\,\Delta x_r.$$

This is a Riemann sum. Proceeding to limit as norm of the partition $\to 0\, and\, n \to \infty$, the volume V of the solid is

$$V = \int_a^b 2\pi\, f(x)\,dx.$$

Example. Find the volume of the solid of revolution generated by revolving the area enclosed by $y - x^2 - 1$ form $x = 1\, to\, x = 3$ about the $y-axis.$

Solution. By shell Method

Required volume

$$= \int_1^3 2\pi x\, f(x)\,dx$$

$$= 2\pi \int_1^3 x\left(x^2 - 1\right)dx$$

$$= 2\pi\left[\frac{x^4}{4} - \frac{x^2}{2}\right]_1^3 = 32\pi.$$

By Washer Method. The revolution is about the $y-axis.$ Here $f(y) = 3g(y)_{=}\ \sqrt{1+y}$ and $f(y) \geq g(y) in$ $_0\ \leq y \leq 8.$ The integration is with respect to y from $y = 0\, to\, y = 8.$ The required volume is

$$V = \pi \int_0^8 \left[\left(\sqrt{2+y} \right)^2 \right] dx$$

$$\pi \left[8y - \frac{y^2}{2} \right]_0^8 = \pi[64 - 32] = 32\pi.$$

2.9.3. Cross-Sectional Method

In the disc method the volume V is given by

$$V = \int_a^b \pi y^2 \, dx$$

The integrand $\pi y^2 = \pi \{f(x)\}^2$ can be interpreted as the area of the cross section of the solid made by a plane perpendicular to the $x - axis$ and at a distance a units from the origin. Thus, we have the following:

If for each $x in[a,b] the area A(x)$ of a cross-section of a solid is known and is continuous on $[a,b]$, then the volume of V of the solid is given by

$$V = \int_a^b A(x) dx.$$

Example. Find the volume of the right cone of height h whose base is an ellipse of major axis $2a$ and minor axis $2b$.

Solution. A section of the cone by a plane parallel to the base is an ellipse whose major axis is 2x and minor axis 2y and height of the smaller cone is $MP = z and AOM$, we have

$$\frac{PC}{OA} = \frac{PM}{OM}$$

or $\frac{x}{a} = \frac{h-z}{h}$

Also from similar triangles $PDM and OBM$, we have

$$\frac{PD}{OB} = \frac{PM}{OM}$$

or $\frac{y}{b} = \frac{h-z}{h}$

The area of the section is $\pi xy = \frac{\pi ab(h-z)^2}{h^2}$

Hence volume of the cone = $\frac{\pi ab}{h^2}\int_0^h (h-z)^2\,dz = \frac{1}{3}\pi\,a\,bh.$

If the base of the cone is a circle, then $a = b$ so that its volume is $\frac{1}{3}\pi a^2\,h.$

Theorem. (First theorem of Pappas). If a plane area is revolved about n axis in its plane and not crossing the area, the volume of the solid generated is equal to the product of the area and the length of the path described by the centroid of the area.

Proof. Let A be the area to be revolved. Let the axis of rotation be taken as $x-axis$ and a straight line perpendicular to it as $y-axis.$ Divide the area into a large numbers of small parts and let ΔA be an elements of area $P.$ if the area rotated through an angle of measure $\theta,$ the length of the arc described by p is $y\theta.$ Thus the volume generated is $y\theta\,\Delta A.$ Hence the volume generated by the whole area is Let the axis of rotation be taken as $x-axis$ and a straight line perpendicular to it as $y-axis.$ Divide the area into a large numbers of small parts and let ΔA be an elements of area $P.$ if the area rotated through an angle of measure $\theta,$ the length of the arc described by p is $y\theta.$ Thus the volume generated is $y\theta\,\Delta A.$ Hence the volume generated by the whole area is

$$= \lim_{\Delta A\to 0} \Sigma\, y\,\theta\,\Delta A = \theta \int y\,dA$$

If $(\overline{x}, \overline{y})$ are coordinates of the centroid, then

$$\overline{y} = \frac{\int y\,dA}{\int dA} = \frac{\int y\,dA}{A}$$

Thus, Volume = $\theta \int y\,dA = A(\bar{y}\theta)$

= area $\times$ length of the path described by the centroid of the area.

Example. Find the volume of the torus generated by revolving the disc $x^2 + y^2 = 4$ about the line $x = 3$.

Solution. The centroid $(\bar{x}, \bar{y})$ of the area enclosed by the disc is the point (0,0) it described a circle of radius 3. Hence, volume V of the torus is

$$V = (\text{area of the disc}) \times (\text{circumference of the circle of radius } 3)$$

$$= \pi(2)^2(6\pi) = 24\pi^2.$$

2.10. AREA OF SURFACE OF REVOLUTION

Definition. If an arc a portion of a plane curve, is rotated about a straight line, the arc generated a surface called a surface a surface of revolution.

In this section we propose to find areas of such surfaces. If a surface is generated by a straight line segment, its area just the surface of a frustum of a cone as shown in figure. In this section we propose to find areas of such surfaces. If a surface is generated by a straight line segment, its area just the surface of a frustum of a cone as shown in figure.

If and are the end points of the line segment that generates the surface when rotated about the x-axis, then its area is the surface of the frustum of a cone with base radii We first derive a formula for the lateral area of the frustum of a cone.

Lemma. We know that the lateral surface area of a right circular come with base radius and slant height

Now area of the frustum of the cone = area of the entire cone with radius – area of the cone with radius r_1. Now area of the frustum of the cone = area of the entire cone with radius $r_2 - r_2 -$ – area of the cone with radius r_1,

Let $|AP| = l_1 \text{ and } |PQ| = l_2$

Then $l = l_1 + l_2$, where l is a slant height of the cone

Area A of the frustum is

$$A = \pi r_2 l - \pi r_1 l_1$$

$$= \pi r_2 (l_1 + l_2) - \pi r_1 l_1$$

$$= \pi (r_2 - r_1) l_1 + \pi r_2 l_2$$

But, we have from the similar triangles AMP and AOQ,

$$\frac{r_1}{l_1} = \frac{r_2}{l} = \frac{r_2}{l_1 + l_2}$$

or $r_1 l_1 + r_1 l_2 = r_2 l_{1s}$

or $(r_2 - r_1) l_1 = r_1 l_2$

Therefore,

$$A = \pi r_1 l_2 + \pi r_2 l_2$$

$2\pi l_2 \left(\frac{r_1 + r_2}{2} \right) = 2\pi \times$ slant height of the frustum $\times$ average radius.

2.10.1. Area of a Surface of Revolution

Let $y = f(x), a \le x \le b$ be a continuous curve such that $f'(x)$ exist and is continuous. Further suppose that $f(x) \ge 0 \, on [a,b].$ Let the arc s of the curve between $a \le x \le b$ be rotated about the $x - axis.A$ formula for the area of the surface thus generated will bc derived.

Let $P = \{ a = x_0, x_1, x_2, \ldots, x_{r-1}, x_r, \ldots x_n = b \}$ bc a partition of $[a,b].$ we first findtheareasofthesurfacesgeneratedbythelinesegments $|P_{r-1}P_r| r = 1,2,\ldots,n.$

Length p_f the segment $P_{r-1}P_r$ is $l_r = \sqrt{(x_r - x_{r-1})^2 [f(x_r) - f(x_{r-1})]^2}$ By the Means value Theorem, we have

$$\frac{f(x_r) - f(x_{r-1})}{x_r - x_{r-1}} = f'(\omega_r), \omega_r \in [x_{r-1,} x_r]$$

Therefore

$$l_r = \sqrt{1+\left[f'(\omega_r)\right]^2}\,\Delta x_r \qquad (1)$$

The conical frustum obtained by revolving the segment $|P_{r-1}P_r|$ about the axis is shown in figure its slant height l_r is given by (1). Average radius R_r of the frustum is

$$R_r = \frac{f(x_{r-1}) + f(x_r)}{2}$$

Now, Δx_r is small, x_{r-1} *and* x_r are close to each other and ω_r lies between them. Due to the continuity of both f *and* f' *on* $[x_{r-1}, x_r]$, the value of R_r is approximately equal to $\dfrac{f(\omega_{r-1}) + f(\omega_r)}{2}$ $f(\omega_r)$

Hence the area of the conical frustum generated by $|P_{r-1}P_r|$ is

$$= 2\pi f(\omega_r)\sqrt{1+\left[f'(\omega_r)\right]^2}.\Delta x_r$$

Adding the areas of the such frustum, for $r = 1,2,3,\ldots,n,$ we have $\sum_{r=1}^{n} 2\pi f(\omega_r)\sqrt{1+\left[f'(\omega_r)\right]^2}.\Delta x_r =$ Area of the n frustums

This is a Riemann sum for the function $f\sqrt{1+f^2}$ *on* $[a,b]$. Proceeding to limit the norm of the partition $P \to 0,$ the desired area A of the surface of revolution

$$A = \int_a^b 2\pi f(x)\sqrt{1+\left[f'(x)\right]^2}\,dx$$

$$= 2\pi \int_a^b y\sqrt{1+\left(\frac{dy}{dx}\right)^2} = dx = 2\pi \int_a^b y\,ds.$$

If the rotation is about the $y-axis$, the formula for the area is

$$A = 2\pi \int_a^b x \, ds.$$

If the curve is defined by the parametric equations

$$x = f(t) = g(t), a \le t \le b$$

and it is rotated bout the x-axis, then the formula for area of the surface of revolution becomes

$$A = 2\pi \int_a^b y \, ds.$$

$$= 2\pi \int_a^b g(t) \sqrt{\left(\frac{dx}{dt}\right)^2 + \left(\frac{dy}{dt}\right)^2} \, dt, since \, ds^2 = dx^2 + dy^2$$

$$= 2\pi \int_a^b g(t) \sqrt{\left[f'(t)\right]^2 + \left[g'(t)\right]^2} \, dt.$$

Example. Find the area of the surface of revolution generated by revolving about the $x-axis$ the hypocycloid $x = a \cos^3 \theta, y = \sin^3 \theta.$

Solution. The required surface is generated by revolving the arc from $\theta = 0 \, to \, \theta = \pi.$

Here $$\frac{dx}{d\theta} = -3a\cos^2\theta\sin\theta, \frac{dy}{d\theta} = 3a\sin^2\theta\cos\theta$$

and $$\left(\frac{dx}{d\theta}\right)^2 + \left(\frac{dy}{d\theta}\right)^2 = 9a^2\cos^2\theta\sin^2\theta.$$

Required area $$A = 2.2\pi \int_0^{\pi/2} y \sqrt{\left(\frac{dx}{dt}\right)^2 + \left(\frac{dy}{dt}\right)^2} \, d\theta$$

$$4\pi \int_0^{\pi/2} (a\sin^3\theta)(3a\cos\theta \, sin\theta) d\theta = \frac{12a^2\pi}{5}.$$

Theorem. (Second Theorem of Pappas). If the area bounded by an arc of a curve is revolved about an axis in its plane but not crossing the arc, the area of the surface generated is equal to the product of the length of the arc and the length described by the centroid of the arc. If the area bounded by an arc of a curve is revolved about an axis in its plane but not crossing the arc, the area of the surface generated is equal to the product of the length of the arc and the length described by the centroid of the arc.

Proof. Let s be the length of the arc. Divide it into a large numbers of small parts, Consider an element of the arc of length $\Delta s.$ when the arc is rotated through an angle of measure $\theta,$ the length of the arc described by a point P of arc AB is $y\theta.\Delta s.$ Therefor, the area A of the surface generated

$$\lim_{\Delta s \to 0} \Sigma\, y\theta\,\Delta s = \theta \lim_{\Delta s \to 0} \Sigma\, y\Delta s = \theta \int y\, ds$$

Now if $(\bar{x}, \bar{y})$ are the coordinates of centroid of the arc, then

$$\bar{y} = \frac{\int y\, ds}{\int ds} = \frac{\int y\, ds}{s}$$

Thus $$A = \theta \int y\, ds = \theta(\bar{y}.s)$$

$= s(\bar{y}\theta) =$ length of the arc $\times$ Length of path described by the centroid of the curve

Example. Find the area of the surface generated by revolving the rectangular area of dimension *a* by *b* about an axis *c* units from the centroid describes a circle of radius $c.$ Therefore, required area

$$= 2(a+b)2\pi = 4\pi(a+b)c.$$

2.11. VOLUME AND AREA IN POLAR COORDINATES

Volume. Suppose a solid is generated by revolving about the initial line the area bounded by the curve $r = f(\theta)$ and the radii $\theta = a \, and \, \beta.$ Suppose $f(\theta) and\, f'(\theta)$ are continuous for $a \le \theta \le \beta \, and \, f(\theta)$ does not change sign in the interval $a \le \theta \le \beta.$ Let the arc AB represent the curve $r = f(\theta) and\, OA, OB$ the radius vectors $\theta = a, \theta = \beta.$ We divide the area into a large number of sectors. Consider a representative sector OPQ. We can consider thus sector to be a triangle with base of length $r\Delta\theta$ and altitude of length r. Area of the sector is then $\frac{1}{2}r^2\,\Delta\theta.$ Let G be the centroid of the sector $OPQ.$ Then $OG=\frac{2}{3}r$ and $G:L=\frac{2r}{3}\sin\theta.$ Thus volume of the solid generated by the sector OPQ

$$=\frac{1}{2}r^2\,\Delta\theta\times 2\pi\left(\frac{2}{3}r\sin\theta\right),$$

$$=\frac{2\pi}{3}r^3\sin\theta\,\Delta\theta.$$

Hence the volume V of the solid generated by revolving the given plane area is

$$V=\int_a^\beta \frac{2\pi}{3}r^3\sin\theta d\theta.$$

The volume of the solid generated by revolving the given area about the straight line through O and perpendicular to the initial line Ox is obtained by changing $\theta\, to\, \frac{\pi}{2}-\theta$ and is $\int_a^\beta \frac{2\pi}{3}r^3\cos\theta\, d\theta.$

Similarly, the volume of the solid generated by revolving about the straight line $\theta=\phi$, the area bounded by the curve $r=f(\theta)$ and the radius vectors

$$\theta=\alpha,\theta=\beta\, is\int_a^\beta \frac{2\pi}{3}r^3\sin(\theta=\phi)\,d\theta.$$

Area. Suppose that a curve defined by $r=f(\theta),a<\theta<\beta$ is revolved about the polar axis. The parametric equation of the curve are

$$x=f(\theta)\cos\theta.y=f(\theta)\sin\theta.a<\theta<\beta$$

$$\frac{dx}{d\theta}=f'(\theta)\cos\theta-f(\theta)\sin\theta$$

$$dy\setminus d\theta=f'(\theta)\sin\theta+f(\theta)\cos\theta$$

$$\left(\frac{ds}{d\theta}\right)^2 = \left(\frac{dx}{d\theta}\right)^2 + \left(\frac{dy}{d\theta}\right)^2 = \left[f(\theta)\right]^2 + \left[f'(\theta)\right]^2.$$

Area A of the surface of revolution is given by

$$A = 2\pi \int_{\alpha}^{\beta} y \frac{ds}{d\theta} d\theta$$

$$= 2\pi \int_{\alpha}^{\beta} f(\theta) sin\theta \sqrt{\left[f(\theta)\right]^2 + \left[f'(\theta)\right]^2 d\theta}.$$

If the rotation is about the line $\theta = \frac{\pi}{2}$, then the formula for area is

$$A = 2\pi \int_{\alpha}^{\beta} f(\theta) \cos\theta \sqrt{\left[f(\theta)\right]^2 + \left[f'(\theta)\right]^2} d\theta$$

Summary and Discussion

In this chapter we discussed about the function of several variables. Another very important topic in Calculus and applied mathematics. Application of this topic is in different fields as well. The basic concept of this chapter is used in different fields of mathematics as well.

2.2. Homogeneous Functions

Definition. A function g defined by $u = f(x, y, z, \ldots)$ of any number of variables is said to be homogenous of degree n in these variables if multiplication of these variables by any number $t(\neq 0)$ result in the multiplication of the function by t^n, i.e.,

$$f(tx, ty, tz, \ldots.) = t^n f(x, y, z, \ldots.)$$

Theorem ('Euler's). If $u = f(x, y)$ is a homogenous function of x, y of degree n, then $x\frac{\partial u}{\partial x} + y\frac{\partial u}{\partial y} = nu.$

2.3. Differentials

Suppose $u = f(x,y)$ processes first partial derivatives $f_x(x,y)$ and $f_y(x,y)$ in a region D, and that $f_x and f_y$ are continuous at some point (x_0, y_0) of D. Furthermore, take a square neighborhood of (x_0, y_0) that lies in D and let $\Delta x and \Delta y$ be numerically less than half the length of a side of this neighborhood, so that the point $x_0 + \Delta x, y_0 + \Delta y)$ lies in D. Then the increment Δu of the dependent variable can be written as follows:

$$\Delta u = f(x_0 + \Delta x, y_0 + \Delta y) - f(x_0, y_0)$$

$$= f(x_0 + \Delta x, y_0 + \Delta y) - f(x_0, y_0 + \Delta y) + f(x_0, y_0 + \theta_2 \Delta y)\Delta y$$

where $0 < \theta_1 < 1, 0 < \theta_2 < 1 and \theta_1 and \theta_2$ depend upon $\Delta x and \Delta y$, respectively.

Theorem. If $u = f(x,y)$ processes first order partial derivatives in a region D, and if these partial derivatives are continuous at some point (x_0, y_0) of D, then (for increments $\Delta x and \Delta y$ sufficiently small numerically,) the increment Δu of the dependent variable can be written as

$$\Delta u = \frac{\partial u}{\partial x}\Delta x + \frac{\partial u}{\partial y}\Delta y \varepsilon_1 \Delta x + \varepsilon_2 \Delta y \tag{1}$$

where the partial derivatives are evaluated at the point (x_0, y_0) and each of $\varepsilon_1, \varepsilon_2$ tends to zero when $(\Delta x, \Delta y) \to (0,0)$.

Definition. The differential (or total differential) du of a differentiable function $u = f(x,y)$ of two variables is defined as

$$du = \frac{\partial u}{\partial x}dx + \frac{\partial u}{\partial y}dy$$

where $dx = \Delta x, dy = \Delta y$ are differentiable of x and y.

2.4. Change Of Variable–The Chain Rule

Let $u = f(x, y)$ be a differential function. Suppose each of x and y is a differentiable function of a single variable y. Suppose t changes to $t + \Delta t(\Delta t \neq 0)$. Then using (1) we can write

$$\frac{\Delta u}{\Delta t} = \frac{\Delta u}{\Delta x}\frac{\Delta x}{\Delta t} + \frac{\Delta u}{\Delta y}\frac{\Delta y}{\Delta t} + \varepsilon_1 \frac{\Delta x}{\Delta t} + \varepsilon_2 \frac{\Delta y}{\Delta t}\sqrt{b^2 - 4ac}$$

Considering u as a function of the single variable t, we let $\Delta t \to 0$ and take limits of both sides of the above equation. Since $x and y$ are differentiable functions of t, and $\varepsilon_1 and \varepsilon_2$ tend to zero), we conclude that

$$\frac{du}{dt} = \frac{\partial u}{\partial x}\frac{dx}{dt} + \frac{\partial u}{\partial y}\frac{dy}{dt}$$

If each of xandy is a differentiable function of variable r and s, we would similarly get,

$$\frac{\partial u}{\partial r} = \frac{\partial u}{\partial x}\frac{\partial x}{\partial r} + \frac{\partial u}{\partial y}\frac{\partial y}{\partial r}$$

$$\frac{\partial u}{\partial s} = \frac{\partial u}{\partial x}\frac{\partial x}{\partial s} + \frac{\partial u}{\partial y}\frac{\partial y}{\partial r}$$

These results can be generalized for a function of several variables. Thus, if $u = f(x, y, z \ldots)$ is a differentiable function and each of $x, y, z, \ldots$ is a differentiable function of several variables $r, s, t, \ldots$ then

$$\frac{\partial u}{\partial r} = \frac{\partial u}{\partial x}\frac{\partial x}{\partial r} + \frac{\partial u}{\partial y}\frac{\partial y}{\partial r} + \frac{\partial u}{\partial z}\frac{\partial z}{\partial r} + \ldots$$

$$\frac{\partial u}{\partial s} = \frac{\partial u}{\partial x}\frac{\partial x}{\partial s} + \frac{\partial u}{\partial y}\frac{\partial y}{\partial s} + \frac{\partial u}{\partial z}\frac{\partial z}{\partial s} + \ldots$$

$$\frac{\partial u}{\partial t} = \frac{\partial u}{\partial t}\frac{\partial x}{\partial t} + \frac{\partial u}{\partial y}\frac{\partial y}{\partial t} + \frac{\partial u}{\partial z}\frac{\partial z}{\partial t} + \ldots$$

These equations are known as the Chain Rule.

2.7. Tangent Planes and Normal Lines

Consider a surface S defined by $f(x,y,z)=0$ Let p be a point of $S.$ Consider a curve through p and lying on $S.$ Then for all points on the curve we have $f(x,y,z)=0$ since each point of the curve is a point of $S.$ Thus, at p the various curves of S give the directions for which the rate of change of f is zero. Perpendicular to all these directions is the direction of grad $f.$ Thus we define normal line to a surface as follows:

Definition. Consider all curves through a point p on a surface $S.$ The straight line perpendicular to all these curves is called the normal line to surface at p and if $\frac{\partial f}{\partial x},\frac{\partial f}{\partial y},\frac{\partial f}{\partial z}$ are all continuous at p and not all of them are zero, then grad f at p is the normal vector to the surface $p.$

Definition. The tangent plane to the surface S at the point p is the planes whose normal vector is grad $fatp.$

2.7.1. Parametric Equation of a Surface

The sphere $x^2+y^2+z^2=a^2$ can be described by the parametric equations. $x=a\sin\varnothing\cos\theta, y=a\sin\varnothing\cos\theta, z=a\cos\varnothing$, where $-\pi<\theta<\pi, 0\le\varnothing\le\pi.$ Here $\theta and\varnothing$ are parameters.

Definition. Let f be a real –valued function of two variables defined and continuous on the closed disc $D=\left\{(x,y)\in R^2 : x^2+y^2\le r^2\right\} ofR^2.f$ is said to have an absolute (global) maximum value D of f on D.

The function f is said to have a relative (local) maximum at some point $(x_0,y_0) ofD$ if there exist an open disc k centered at (x_0,y_0) and of radius a i.e.

$$K=\left\{(x,y)\in R^2 : (x-x_0)^2+(y-y_0)^2<a^2\right\}$$

with $K\subset D$ such that $f(x_0,y_0)\ge f(x,y) froall(x,y)\in K$ i.e., f has an absolute maximum on $K.$

Definition. A function f of the variables as defined in is said to have an absolute (global) minimum value on D if there is some point (x_0, y_0) of D such that $f(x_0, y_0) \leq f(x, y)$ for all $(x, y) \in D.$ in such a case $f(x_0, y_0)$ is the absolute minimum value of $fonD.$

The function f is said to have a relative (local) minimum at some point $(x_0, y_0) ofD$ if there exists an open disc K centered at (x_0, y_0) and of radius r with $K \subset D$ such that $f(x_0, y_0) \leq f(x, y) forall (x, y) \in K.$ i.e., f has an absolute minimum value on $K.$

The maximum and minimum values of are referred to as the extreme values or extrema of f.

Definition. A point (x_0, y_0) in the domain of a function f of two variables is called a critical point of f if either $f_x(x_0, y_0) = f_y(x_0, y_0) = 0$ or one (or both) of the first order partial derivatives of f at (x_0, y_0) does not exist.

Theorem. Let f be a function of two variables x, y defined in a neighborhood of z (x_0, y_0) and have continuous second order partial derivatives in this neighborhood. Suppose (x_0, y_0) is a critical point of f. Set $A = f_{xx}(x_0, y_0), B = f_{xx}(x_0, y_0) = f_{yx}(x_0, y_0), C = f_{yy}(x_0, y_0)$ and

$$D = \begin{vmatrix} f_{xx} f_{xy} \\ f_{yx} f_{yy} \end{vmatrix} = \begin{vmatrix} AB \\ BC \end{vmatrix} = AC - B^2.$$

Then

(i) f has a local minimum at $(x_0, y_0) ifD > 0 andA > 0$

(ii) f has a local maximum at $(x_0, y_0) ifD > 0 andA < 0$.

(iii) f has a saddle point if $D < 0.$

(iv) The test fails if $D = 0$.

2.9. Extrema with Constraints

The Method of Lagrange's Multipliers. In order to find the extrema of the function $w = f(x,y,z)$ subject to the constraints $g(x,y,z) = 0,$ we introduced a new variables $\lambda -$ called Lagrange's Multiplier- and form the function

$$F(x,y,z,\lambda) = f(x,y,z) + \lambda g(x,y,z)$$

The function $F(x,y,z\lambda)$ is called Lagrange Function. Thus, for critical points, we have

$$F_x = \frac{\partial f}{\partial x} + \lambda \frac{\partial g}{\partial x} = 0$$

$$F_y = \frac{\partial f}{\partial x} + \lambda \frac{\partial g}{\partial y} = 0$$

$$F_z = \frac{\partial f}{\partial z} + \lambda \frac{\partial g}{\partial z} = 0$$

$$f_\lambda = g(x,y,z) = 0.$$

The solution of these four equations give the critical points of $f.$ The extrema of f may be found by listing the values of f at each critical point and considering the largest and the smallest values.

2.9.1. Washer Method

Consider the plane area enclosed by the graph of $x = a, x = b$ and of two continuous function $f \; and \; g$ where $f(x) \geq 0$ for all $x \dot{o} [a,b].$

Let $p = \{a = x_0, x_1, \ldots, x_n = b\}[a,b].$ Let ω_r be a point of $[x_{n-1}, x_r]$. The area obtained by cutting the solid at $x = \omega_r$ is the difference of the area of the outer circle and the area of the inner circles, namely

$$\pi\left[f(\omega_r)\right]^2 - \pi\left[g\ (\omega_r\right]^2.$$

The solid of revolution obtained from the rth subinterval is called a washer and its volume is $\left\{\pi\left[f\left(\omega_r\right)\right]^2-\pi\left[g\left(x_r\right)\right]^2\right\}\Delta x_r$ Adding up the volumes of all washers, we obtain a Riemann sum

$$\sum_{r=1}^{\pi}\left\{\pi\left[f\left(\omega_r\right)\right]^2-\pi\left[g\left(\omega_r\right)\right]^2\right\}\Delta x_r$$

If we take the limit of this sum as the norm of the partition $p \to 0$ and $n \to \infty$, we get the volume V of the solid of revolution as

$$V=\int_a^b\left(x\left[f(x)\right]^2-\pi\left[g(x)\right]^2\right)dx.$$

If $g(x)>0$ for all x in $[a,b]$, there is a hole through the solid.

2.9.2. Shell Method

In this case, the volume of a solid of revolution will be found by using cylindrical shell. A cylindrical shell is the solid region between two concentric cylinders if the same height. If the same height. If the inner radius is r_1 and the outer radius is r_2, then the volume of a cylindrical shell of altitude h is

$$\pi r_2^2 h-\pi r_1^2 h$$

$$\pi h\left(r_2+r_1\right)\left(r_2-r_1\right)=2\pi h\left(\frac{r2-r_1}{2}\right)\left(r_2-r_1\right)$$

$=2\pi$ altitude $\times$ average radius thickness.

This formula will be used in evaluating the volume of a solid of revolution.

2.9.3. Cross-Sectional Method

In the disc method the volume V is given by

$$V=\int_a^b \pi y^2\,dx$$

The integrand $\pi y^2 = \pi\{f(x)\}^2$ can be interpreted as the area of the cross section of the solid made by a plane perpendicular to the $x-axis$ and at a distance a units from the origin.

Theorem. (First theorem of Pappas). If a plane area is revolved about n axis in its plane and not crossing the area, the volume of the solid generated is equal to the product of the area and the length of the path described by the centroid of the area.

2.10. Area of Surface of Revolution

Definition. If an arc s, a portion of a plane curve, is rotated about a straight line, the arc generated a surface called a surface a surface of revolution.

Theorem. (Second Theorem of Pappas). If the area bounded by an arc of a curve is revolved about an axis in its plane but not crossing the arc, the area of the surface generated is equal to the product of the length of the arc and the length described by the centroid of the arc. If the area bounded by an arc of a curve is revolved about an axis in its plane but not crossing the arc, the area of the surface generated is equal to the product of the length of the arc and the length described by the centroid of the arc.

CHAPTER 3

APPLICATION OF FRICTION

CONTENTS

3.1. INTRODUCTION

When two smooth surfaces are in contact with each point P, the reaction of one body on the other body is along common normal. But if the bodies are not smooth the reaction of one body on the other is not along the common normal but it is inclined at an angle say to the common normal at the point of contact P. Let R has component R along the common normal at P and F along the tangent plane at P Therefore,

$$R = R' \cos\theta \; and \; F = R' \sin\theta$$

where R is called normal reaction or normal pressure and F is called the force of friction or simply friction. The force of friction prevents of sliding of one body on the surface of the other. is resultant of normal reaction R and force of friction F and is also called *resultant friction.*

3.2. TYPES OF FRICTION

3.2.1. Statically Friction

When one body in contact with another is in equilibrium and is not on the point of sliding along the other, the force of friction in this case is called The equilibrium in this case is non-limiting.

3.2.2. Limiting Friction

When one body is just at the point of sliding on the other; the friction in this case is maximum and is called limiting friction. The equilibrium in the case is called *limiting equilibrium.*

3.2.3. Dynamical Friction

When one body is sliding on the other the friction in this case is slightly less than the limiting friction and is called or

3.3. LAW OF FRICTION

1. The direction of friction is opposite to the direction in which the body moves of tends to move.
2. The magnitude of friction is, up to certain extent. Equal to the force tending to produce motion.
3. The magnitude of the limiting friction bears a constant ratio

reaction between the surfaces. The contain is called

4. The amount of friction is independent of the area and shape of the surface in contact provided the normal pressure remains unaltered.
5. When the motion takes place, the friction still opposes the motion, it is independent of the velocity and is proportional to the normal, it is independent of the velocity and is proportional to the normal reaction but is slightly less than the limitingfriction.

3.4. COEFFICIENT OF FRICTION

The magnitude of the limiting friction F bears a constant ration μ to the normal reaction R between the surfaces. The constant μ is called *coefficientoffriction* and

$$\frac{F}{R} = \mu$$

3.5. ANGLE OF FRICTION

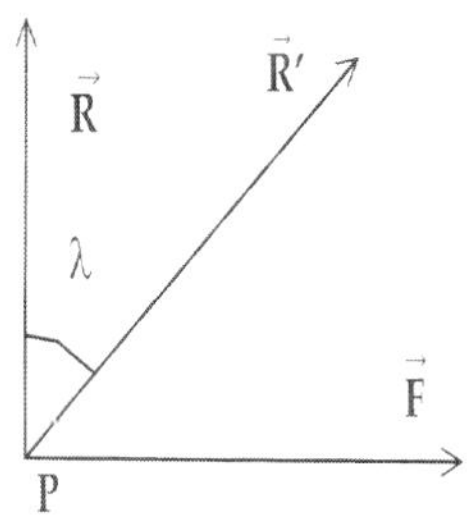

The angle λ, which the direction of the resultant reaction R' when the body is just on the point of motion, is called *angleoffriction.*

Now $R = R'\cos\lambda$ and $F = R'\sin\lambda$

Since $\frac{F}{R} = \mu$, therefore

$$\mu = \frac{R'\sin\lambda}{R'\cos\lambda} \text{ or } \mu = \tan\lambda$$

Also $R^2 + F^2 = R'^2$

or $R' = \sqrt{R^2 + F^2} = \sqrt{R^2 + \mu^2 R^2}$

or $R' = R\sqrt{1+\mu^2}$.

3.6. CONE OF FRICTION

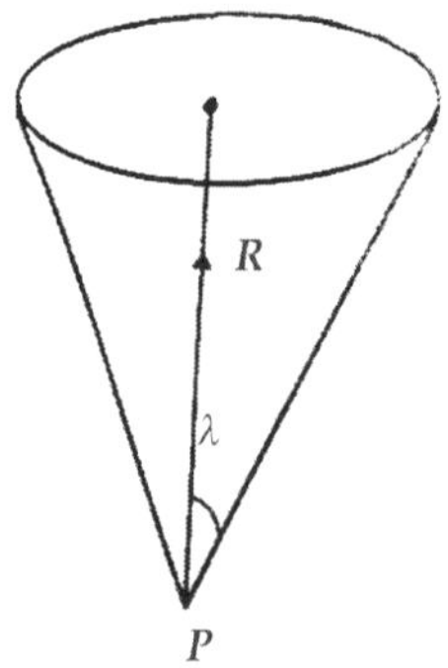

The cone of friction is a cone whose vertex is the point of contact P, its axis along the common normal at P and whose semi – vertical angle is the angle of friction.

3.7. EQUILIBRIUM OF A PARTICULAR ON ROUGH HORIZONTAL PLANE

3.7.1. To find the Least Force Which Will Sent into Motion a Particle at Rest on a Rough Horizontal Plane

Let P be the required least force which makes angle θ with the horizontal plane. Let m be the mass of the particle. For limiting equilibrium

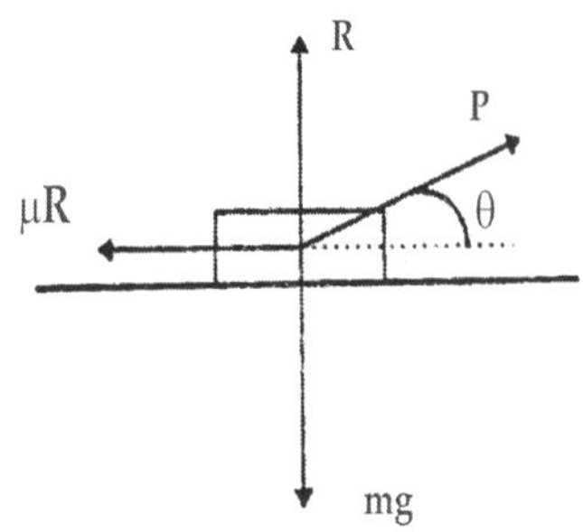

$P\,cos\theta = \mu R$ (1)

$R + P\,sin\theta = mg$ (2)

From (2)

$$R = mg - p\sin\theta$$

Putting in eq. (1), we get

$$P(cos\theta = \mu\left(mg - psin\theta\right)$$

$$P\left(cos\theta + \mu sin\theta\right) = \mu mg$$

$$P = \frac{\mu mg}{cos\theta + \mu\,sin\theta}$$

Since $\mu = \tan\lambda,\ therefore$

$$P = \frac{mgtan\lambda}{cos\theta + tan\lambda\,sin\theta}$$

$$= \frac{mgsin\lambda}{cos\lambda + \tan\lambda\,sin\theta}$$

$$= \frac{mgsin\lambda}{\cos\left(\theta - \lambda\right)}$$

P is least when $\cos\left(\theta - \lambda\right)$ is maximum i.e

$$\cos\left(\theta - \lambda\right) = 1\,or\,\theta - \lambda = 0\,or\,\theta - \lambda$$

Thus

$P_{least} = mg\sin\lambda$.

Equilibrium of a Particular on Rough Horizontal Plane

Show that if a particular is in equilibrium on an inclined plane under its own weight, the inclination of the plane is equal to the angle of friction.

Proof.

Let α be the inclination of the plane and m be the mass of the particle. For limiting equilibrium

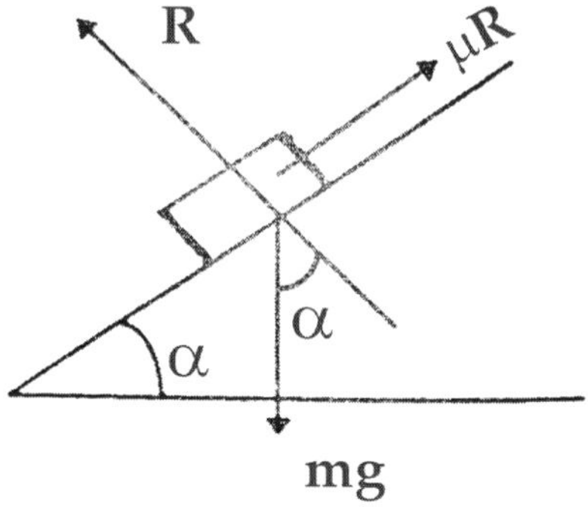

$R = mg\cos\alpha$ (1)

$\mu R = mg\sin\alpha$ (2)

From (1) and (2) we have

$\mu = \dfrac{mg\sin\alpha}{mg\cos\alpha} = \tan\alpha$

Since $\mu = \tan\lambda$ therefore

$\tan\lambda = \tan\alpha$

$\alpha = \lambda$

where λ is the angle of friction

3.7.2. To Find the Least Force Which Will Sent into Motion a Particle at Rest on a Rough Horizontal Plane

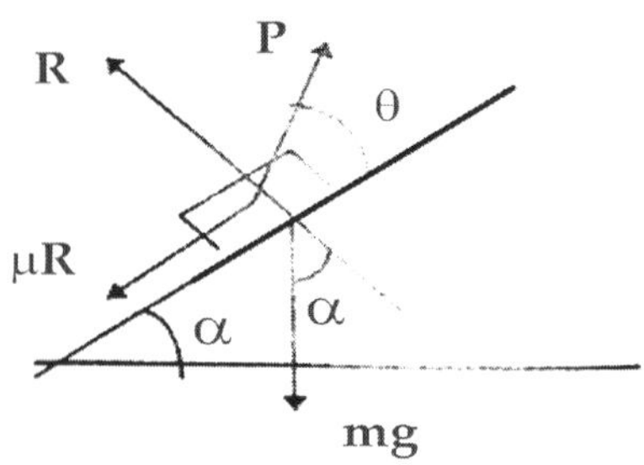

Let p be the required least force which makes angle θ with the inclined plane. Let α be the inclination of the plane and m be the mass of the particle. For limiting equilibrium,

$Pcos\theta = \mu R + mgsin\alpha$ (1)

$Psin\theta + R = mgcos\alpha$ (2)

From (2)

$$R = mgcos\alpha - psin\theta$$

Then (1) becomes

$$P(\cos\theta) = \mu(mgcos\alpha - Psin\theta) + mgsin\alpha$$

$$P(cos\theta + \mu sin\theta) = mg(cos\alpha + sin\alpha)$$

$$P = \frac{mg\ (\mu\cos\alpha + sin\alpha)}{cos\theta + \mu sin\theta}$$

Since $\mu = \tan\lambda$; λ is an angle of friction, therefore

$$P = \frac{mg(sin\lambda cos\alpha + cos\lambda sin\alpha)}{cos\lambda cos\theta + sin\lambda sin\theta}$$

$$= \frac{mg\sin(\alpha + \lambda)}{\cos(\theta - \lambda)}$$

For least P $cos(\theta - \lambda)$ is maximum $i.e cos(\theta - \lambda) = 1$

or $\theta - \lambda = 0$ or $\theta = \lambda$

Thus

$$P_{least} - mg\sin(\alpha + \lambda)$$

Example

The least force which will move a weight up an inclined plane is of magnitude P. Show the least force, acting parallel to plane, which will move the plane upward is $P\sqrt{1+\mu^2}$, where μ is coefficient of friction.

Solution

We know that the least force P is $P_{least} = mg\sin(\alpha + \lambda)$ (1)

where αis the inclination of the plane Let P' be the magnitude of the least force acting parallel to plane. For limiting equilibrium,

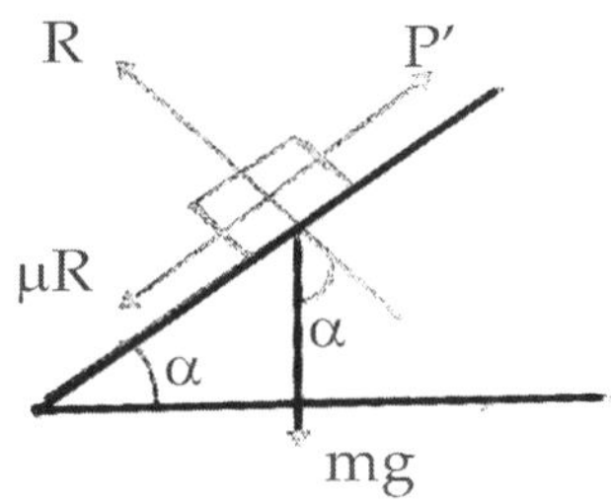

$P' = \mu R + mg \sin\alpha$ (2)

$R = mg cos\alpha$ (3)

From (2) and (3)

P' = $\mu mg cos\alpha + mg \sin\alpha$

P' = mg ($\mu \cos\alpha + \sin\alpha$)

P' = mg $(tan\lambda cos\alpha + \sin\alpha)$

=

= $mg\left(\dfrac{sin\lambda cos\alpha + cos\lambda sin\alpha}{cos\lambda}\right)$

= $\dfrac{mgsin(\alpha+\lambda)}{cos\lambda}$

= $\dfrac{P}{cos\lambda} = Psec\lambda = P\sqrt{1+\tan^2\lambda}$ = P$\sqrt{1+\mu^2}$

3.7.3. To Find the Least Force to Drag a Particle on a Rough Inclined Plane in Downward Direction

Let P be the required force which make angle θ with the plane. Let be the mass of the particle. For limiting equilibrium

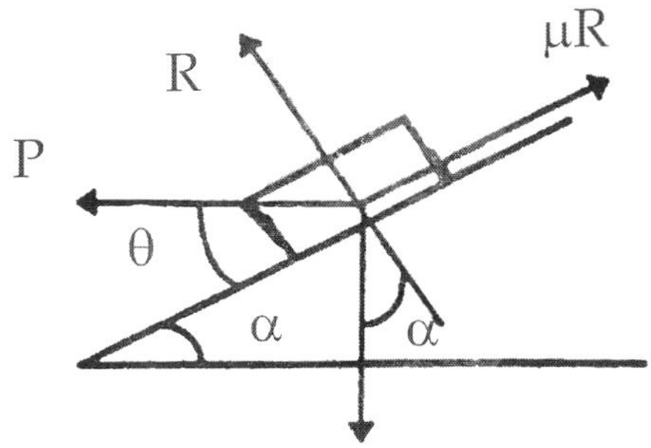

$$Pcos\theta + mgsin\alpha = \mu R$$

$$Psin\theta + R = mgcos\alpha$$

From eq. (2) we have

$$R = mgcos\alpha - psin\theta$$

Put in eq. (1), we get

$$Pcos\theta + mg\sin\alpha = \mu\left(mgcos\alpha - Psin\theta\right)$$

$$P\left(cos\theta + \mu sin\theta\right) = mg\left(\mu\cos\alpha - sin\alpha\right)$$

$$P = mg\frac{\mu\cos\alpha - sin\alpha}{cos\theta + \mu sin\theta}$$

$$P = mg\frac{tan\lambda cos\alpha - sin\alpha}{cos\theta + tan\lambda sin\theta}$$

$$P = mg\frac{sin\lambda cos\alpha - cos\lambda sin\alpha}{cos\lambda cos\theta + sin\lambda sin\theta}$$

$$P = mg\frac{\sin\left(\lambda - \alpha\right)}{\cos\left(\theta - \lambda\right)}$$

For least P $\cos\left(\theta - \lambda\right)$ is maximum, i.e.,

$$\cos\left(\theta - \lambda\right) = 1$$

or $\theta - \lambda = 0 or \theta = \lambda$

$$P_{least} = mg\sin\left(\lambda - \alpha\right)$$

3.7.4. To Find the Force Necessary Just to Support a Heavy Body on an Inclined Plane with Inclination α ;(a > λ)

Let particle be at point A and the required force P acts along AB making angle θ with plane Let n be the mass of the particle. For limiting equilibrium,

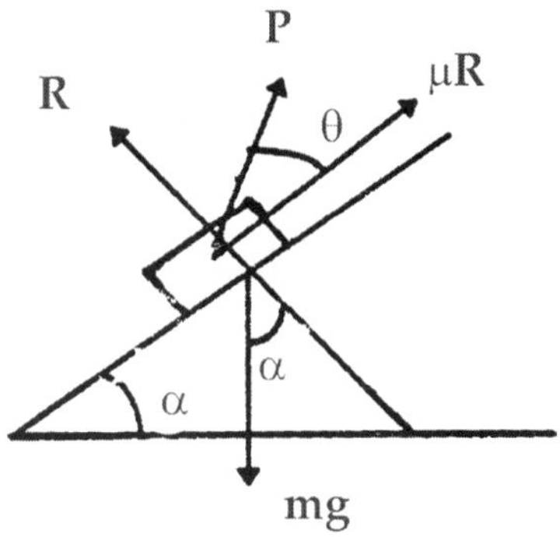

$$Pcos\theta + \mu R = mg\sin a$$

$$Psin\theta + R = mg\ \cos a$$

From eq. (2), we have

$$R = mg\cos a - Psin\theta$$

Put in eq. (1), we get

$$P(cos\theta + \mu\ (mg\cos a - psin\theta\ ) = mgsin\alpha$$

$$P\left(cos\theta - \mu sin\theta\right) = mg(sin\alpha - \mu\cos\alpha\ )$$

$$P = mg\frac{sin\alpha - \mu cos\alpha}{cos\theta - \mu\sin\theta}$$

$$p = mg\frac{sin\alpha - \ \tan\lambda\cos\alpha}{cos\theta - tan\lambda sin\theta}$$

$$P = mg\frac{\cos\lambda\sin\alpha - \ \sin\lambda cos\alpha}{\cos\lambda cos\theta + sin\lambda\sin\theta}$$

$$P = mg\frac{\sin\left(\alpha - \lambda\right)}{\cos\left(\theta + \lambda\right)}$$

For least P $\cos(\theta+\lambda) is$ maximum

$$\cos(\theta+\lambda)=1$$

or $\theta+\lambda=0$ or $\theta=-\lambda$

Thus

$$P_{least}=mgsin(\alpha-\lambda)$$

Example

A light ladder is supported on a rough floor and leans against an equally rough wall. How far up ladder can a man climb without slipping taking place.

Solution

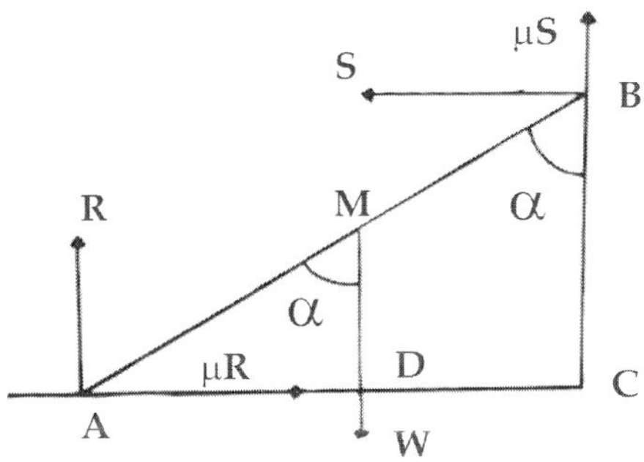

Let AB be the light ladder which makes angle α with the wall. Let man be at point. M when ladder is limiting equilibrium. Let W be weight of the man. Suppose AM = x, R and S be the reaction of ground and wall, respectively and $\mu Rand\, \mu Sbe$ the force of friction. For limiting equilibrium

$$R+\mu S-W$$

S = μR

From equation (1), we have

R = W – $\mu S, puton(2), weget$

S $\mu(W-\mu S)$

S + $\mu^2 S=\mu W$

$$S = \frac{\mu W}{1+\mu^2}$$

Taking moment about A, then

$$W(AD)=S(BC)+\mu S(AC)$$

$$W(AMsin\alpha)=S(ABcos\alpha)+\mu S(ABsin\alpha)$$

$$W(xsin\alpha)=S(lcos\alpha)+\mu S(l\sin\alpha)$$

$$W(xsin\alpha)=Sl(cot\alpha+\mu)\ AB=l=lenghtofladder$$

$$Wx=\frac{\mu Wl}{1+\mu^2}=(cot\alpha+\mu)$$

$$x = \frac{\mu l}{1+\mu^2}(cot\alpha+\mu)$$

Example

A rod, 4 ft. long rests on a rough floor against the smooth edge of a table of height 3 ft. If the rod is on the point of slipping when inclined at an angle of $60^0 to$ the horizontal, find the coefficient of friction.

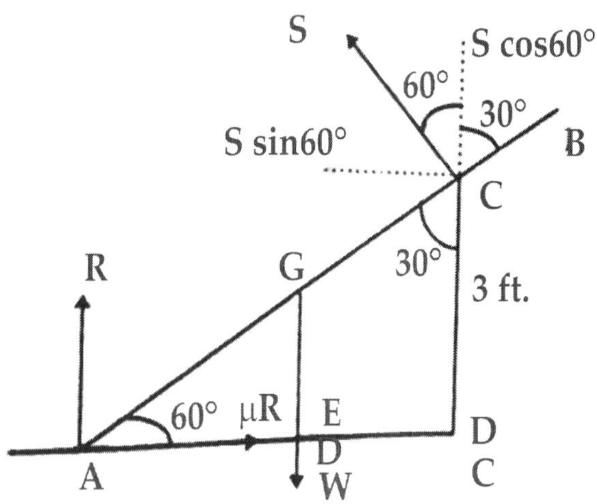

Solution

Let AB be the rod of weight W which acts at its mind point G. Then

AB = 4 ft,; AG = 2 ft.; CD = 3 ft.

Reaction R at A is perpendicular to ground and reaction S act at C is perpendiculars to the rod. For limiting equilibrium,

$$R + Scos60^0 = W$$

$$Ssin60^0 = \mu R$$

From Eq. (1), we have

$R = W - Scos60^0$, put in (2), then

$$Ssin60^0 \mu\left(W - Scos60^0\right)$$

$$S\frac{\sqrt{3}}{2} = \mu\left(W - \frac{S}{2}\right)$$

$$S = \frac{2\mu W}{\sqrt{3+\mu}}$$

Let AC = x, then in $\Delta ACD,$

$$\frac{DC}{AC} = \sin 60^0$$

$\frac{3}{x} = \frac{\sqrt{3}}{2}$ or x = $2\sqrt{3}$ (4)

Taking moment about A, we get

W (AE) = S(AC)

W(AG $cos60^0) = SX$

Using (3) and (4), we have

$$W(2)\left(\frac{1}{2}\right) = \frac{2\mu W}{\sqrt{3}+\mu}(2\sqrt{3})$$

$$1 = \frac{2\mu}{\sqrt{3}+\mu}(2\sqrt{3})$$

$$\sqrt{3} + \mu = 4\sqrt{3}\mu$$

$$\mu = \frac{\sqrt{3}}{4\sqrt{3}-1} = 0.292$$

Example

A uniform ladder rests in limiting equilibrium with one end on rough horizontal plane and the other against a smooth vertical wall. A man ascend the ladder. Show that he cannot go more than half way up.

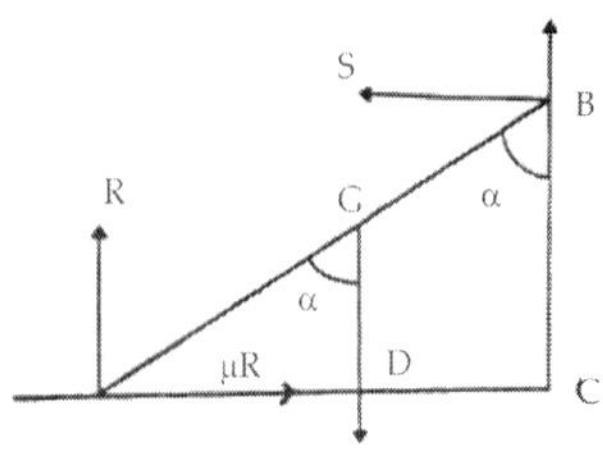

Solution

Let AB be ladder of length I and weight W. Let ladder is limiting equilibrium, then

$$R = W$$

$$S = \mu$$

From (1) and (2), we have

$$S = \mu W$$

Taking moment about A, we get

$$W(AD) = S(BC)$$

$W(AGsin\alpha) = \mu W(ABcos\alpha)$ by using (3)

$$\frac{1}{2} = sin\alpha\mu l\cos\alpha$$

$\tan\alpha = 2\mu$ (3a)

Let W_l be the weight of man who ascend the ladder. Let man be at point M when the ladder is in limiting equilibrium Let AM = x then

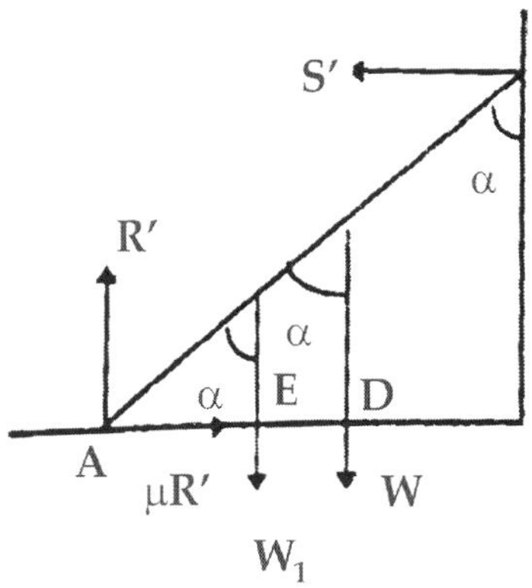

$R' = W + W_l$ (4)

$S' = \mu R'$ (5)

From (4) and (5)

$S' = \mu R'$ (6)

From (4) and (5)

$$S' = \mu(W + W_l)$$

Taking moment about A,

W(AD) + $W_l(AE) = S'(BC)$

W (AG sinα) + $W_l\ (AM \sin\alpha) = S'(ABcos\alpha)$

By using (6), we have

$$W(\frac{1}{2}\sin\alpha) + W_l\ (AM\sin\alpha) = \mu(W + W_l)\ (l\cos\alpha)$$

W $(\frac{l}{2}\ \tan\alpha) + W_l\ (x\tan\alpha\ ) = \mu l(W + W_l)$

Using (3a), we get

$$W\left(\frac{l}{2}2\mu\right) + W_l(2\,\mu x) = \mu lW + \frac{\mu}{W_l}$$

$$W(l\mu) + 2W_l(x\mu) = \mu lW + \mu lW_1$$

$$2W_l(x\mu)=\mu l w_1$$

$$2x = lorx = \frac{1}{2}$$

which shows that the man cannot go more than half way up.

Example

A uniform rod of weight W is placed with its lower end on a rough horizontal floor and its upper and against an equally rough vertical wall. The rod makes an angle α with the wall and is just prevented from slipping down by a horizontal force P applied at its middle point. Prove that

$$P = W\tan(\alpha - 2\lambda).$$

where λ is the angle of friction and $\lambda < \frac{1}{2}\alpha.$

Solution

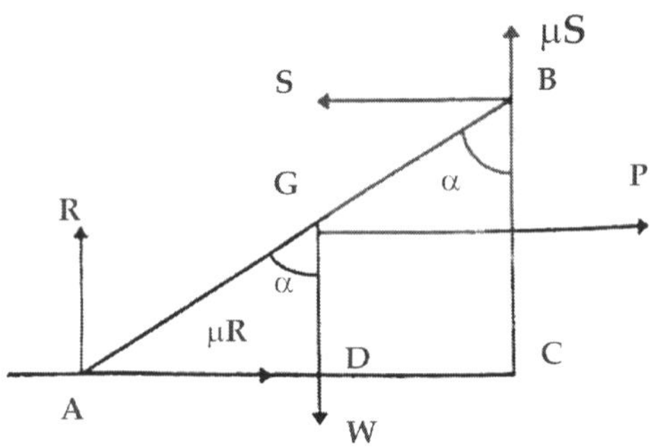

Let AB be rod of length l. Its weight W act at its mid-point G. Then,

$$AB = l$$

$$AG = \frac{l}{2} = BG$$

For limiting equilibrium, we have

$$P + \mu = S$$

$$R + \mu S = W$$

From (1) and (2), we have

$$\frac{P}{W}=\frac{S-\mu R}{R+\mu S}$$

Taking moment about G, we get

$$\mu S(GE)+S(BE)+\mu R(GD)=R(DA)$$

$$\Rightarrow \mu S(BG\sin\alpha)+S\ (BG\cos\alpha)+\mu R(AG\cos\alpha)=R(AG\sin\alpha)$$

$$\Rightarrow \mu\left(\frac{1}{2}\right)\sin\alpha+S\left(\frac{1}{2}\right)\cos\alpha+\mu R\left(\frac{1}{2}\right)cos\alpha=R\left(\frac{1}{2}\right)sin\alpha$$

$$\Rightarrow \mu S\sin\alpha+S\cos\alpha+\mu Rcos\alpha=R\sin\alpha$$

$$\Rightarrow S(\mu\sin\alpha+\cos\alpha)=R(\sin\alpha-\mu cos\alpha)$$

$$\Rightarrow S(\tan\lambda sin\alpha+cos\alpha)=R\ (sin\alpha-tan\lambda cos\alpha)$$

$$\Rightarrow S(\sin\lambda\sin\alpha+cos\lambda cos\alpha)=R(\cos\lambda sin\alpha-sin\lambda cos\alpha)$$

$$\Rightarrow \pm\cos(\ \ -\aleph)=\ \ \sin(\ \ -\ \ )$$

$$\Rightarrow S=Rtan(\alpha-\lambda)$$

From (3) and (4), we get

$$\frac{P}{W}=\frac{Rtan(\alpha-\lambda)-\mu R}{R+\mu\tan(\alpha-\lambda)}$$

$$\Rightarrow\frac{P}{W}=\frac{tan(\alpha-\lambda)-\mu}{1+\mu\tan(\alpha-\lambda)}$$

$$\Rightarrow\frac{P}{W}=\frac{tan(\alpha-\lambda)-tan\mu}{1+tan\lambda\tan(\alpha-\lambda)}=\tan(\alpha-\lambda-\lambda)$$

$$\Rightarrow P=\tan(\alpha-2\lambda)$$

Since P is positive, therefore, $(\alpha-2\lambda)>0, thus\,\lambda<\frac{1}{2}\alpha$

Example

A uniform semi- circular wire hangs on a rough peg, the line joining its extremities makes angle of 45 with horizontal. If it is just on the point of slipping, find the coefficient of friction between the wire and the peg.

Solution

Let P be the peg and APB be the semi-circular wire with center at O and radius r, then

OR = r

Let G be the center of gravity of wire, then

$$OG = \frac{2r}{\pi}$$

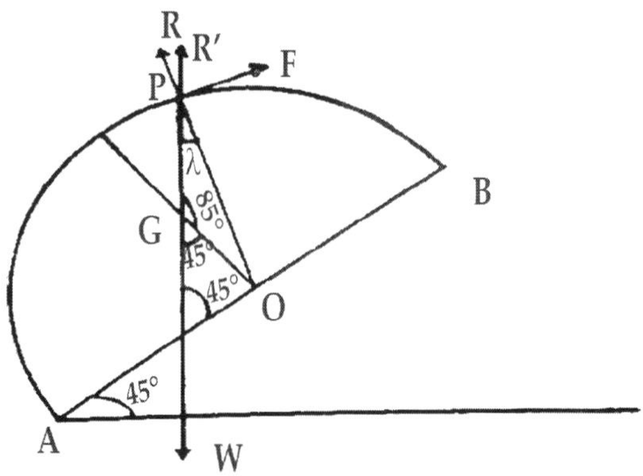

For equilibrium, the resultant reaction R' and W must be equal and opposite and act along the same straight line, therefore

$$< OPG = \lambda and < OPG = 135^0$$

For limiting equilibrium, we have

$$R = Wsin\lambda$$

$$\pm = \mu \quad = \quad \sin\lambda$$

Taking moment about O, we get

$$OG\left(Wsin45^0\right) = OP\left(\mu R\right)$$

$$\frac{\ddot{u}r}{\pi} \pm \sec\lambda\left(\frac{}{\sqrt{2}}\right) = \ .\mu.$$

Using (1), we get

$$\frac{\pm\ \lambda}{\mu}=\frac{\pi}{\sqrt{2}}$$

Squaring both sides, we get

$$\frac{\text{ü}\ ^{2}\lambda}{\mu^{2}}=\frac{\pi^{2}}{2}$$

$$\frac{1+\mu^{2}}{\mu^{2}}=\frac{\pi^{2}}{2}$$

where $\sec^{2} 1+\mu^{2}=1+\tan^{2}=1+\mu^{2}$

$$\frac{1}{\mu^{2}}+1=\frac{\pi^{2}}{2}$$

$$\mu^{2}=\frac{2}{\pi^{2}-2}$$

$$\mu=\sqrt{\frac{2}{\pi^{2}-2}}=0.504$$

Example

A hemi-spherical shell rests on a rough inclined plane whose angle of friction is λ show that the inclination of the plane base to the horizontal cannot be greater than $\sin^{-1}(2 sin\lambda)$.

Solution

Let G be the center of gravity of hemi- sphere, its weight W acts along GA. For limiting equilibrium, the resultant of R and μR must balance the weight W and so it acts along AG. Therefore,

$$< OAG=\lambda and < AGB=\theta$$

Since the angle between two lines is equal to the angle their perpendiculars, therefore

$$< OGA=180^{0}-\theta$$

Let r be the radius of sphere, then

$$OA=r and OG=\frac{r}{2}$$

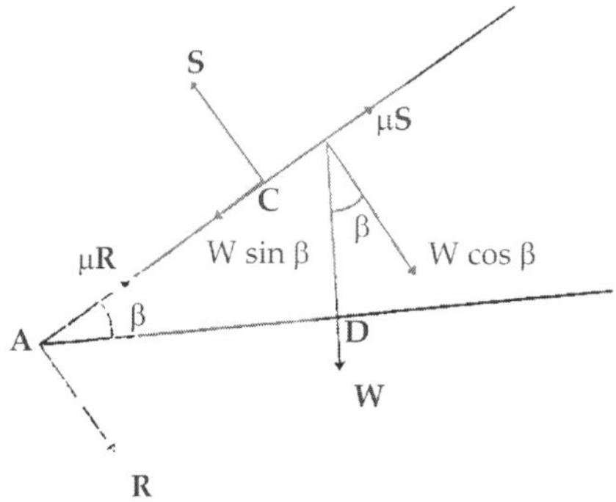

For limiting equilibrium

$$\mu r R = W sin\lambda$$

$$R = W cos\lambda$$

Taking moment about 0

$$(OG) W sin(960^0 - \theta) + (OA)\mu R \sin(90^0) = 0$$

$$\aleph \frac{r \;\; \mu R}{2 \sin\lambda} sin\theta \quad r(\mu R) \quad 0$$

$$\pm \;\; \theta = 2\sin\lambda$$

$$\theta = \sin^{-1}(2 sin\lambda)$$

Example

A thin uniform rod passes over one peg and under another, the coefficient of friction between each peg and the rod being μ. The distance between the pegs is ‘a’ and the straight line joining them makes an angle β with the horizontal. Show that equilibrium is not possible unless the length of the rod is greater than $\frac{a}{\mu}(\mu + \tan\beta)$.

Solution

Let AB be the rod of length x. Let the rod rests over the peg C and under the peg A. The rod will be of least length when its one is in contact with the lower peg. Given AC = a. Let the normal reaction at A and C is R and S, respectively. Resolving components along and perpendicular to the rod.

$$\mu.R + \mu.S = W \sin\beta \qquad (1)$$

$S = R + W\cos\beta$ (2)

Eliminating R from (1) and (2), we have

$$\mu(S - W\cos\beta) + \mu S = W\sin\beta$$

$$2\mu S = W(\sin\beta + \mu\cos\beta)$$

$$S = \frac{W}{2\mu}(\sin\beta + \mu\cos\beta)$$

Taking moment about A,

$$S(AC) \quad W(AD)$$

$$S(a) = W(AG\cos\beta)$$

$$\frac{aW}{2\mu}\sin\beta + \mu\cos\beta) = W\left(\frac{x}{2}\right)\cos\beta$$

$$x = \frac{a}{\mu}(\tan\beta + \mu)$$

which gives the least length of the rod. For equilibrium

$$S\left|\overline{AC}\right| < W\left|\overline{AD}\right|$$

$$x > \frac{a}{\mu}(\tan\beta + \mu).$$

Example

A uniform rod slides with its end on two fixed equality rough rods. One being vertical and the other inclined at an angle α to the horizontal. Show that the angle θ to the horizontal of the movable rod, when it is on the point of sliding is given by

$$\tan\theta = \frac{1 - 2\mu\tan\alpha - \mu^2}{2(\mu + \tan\alpha)}$$

Solution

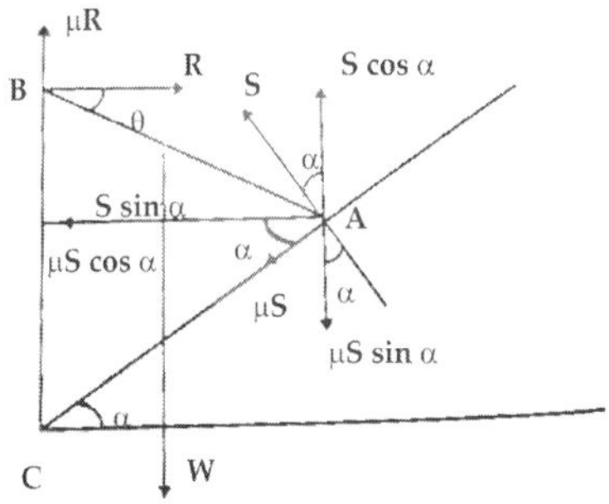

Let the movable rod be the AB under the action of three forces, the reactions of fixed rods and its weight.

Resolving the force horizontally and vertically as shown in the figure. Since the system is in equilibrium therefore, we get

$$R = \mu.S\cos\alpha + S\sin\alpha \qquad (1)$$

$$\mu.R + S\cos\alpha = W + \mu\sin\alpha \qquad (2)$$

From (1) and (2), we have

$$W = \frac{R\left(1+\mu^2\right)\cos\alpha}{\mu\cos\alpha + \sin\alpha} \qquad (3)$$

Taking moment about A, we get

$$\left(\frac{l}{2}\right)W\sin\left(90^0 - \theta\right) = l R\sin\theta + l\mu R\sin\left(90^0 - \theta\right)$$

where l is the length of the movable rod AB,

$$\frac{W\cos\theta}{2} = R\sin\theta + \mu R\cos\theta$$

$$R\tan\theta = \frac{W}{2} - \mu R$$

By using (3), we get

$$R\tan\theta = \frac{R\left(1+\mu^2\right)\cos\alpha}{2\left(\mu\cos\alpha + \sin\alpha\right)} = \mu R$$

$$\tan\theta = \frac{1 - 2\mu\tan\alpha - \mu^2}{2\left(\mu + \tan\alpha\right)}$$

Example

A weight W is laid upon a rough plane $\mu = 1/\sqrt{3}$, inclined at 45 to the horizon and is connected by string passing through a smooth ring A at the top of the plane with the weight P hanging vertically. If W = 3P, show that if θ be the greatest possible inclination of the string AW to the line of greatest possible inclination of the string AW to the line of greatest slope in the plane, them

$$\cos\theta = \frac{2\sqrt{2}}{3}$$

Find also the direction in which w would commence to move

Solution

The line of greatest slope in a plane is line perpendicular to the line if the intersection of the given plane with the horizontal plane. In figure BC is the line of the greatest slope.

$$R = W cos 45^0 = \frac{w}{\sqrt{2}}$$

$$\mu R = \frac{1}{\sqrt{3}}\frac{w}{\sqrt{2}} = \frac{w}{6}$$

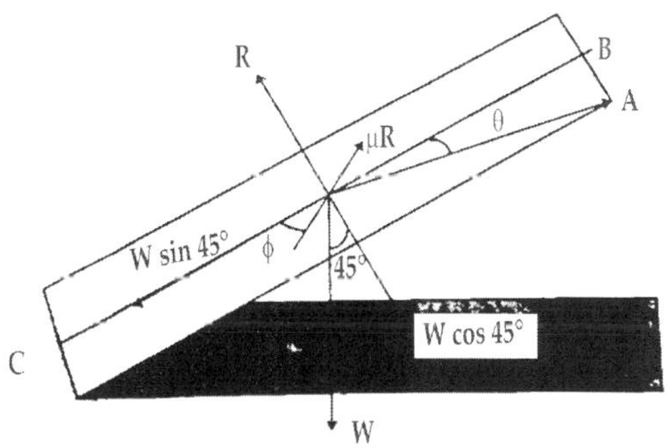

Suppose that the direction of motion of W makes an angle $\varnothing$ with the line of greatest slope. So friction $\mu R acts$ in opposite direction. Now weight W is in equilibrium under the action of three forces P, $\mu R and$ W $sin 45^0$. By Lami's theorem,

$$\frac{P}{\sin\left(180^0-\varnothing\right)}=\frac{\mu R}{\sin\left(180^0-\theta\right)}=\frac{W\sin 45^0}{\sin(\theta+\varnothing)}$$

$$\frac{\frac{W}{3}}{sin\varnothing}=\frac{\frac{W}{\sqrt{6}}}{sin\theta}=\frac{\frac{W}{\sqrt{2}}}{\sin(\theta+\varnothing)}$$

The relation $\frac{\frac{W}{3}}{sin\varnothing}=\frac{\frac{W}{\sqrt{6}}}{sin\theta}$ implies that

$$\sin\varnothing=\frac{\sqrt{6}}{3}=sin\theta$$

And the relation $\frac{\frac{W}{\sqrt{6}}}{sin\theta}=\frac{\frac{W}{\sqrt{2}}}{\sin(\theta+\varnothing)}$ gives

$$\sqrt{6}sin\theta=\sqrt{2}\ (\ \sin\theta\cos\varnothing+\cos\theta sin\varnothing)$$

Using (2), we get

$$\sqrt{3}sin\theta=\left(sin\theta\cos\varnothing+\cos\theta\left(\frac{\sqrt{6}}{3}\right)\right)$$

$$cos\theta=\sqrt{3}-cos\theta\left(\frac{\sqrt{6}}{3}\right)$$

Squaring and adding (2) and (3), we get

$$=\frac{2}{3}\sin^2\theta+3+\frac{2}{3}\cos^2\theta-2\sqrt{2}\cos\theta$$

$$=\frac{2}{3}+3-2\sqrt{2}\cos\theta$$

$$=2\sqrt{2}cos\theta=\frac{4}{3}\text{ or }cos\theta=\frac{2\sqrt{2}}{3}$$

Putting this value in eq. (3), we get

$$cos\varnothing=\sqrt{3}-\frac{2\sqrt{2}}{3}.\frac{\sqrt{6}}{3}=\frac{5\sqrt{3}}{9}$$

$$\varnothing=\cos^{-1}(\frac{5\sqrt{3}}{9})$$

which gives the direction in which W moves.

Summary and Discussion

In this stage, we have discussed closely an important topic of friction which amusement a very important role in mechanics, physics and engineering. Some actuality life problems are given on which we apply different technique to solve them.

3.2.1. Statically Friction

When one body in contact with another is in equilibrium and is not on the point of sliding along the other, the force of friction in this case is called *staticalfriction.* The equilibrium in this case is non-limiting.

3.2.2. Limiting Friction

When one body is just at the point of sliding on the other; the friction in this case is maximum and is called limiting friction. The equilibrium in the case is called *limiting equilibrium*

3.2.3. Dynamical Friction

When one body is sliding on the other, the friction in this case is slightly less than the limiting friction and is called dynamical friction or Kinetic friction

3.3. Law of Friction

1. The direction of friction is opposite to the direction in which the body moves of tends to move.
2. The magnitude of friction is, up to certain extent, equal to the force tending to produce motion.
3. The magnitude of the limiting friction bears a constant ratio $\mu to the normal$ reaction between the surfaces. The contain μ is called coefficient of friction

4. The amount of friction is independent of the area and shape of the surface in contact provided the normal pressure remains unaltered.

5. When the motion takes place, the friction still opposes the motion, it is independent of the velocity and is proportional to the normal, it is independent of the velocity and is proportional to the normal reaction but is slightly less than the limiting friction.

3.4. Coefficient of Friction

The magnitude of the limiting friction F bears a constant ration μ to the normal reaction R between the surfaces. The constant μ is called *coefficientoffriction* and $\frac{F}{R} = \mu$

3.5. Angle of Friction

The angle λ, which the direction of the resultant reaction R' when the body is just on the point of motion, is called *angleoffriction.* Now

$$R = R' cos\lambda and F - R' sin\lambda$$

Since

$$\frac{F}{R} = \mu,$$

therefore

$$\mu = \frac{R' \sin\lambda}{R' \cos\lambda}$$

or

$$\mu = tan\lambda$$

Also

$$R^2 + F^2 = R'^2$$

or

$$R' = \sqrt{R^2 + F^2} = \sqrt{R^2 + \mu^2 R^2}$$

or

$$R' = R\sqrt{1+\mu^2}$$

3.6. Cone of Friction

The cone of friction is a cone whose vertex is the point of contact P, its axis along the common normal at P and whose semi – vertical angle is the angle of friction.

CHAPTER 4

APPLICATIONS OF KINEMATICS, VIRTUAL WORK, AND RECTILINEAR MOTION

CONTENTS

4.1. WORK DONE BY A FORCE

If a constant force F acts on a particle and particle is displaced from point A to point B. Let the displacement AB = S, then work done W by the force F is defined as

$$W = F.S = FS\cos\theta$$

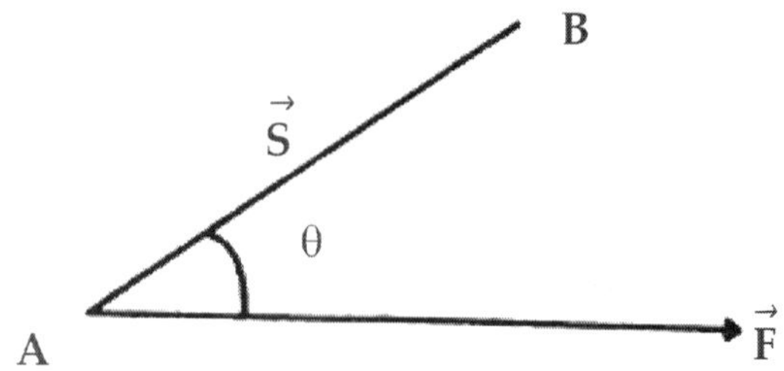

where θ is the angle between F and S

Case (i) $\theta = 0, then$

Work done $= W = FS\cos\theta = FS$

Case (ii) $\theta = \frac{\pi}{2}, then$

Work done $= W = FS\cos\frac{\pi}{2} = FS(0) = 0$

Therefore when the applied force is perpendicular to the displacement the work done will be zero.

Case (iii) $\theta = \pi, then$

Work done $= W = FS\cos\pi = FS(-1) = -FS$

Case (iv) $0 \le \theta \le \frac{\pi}{2}, then$

Work done $= W$ is positive, because cos $\theta \ge 0$.

Case (v) $\frac{\pi}{2} \le \theta \le \pi, then$

Work done = W is negative, because $\cos\theta \le 0$.

4.2. WORK DONE BY THE WEIGHT W

If body of weight W falls down vertically through a distance h, the work done by the gravity is positive and is Wh . If the body rises vertically throughout the height h. the work done is negative and is equal to $-Wh$ and we say that work is being done against gravity.

4.3. WORK DONE BY TENSION OF AN EXTENSIBLE LIGHT STRING

Let AB be a string whose tension is T. in displaced position, let $A'B'$ be the string Let θ be the angle which $A'B'$ makes with AB, as shown in figure

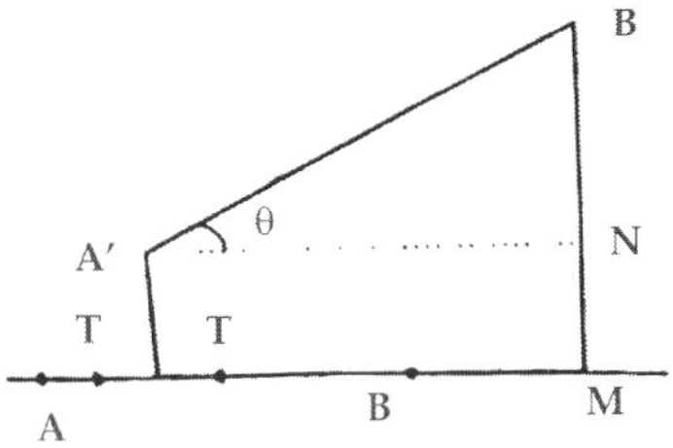

From figure

Work done $=T(AL)-T(BM)$

$$=T(AB-LB)-T(LM-LB)$$

$$=T(AB-LB-LM+LB)$$

$$=T(AB-LM)$$

$$=T(AB-A'N)$$

$$=T(AB-A'B'cos\theta)$$

$$=T\left\{AB-A'B'\left(1=\frac{\theta^2}{2!}+\frac{\theta^4}{4!}-\dots\right)\right\}$$

Neglecting θ^2 and higher powers of θ, as being small, then

Work done $=T\{AB-A'B'\}$

$$= -T\left(A'B' - AB\right)$$

$$= -T\,\delta l$$

where δl is the extension in the length of the string. Similarly, work done by thrust T of the light $T\,\delta l.$

4.4. WORK DONE BY A FORCE DURING A ROTATION

Suppose a forces F acts on a body and rotates it about point O in $xy-$ plane through an angle θ. Let 'a' be distance a force F from point O. Then

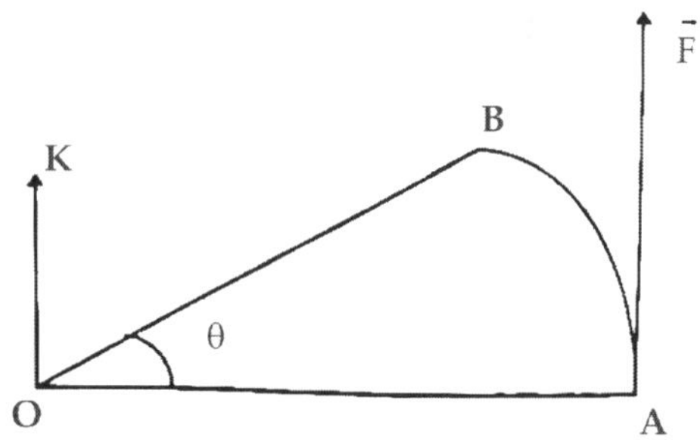

$G =$ moment of F about O $= Fak$

where k is a unit vector along z-axis.

Displacement $= arc\ AB = \theta\left(s = r\theta\right)$

Work done $= F\,a\,\theta = F\,a\,\theta k.k$

$$= Fak.\theta k = G.\theta$$

where $\theta = \theta K$ is angular displacement.

4.5. WORK DONE BY A COUPLE

$Let\left(F, F\right)$ be a couple acting on a body in xy plane whose forces F and – F act at A and B

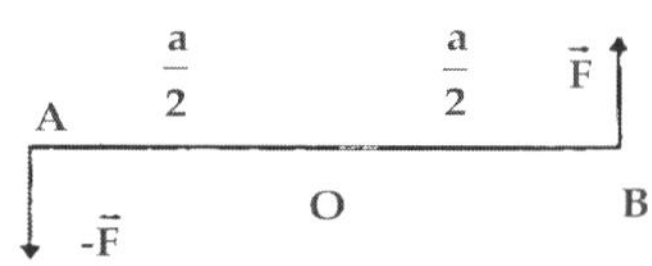

Let AB = a, let axis of rotation passes through O where O is the midpoint of AB, then

$$G = momnet\ of\ couple = Fa\,k.$$

Work done for a small angular displacement θ will be

$$= F\left(\frac{a}{2}\right)\theta + F\left(\frac{a}{2}\right)\theta = F\,a\,\theta = Fa\,\theta k.k$$

$$= Fa\,k.\theta K = G.\theta$$

The Total Work done by all the Forces is Equal to the Work done by their Resultant

Let R be the resultant of forces $F_1.F_2, F_3, \ldots .F_n$ acting on a particle then

R = $F_1 + F_2 + F_3 + \ldots\ldots. + F_n$

Let S be the displacement, then

$$R.S = (F_1 + F_2 + F_3 + \ldots\ldots F_n).S$$

$$= F_1.S + F_2.S + F_3.S + \ldots\ldots. + F_n.S$$

4.6. VIRTUAL DISPLACEMENT AND VIRTUAL WORK

If a set of particles or a body is in equilibrium under the action of forces then there is no motion and consequently there is no actual displacement.

Suppose that the set of particles or the body receives an imaginary displacement, the forces acting there on being regarded as constant during the displacement which is called *virtual displacement* and the work done by the forces during such displacement is called *virtual work*

4.7. CONSTANT

A constraint is a restriction imposed on a mechanical system to get a desired position or motion. In case of a particle moving down an inclined plane, the constraint on its motion is its motion is its contact with the plane. Any displacement along the plane is consistent with the constraint and the displacement away from the plane is not consistent with the constraint. Due to imposition of constraint we get unknown forces acting on the system. These forced are called reaction of constraint

4.7.1. Ideal Constraint or Workless Constraints

If the forces of constraint do not virtual work, the constraints are said to be *ideal* or *workless.*

4.7.1.1. Example of Workless Constraints

(1) The reaction $\bar{R}$ *of any smooth surface with which the body is in contact is workless.*

The surface is smooth, the reaction $\bar{R}$ is normal to the surface at the point of contact p. if p moves to neighboring point p'/, the displacement $\overline{pp'}$ is at right angle to $\bar{R}.S$ o its virtual work is zero.

(2) The tension on an inextensible string is workless.

Let AB be a string whose tension is T. Let A 'B' be the displaced position of the string. The string is inextensible, therefore AB = A'B'/ Let θ be the angle which A'B' makes AB, as shown in figure

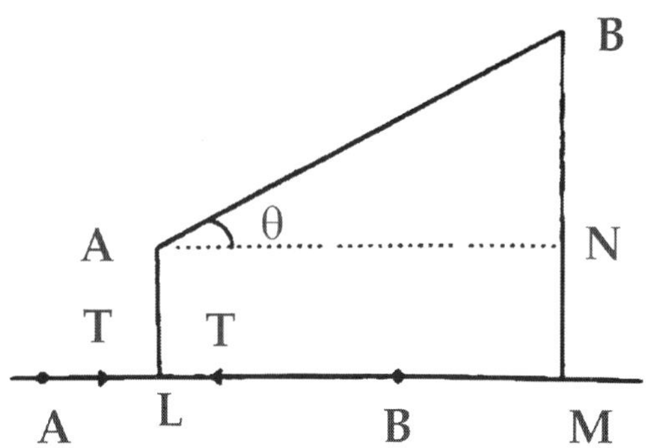

From figure,

Work done = $T(AL)-T(BM)$

$$=T(AB-LB)-T(LM-LB)$$

$$=A(AB-LB-LM+LB)$$

$$=T(AB-LM)$$

$$=T(AB-A'N)$$

$$= T\left(AB - A'B'cos\theta\right)$$

$$= T\left\{AB - A'B'(1 - \frac{\theta^2}{2!} + \frac{\theta^4}{4!} - ....\right\}$$

Neglecting θ^2 and higher powers of $\theta,$ as being small, then

Work done = $T\left\{AB - A'B'\right\}$

$$= -T\left(A'B' - AB\right) = 0; AB = A'B'.$$

Similarly, the thrust of an inextensible rod is workless constraint.

(3) Mutual reaction of two rigidly connected particles Let $\bar{r}_1$ *and* $\bar{r}_2$ *be the position vectors of A and B, respectively, as shown below.*

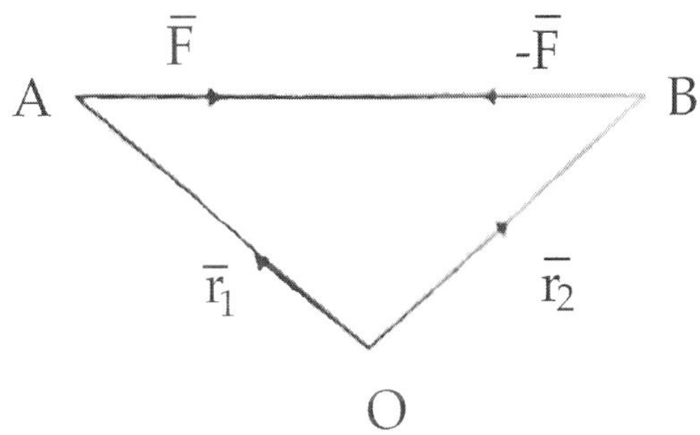

Let $\bar{F}$and $-\bar{F}$ be forces of mutual reaction acting along AB and BA, respectively, then $\bar{F}$ *and AB are collinear*. Therefore,

$$\bar{F} = k\overline{AB}$$

Where k is constant.

$$\bar{F} = k\left(\bar{r}_2 - \bar{r}_1\right)$$

Since the particle are rigidly connected, therefore

$$\left|\overline{AB}\right| = constant$$

$$\left|\bar{r}_2 - \bar{r}_1\right| = constant$$

$$\left|\bar{r}_2 - \bar{r}_1\right|^2 = constant$$

$$\left(\overline{r_2}-\overline{r_1}\right).\left(\overline{r_2}-\overline{r_1}\right)=constant$$

$$\delta\left\{\left(\overline{r_2}-\overline{r_1}\right).\left(\overline{r_2}-\overline{r_1}\right)\right\}=0$$

$$\left(\overline{r_2}-\overline{r_1}\right).\left(\delta\overline{r_2}-\delta\overline{r_1}\right)+\left(\delta\overline{r_2}-\delta\overline{r_1}\right)+\left(\overline{r_2}-\overline{r_1}\right)=0$$

$$2\left(\overline{r_2}-\overline{r_1}\right).\left(\delta\overline{r_2}-\delta\overline{r_1}\right)=0$$

Multiplying by $\frac{k}{2}, then$

$$K\left(\overline{r_2}-\overline{r_1}\right).\left(\delta\overline{r_2}-\delta\overline{r_1}\right)=0$$

$$\overline{F}.\left(\delta\overline{r_2}-\delta\overline{r_1}\right)=0$$

$$\overline{F}.(\delta\overline{r_2})-\overline{F}.(\delta\overline{r_1})=0$$

$$\overline{F}.(\delta\overline{r_2})+(-\overline{F}).(\delta\overline{r_1})=0$$

Therefore the sum of work done $\overline{F}$ and $-\overline{F}$ is zero. So this constraint is workless.

(4) The reaction at a point or a fixed axis of rotation is workless because the displacement of the point of application of the force is zero.

4.8. THE PRINCIPLE OF VIRTUAL WORK FOR A SINGLE PARTICLE

A particle subject to workless constraints is in equilibrium if and only if zero virtual work is done by the applied forces in any arbitrary infinitesimal displacement consistent with the constraints.

Proof

Let the total force applied on the particle be $\overline{F}$ and the total forces applied by the constraint be $\overline{F}_c$. Let the particle be in equilibrium, then

$$\overline{F}_a+\overline{F}_c=0$$

Let $\delta\overline{r}$ *be an* infinitesimal displacement consistent with the constraints, then

$$\left(\overline{F}_a + \overline{F}_c\right).\delta\overline{r} = 0$$

$$\overline{F_a}.\delta\overline{r} + \overline{F_c}.\delta\overline{r}. = 0$$

Since the constraints, is workless, then

$$\overline{F_c}.\delta\overline{r} = 0$$

Thus $\overline{F_a}.\delta\overline{r} = 0,$ i.e., virtual work done by the applied force is zero.

Conversely, let the virtual work done by the applied force be zero, then

$$\overline{F_a}.\delta\overline{r} = 0$$

Also since the constraints is workless, therefore

$$\overline{F_c}.\delta\overline{r} = 0$$

Adding (1) and (2), we get

$$\overline{F_a}.\delta\overline{r} + \overline{F_c}.\delta\overline{r} = 0$$

$$(\overline{F_a} + \overline{F_c}).\ \delta\overline{r} = 0$$

Since $\delta\overline{r}$ is arbitrary, therefore $\delta\overline{r} \neq 0$ and hence

$$\overline{F_a} + \overline{F_c} = 0$$

which shown that the particle is in equilibrium

4.9. THE PRINCIPLE OF VIRTUAL WORK FOR SET PARTICLES

A set of particles, subject to workless constraints is in equilibrium if and only if zero virtual works is one by the applied forces in an arbitrary infinitesimal displacement consistent with constraints.

Let there be n particle in the set. Let $\overline{F_{ia}}$ be total applied force and $\overline{F_{ic}}$ be total force of constraints on the ith particle. Total force on the ith particle will be $\overline{F_{ia}} + \overline{F_{ic}}$. Thus

Total force on all particles $= \sum_{i=1}^{n} \overline{F_{ia}} + \overline{F_c} = 0$

Let the set of particles be in equilibrium, then

$$\sum_{i=1}^{n}\overline{F_{ia}} + \overline{F_{ic}} = 0$$

Let $\delta\overline{r}$ be arbitrary infinitesimal displacement, then

$$\sum_{i=1}^{n}(\overline{F_{ia}} + \overline{F_{ic}}).\delta\overline{r} = 0$$

$$\sum_{i=1}^{n}(\overline{F_{ia}}.\delta\overline{r} + \overline{F_{ic}}.\delta\overline{r}) = 0$$

Since the constraint is workless, therefore

$$\sum_{i=1}^{n}\overline{F_{ic}}.\delta\overline{r} = 0$$

Hence

$$\sum_{i=1}^{n}\overline{F_{ia}}.\delta\overline{r} = 0$$

That is zero virtual work is done by applied forces. Conversely, let virtual work is done by the applied forces be zero, then

$$\sum_{i=1}^{n}\overline{F_{ia}}.\delta\overline{r} = 0$$

Also given that constraints are workless, therefore

$$\sum_{i=1}^{n}\overline{F_{ic}}.\delta\overline{r} = 0$$

Adding (1) and (2), we get

$$\sum_{i=1}^{n}\overline{F_{ia}}.\delta\overline{r} + \sum_{i=1}^{n}\overline{F_{ic}}.\delta\overline{r} = 0$$

$$\sum_{i=1}^{n}(\overline{F_{ia}}.\delta\overline{r} + \overline{F_{ic}}.\delta\overline{r}) = 0$$

$$\sum_{i=1}^{n}(\overline{F_{ia}} + \overline{F_{ic}}).\delta\overline{r} = 0$$

Since $\delta\overline{r}$ is arbitrary, therefore $\delta\overline{r} \neq 0, hence$

$$\sum_{i=1}^{n}(\overline{F_{ia}} + \overline{F_{ic}}) = 0$$

4.10. THE PRINCIPLE OF VIRTUAL WORK OF A SINGLE RIGID BODY

A rigid body, subject to workless constraints is equilibrium if and only if zero virtual work is done by the applied forces and applied torques in any arbitrary infinitesimal displacement consistent with constraints.

Proof

A given system of forces acting on rigid body can be reduced to a single $\overline{R}$ at any arbitrary point in the plane of forces together with a couple $\overline{G}.Let$

$\overline{R} = \overline{R}_a + \overline{R}_c$

where $\overline{R}_a\, is$ sum of applied forces and $\overline{R_c}\, is$ the sum of forces of constraints. Let

$$\overline{G} = \overline{G_a} + \overline{G_c}$$

where $\overline{G_a}\, is\, the$ sum of applied torque and $\overline{G}_c$ is the sum of torque of constraints. Let the body be in equilibrium, then

$$\overline{R} = 0\, and\, \overline{G} = 0$$

Any arbitrary virtual displacement of the body will consist of a translation $\delta\overline{r}$ and rotation $\overline{\delta\theta}$. Thus

$$\overline{R}.\delta\overline{r} = 0\, and\, \overline{G}.\delta\overline{\theta} = 0$$

By adding, we have

$$\overline{R}\,.\,\delta\overline{r} + \overline{G}.\delta\overline{\theta} = 0$$

By using $(1)\, and\, (2),$ we get

$$(\overline{R_a} + \overline{R_c}).\delta\overline{r} + \left(\overline{G_a} + \overline{G_C}\right).\delta\theta = 0$$

$$\overline{R_a}.\delta\overline{r} + \overline{R_c}.\delta\overline{r} + \overline{G_a}.\delta\theta + \overline{G_c}.\delta\theta = 0$$

Since constraints are workless, therefore,

$$\overline{R_c}.\delta\overline{r} + \overline{G_c}.\delta\theta = 0$$

And therefore

$$\overline{R_a}.\delta\overline{r} + \overline{G_a}.\delta\theta = 0$$

So, the work done by applied forces and applied torque is zero. Conversely Let work done by applied forces and applied torque be zero, then

$$\overline{R_a}.\delta\overline{r} + \overline{G_a}.\delta\theta = 0 \qquad (3)$$

Also given constraints are workless, therefore

$$\overline{R_c}.\delta\overline{r} + \overline{G_c}.\delta\theta = 0 \qquad (4)$$

Adding (3) *and* (4). We get

$$\overline{R_a}.\delta\overline{r} + \overline{R_c}.\delta\overline{r} + \overline{G_a}.\delta\theta + \overline{G_c}.\delta\theta = 0$$

$$(\overline{R_a} + \overline{R_c}).\delta\overline{r} + \left(\overline{G_a} + \overline{G_C}\right).\delta\theta = 0$$

$$\overline{R_c}.\delta\overline{r} + \overline{G_c}.\delta\theta = 0$$

Since $\delta\overline{r}$ and $\delta\theta$ are arbitrary, therefore

$\overline{R} = 0$ And $\overline{G} = 0$

That is the body is in equilibrium.

4.11. THE PRINCIPLE OF VIRTUAL WORK FOR A SET OF RIGID BODIES

A set of rigid bodies. Subject to workless constraints, is in equilibrium if and only if zero virtual work is done by the applied forces and applied torques in any arbitrary infinitesimal displacement consistent.

Proof

Let there be n rigid bodies. Given system of forces acting on the ith rigid body can be reduced to a single force $\overline{R_i}$ at an arbitrary point in the plane of forces together with a couple $\overline{G_i}$.

Let $\overline{R_i} = \overline{R_{ia}} + \overline{R_{ic}}$

where $\overline{R_{ia}}$ is the sum of the applied forces and $\overline{R}\_ic$ is the sum of the forces of constraints on the ith rigid body. Let

$$\overline{G_i} = \overline{G_{ia}} + \overline{G_{ic}}$$

where $\overline{G_{ia}}$ the sum of is applied torques and $\overline{G_{ic}}$ is the sum of the torque of constraints on the ith rigid body. For equilibrium

$$\sum_{i=1}^{n} \overline{R_i} = 0 \, and \sum_{i=1}^{n} \overline{G_i} = 0$$

Any arbitrary virtual displacement of the body will consist of a translation $\delta\overline{r}$ *and* rotation $\delta\theta$. Thus

$$\sum_{i=1}^{n} \overline{R_i}.\delta\overline{r} = 0 \quad \text{and} \quad \sum_{i=1}^{n} \overline{G_i}.\delta\overline{\theta} = 0$$

Adding

$$\sum_{i=1}^{n} \overline{R_i}.\delta\overline{r} = 0 + \sum_{i=1}^{n} \overline{G_i}.\delta\overline{\theta} = 0$$

By using (1) and (2), we get

$$\sum_{i=1}^{n} (\overline{R_{ia}} + \overline{R_{ic}}).\delta\overline{r} = 0 \sum_{i=1}^{n} (\overline{G_{ia}} + \overline{G_{ic}}).\delta\overline{\theta} = 0$$

$$\sum_{i=1}^{n} \overline{R_{ia}}.\delta r + \sum_{i=1}^{n} \overline{R_{ic}}.\delta r + \sum_{i=1}^{n} \overline{G_{ia}}.\delta\overline{\theta} + \sum_{i=1}^{n} \overline{G_{ic}}.\delta\overline{\theta} = 0$$

Since constraints are workless

$$\sum_{i=1}^{n} \overline{R_{ic}}.\delta\overline{r} + \sum_{i=1}^{n} \overline{G_{ic}}.\delta\overline{\theta} = 0$$

and therefore

$$\sum_{i=1}^{n} \overline{R_{ia}}.\delta\overline{r} + \sum_{i=1}^{n} \overline{G_{ia}}.\delta\overline{\theta}$$

That is virtual work done by the applied forces and torque is zero. Conversely . Let virtual work done by the applied forces and applied torque be zero, then

$$\sum_{i=1}^{n} \overline{R}_{ia}.\delta\overline{r} + \sum_{i=1}^{n} \overline{G}_{ia}.\delta\overline{\theta}$$

Also given constraints are workless, then

$$\sum_{i=1}^{n} \overline{R}_{ic}.\delta\overline{r} + \sum_{i=1}^{n} \overline{G}_{ic}.\delta\overline{\theta} = 0$$

Adding (3) and (4), we get

$$\sum_{i=1}^{n} \overline{R}_{ia}.\delta\overline{r} + \sum_{i=1}^{n} \overline{R}_{ic}.\delta\overline{r} + \sum_{i=1}^{n} \overline{G}_{ia}.\delta\overline{\theta} + \sum_{i=1}^{n} \overline{G}_{ic}.\delta\overline{\theta} = 0$$

$$\sum_{i=1}^{n} (\overline{R}_{ia} + \overline{R}_{ic}).\delta\overline{r} = 0 \sum_{i=1}^{n} (\overline{G}_{ia} + \overline{G}_{ic}).\delta\overline{\theta} = 0$$

$$\sum_{i=1}^{n} \overline{R}_{i}.\delta\overline{r} + \sum_{i=1}^{n} \overline{G}_{i}.\delta\overline{\theta} = 0$$

Since $\delta\overline{r}$ and $\delta\overline{\theta}$ are arbitrary. Thus

$$\sum_{i=1}^{n} \overline{R}_{i}.\delta\overline{r} = 0 \, and \sum_{i=1}^{n} \overline{G}_{i}.\delta\overline{\theta} = 0$$

i.e., the Set of rigid body is in equilibrium.

Example

Four equal heavy uniform rods are freely jointed to form a rhombus $ABCD$ which is freeky suspended from A and kept in shape of a square by an inextensible string connecting A and C. show that tension in the string is 2w, where W is the weight of one rod

Solution

Since A is a fixed point so we measure the distance from A. The weight of four rod is 4W which act at center of gravity G, where G is point of intersection of Diagonals AC and BD.

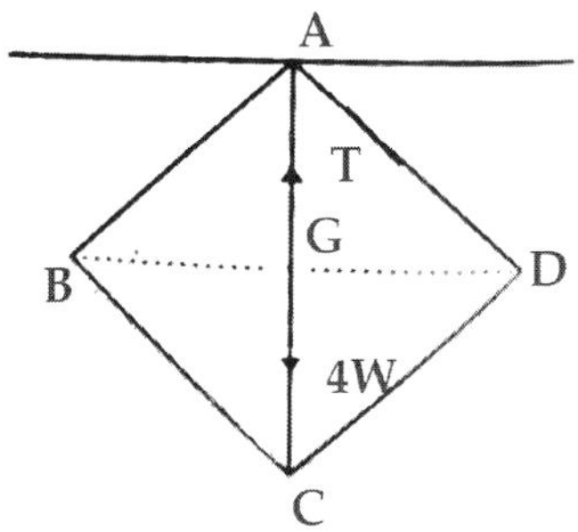

Suppose the string is removed and a force equal to tension T is applied at C along AC. Let

AG = y then AC =2y

Equation of virtual work

$$4W\,\delta(AG) - T\,\delta(AC) = 2y$$

$$4w\delta(y) - T\delta(2y) = 0$$

$$4W - 2T = -$$

$$T = 2W$$

Example

A frame work ABCD consist of four equal light rods are smoothly jointed together to form a square; it is suspended from a peg at A and a weight W is attached to c, the framework being kept in shape by a light rod connecting B and D. Determine the thrust in the rod

Solution

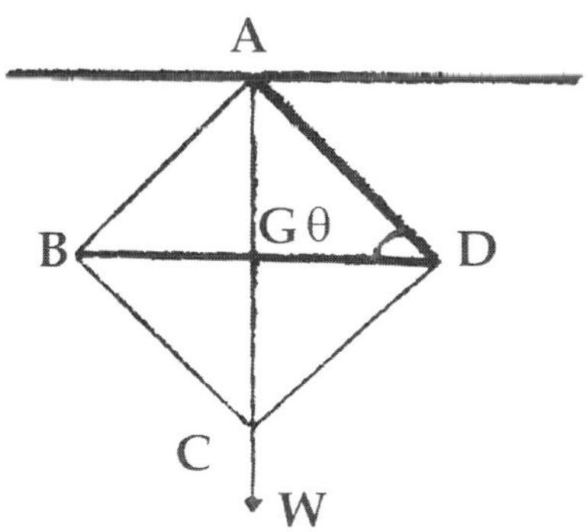

Let $\angle ADB = \theta$

Let l be the length of each rod, then

AD = l

Thus
$$AG = AD\ \sin\theta = l\ sin\,\theta$$
$$DG = AD\, cos\theta = l\ cos\theta$$
$$BD = 2(DG) = 2l\ cos\theta$$
$$AC = 2(AG) = 2l\sin\theta$$

Let T be the thrust in rod BD. The equation of the virtual work
$$W\,\delta(AC) + T\,\delta(2lcos\theta) = 0$$
$$W\,\delta(2lsin\delta) + T\,\delta(2lcos\theta) = 0$$
$$W\ 2lcos\theta\delta\theta - T\ 2l\ sin\theta\delta\theta = 0$$
$$W\cos\theta - T\sin\theta = 0$$
$$T = W\ cot\theta$$

Put $\theta = 45^0$, because $ABCD$ is a square, then
$$T = W$$

Note: For the thrust in the rod T is positive whereas for tension the string T is negative.

Example

Four uniform rods are freely jointed at their extremities and from a parallelogram ABCS, which is suspended from the joint A and is kept in shape by an inextensible string AC. Prove that the tension of the string is equal to half of the whole weight.

Solution

Let W be the weight of the four rods. It acts at its center of gravity D where G point of intersection is ofAC andBD. Suppose the string is removed and forces equal to tension T is applied at C along AC.

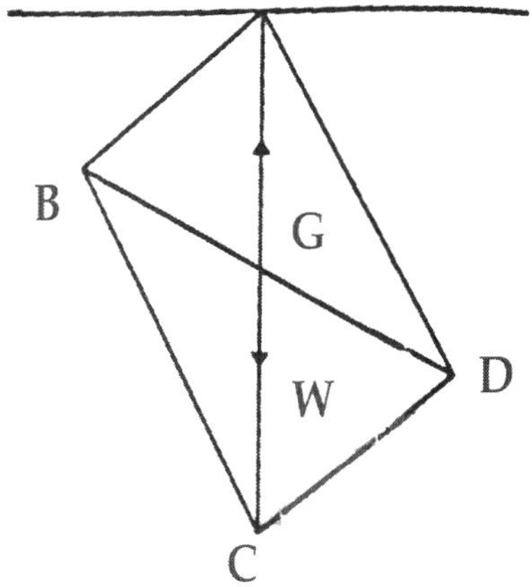

Let $AG = y$, then

$$AC = 2y$$

Equation of virtual work us

$$W\delta(AG) - T\delta(AC) = 0$$

$$W\delta(y) - T\delta(2y) = 0$$

$$W\delta - 2T\delta y = 0$$

$$W - 2T = 0$$

$$T = \frac{W}{2}$$

Example

Six equal bars are smoothly hinged at their extremities, forming a regular hexagon ABCDEF which is kept in shape by vertical strings joining the middle points of BC, CD and AF, FE, respectively. The side AB being held horizontal and upper most. Prove that the tension of each string is three times the weight of the bar.

Solution

Let W be the weight of each bar, then weight of six bars in 6W which acts at the center of gravity G. Let P, Q, L, M and N be mid points of bars BC, CD, AB, AF and EF, respectively. Let θ be the angle which bar AF

makes with horizontal. Let l be the length of each bar. Let $PQ \, and \, MN$ intersect CF at K and H, respectively.

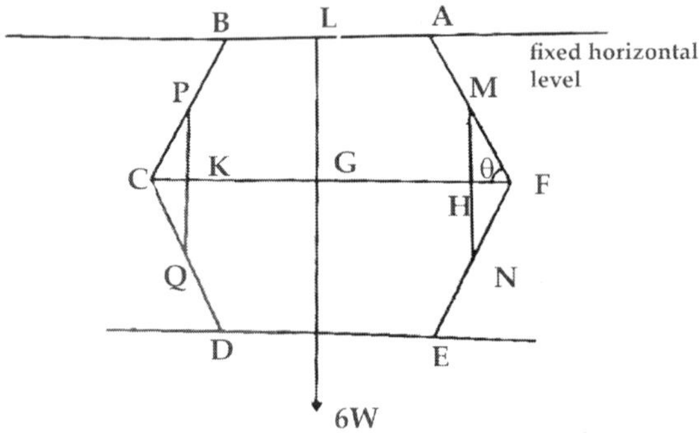

In triangle MFH

$$HF = MF \, c0s\,\theta = \frac{1}{2}(AF)\theta = \frac{1}{2} l \cos\theta$$

$$MH = MF \sin\theta = \frac{1}{2}(AF)\sin\theta = \frac{1}{2} l \sin\theta$$

Also,

$$MN = 2(MH) = l \, sin\theta$$

$$PQ = MN = l \sin\theta$$

$$LG = 2(MH) = l \, sin\theta$$

Let Tbe ether tension in each string, then the equation of virtual work is

$$6W\,\delta(LG) - T\delta(PQ) - (MN) = 0$$

$$6W\,\delta(lsin\theta) - T\delta(lsin\theta) - T\delta(lsin\theta) = 0$$

$$6W\,cos\theta - T\,cos\theta - T\,cos\theta = 0$$

$$6W - 2T = 0, T = 3W$$

Example

Six equal uniforms rods $AB, BC\,SD, SE, EF, FA$, each of weight W are freely jointed to form a regular hexagon. The rod AB is fixed in horizontal position and the slope of the hexagon is maintained by a light rod joining C and F. Show that the thrust in thus rod is $\sqrt{3}$ W.

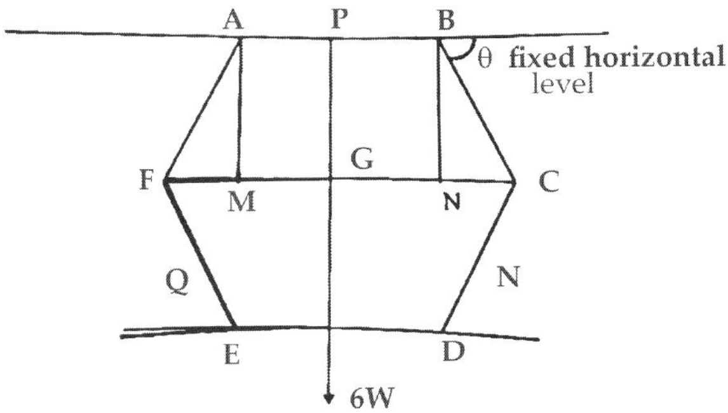

Solution

Let θ be the angle which rod BC makes with horizontal. Let Am and BN are perpendicular to CF. Let p be mined point of AB. *Let P* be mind point of AB. Let l be the length of each rod, weight of six rods is 6w which acts at G, the center of gravity of six rods.

In ΔBCN

$$BN = BC \sin\theta = l\,sin\theta$$

$$CN = BC \cos\theta = l\,cos\theta$$

Therefore,

$$PG = BN = l\,sin\theta$$

$$CF = CN + MN + MF$$

$$CF = l\,cos\theta + l + lcos\theta = 1 + 21\,cos\theta$$

Let be thrust in rod CF, then equation of virtual work is

$$6W\,\delta(PG) + T\delta(l + 2l\,cos\theta) = o$$

$$6w\delta(l\sin\theta) + T\delta(l + 2lcos\theta) = 0$$

$$6wl\,cos\theta\delta\theta - T2l\,sin\theta\delta\theta = 0$$

$$6W\,cos\theta - 2T\,sin\theta = 0$$

$$T = 3W\,cot\theta$$

For regular hexagon, put $\theta = 60^0, then$

$$T = \sqrt{3}W$$

Example

A string of length 'a' forms the shorter diagonal of a rhombus formed by four uniform rods, each of length 'b' and weight W, which are hinged together. If one of the rods be supported in a horizontal position, prove that the tension in the string is

$$\frac{2W\left(2b^2 - a^2\right)}{b\sqrt{4b^2 - a^2}}$$

Solution

Given length of each rod = b, so that

$$AB = BC = CD = DA = b$$

Also given $AC = a$. Let $\angle ADC = 2\theta, then \angle ABD = \theta$ and $\angle BDC = \theta. since W$ is the weight of each rod, therefore weight of four rods is 4W, which act at point G, the point of intersection $of\ AC\ and\ BD$

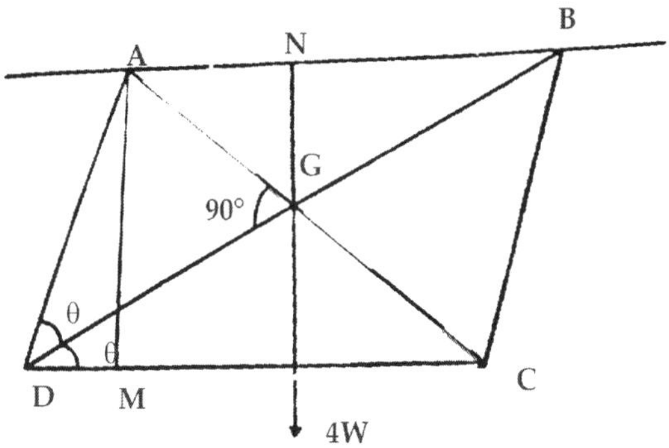

In triangle ΔADM ,AM = AD $\sin\theta = b\sin 2\theta$

Also

$NG =$ Depth of G below AB = $\frac{1}{2}(AM) = \frac{1}{2}b\sin 2\theta$

In triangle AGD

$$AG = AD\,sin\theta = b\,sin\theta$$

Thus

$$AC = 2(AG) = 2b\sin\theta$$

But $AC = a$, therefore

$$sin\theta = \frac{a}{2b}$$

Let T be the tension is the string, the equation of virtual work is

$$4w\delta(NG) - T\delta(AC) = 0$$

$$4W\delta\left(\frac{1}{2}b\sin 2\theta\right) - T\delta(2bsin\theta) = 0$$

$$4W\, cos2\theta - 2T\cos\theta = 0$$

$$T = \frac{2W\, cos2\theta}{cos\theta} = \frac{2W\,(1-\sin^2\theta)}{\sqrt{1-\sin^2\theta}}$$

Put $\sin\theta = \frac{a}{2b}$. Then

$$= \frac{2\left(1 \quad \frac{}{}\right)}{\sqrt{\quad \frac{a^2}{}}} = \frac{2W(2b^2 \quad a^2)}{b\sqrt{b^2 \quad a^2}}$$

Example

A uniform rod of length 2 rest in equilibrium against a smooth vertical wall and upon a smooth peg at a distance b from the wall. Show that, in position of equilibrium, the beam is inclined to the wall at an angle $\sin^{-1}\left(\frac{b}{a}\right)^{\frac{1}{3}}$

Solution

Let the rod be inclined at an angle θ to the wall. Given that

$$Ab = 2a$$

Therefore'

$$AG = a$$

$$PN = b.$$

Height of center of gravity G above the peg is MN. So that

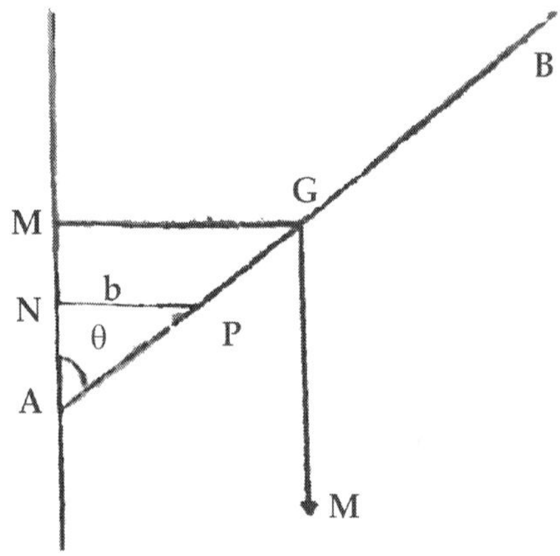

$$MN = AM - AN = AG\cos\theta - NP\cot\theta$$

$$MN = a\cos\theta - b\cot\theta$$

Let $z = a\cos\theta - b\cot\theta$, then equation of virtual work is

$$-W\,dz = 0$$

$$a\sin\theta\, d\theta + b\,\mathrm{cosec}^2\theta\, d\theta = 0$$

$a\sin^3\theta = b$ or $\theta = \sin^{-1}\left(\frac{b}{a}\right)^3$

Example

A heavy elastic string whose natural length is $2\pi a$ is placed round a smooth cone whose axis is vertical and whose semi vertical angle has measure α. If W be the weight and λ the modules of the string, prove that it will be in equilibrium in the form of a circle when radius is

$$a\left(1 + \frac{W}{2\pi a}\cot\alpha\right).$$

Solution

Let the string be in equilibrium at a depth y below the vertex of the cone, the radius of circle be

$$x, then\, tan\alpha = \frac{x}{y}$$

$$x = y \tan \alpha$$

Circumference = $2\pi x$

By Hook's law the tension in the string = T

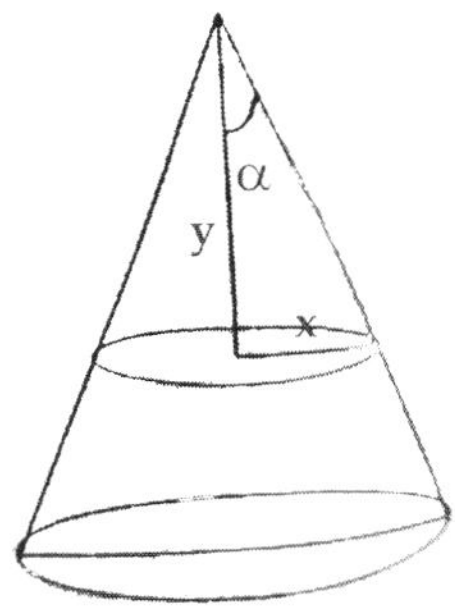

or

$$T = \frac{\lambda(\text{change in lenght})}{\text{orignal lenght}} = \frac{\lambda(2\pi x - 2\pi a)}{2\pi a} = \frac{\lambda(x - a)}{a}$$

Let the ring be displaced downward by displacement δy, the equation of virtual work is

$$W\,\delta y - T\delta(2\pi x) = 0$$

$$W\,\delta - T\,\delta(2\pi \tan \alpha) = 0$$

$$W\,\delta y - 2\pi T \tan \alpha\, \delta y = 0$$

$$W = 2\pi T\alpha = 2\pi \frac{\lambda(x - a)}{a} tan\alpha$$

$$x = a\left(1 + \frac{w}{2\pi a} cot\alpha\right)$$

or

Example

A rhombus is formed of four uniforms rods jointed together and suspended from A. it is kept in position by a light rod joining the middle points of BN and CD; prove that if T be the thrust in this rod and W the weight of the rhombus, then

$$T = W tan\left(\frac{A}{2}\right)$$

Solution

The weight W of the rhombus acts at its center of gravity G. Let P and Q be midpoint of BC and CD respectively. Let l be the length of each rod and

$$\angle CAD = \theta \, and \, \angle ACD = \theta . In \, triangle \, ADG, we \, have$$

$$AG = AD\,cos\theta = l\,cos\theta$$

$$Let\, PQ\, interest\, AC\, at\, N, then\, in\, \Delta CNQ$$

$$QN = CQ\,sin\theta$$

$$PQ = 2QN = 2CQ\,sin\theta = CD\,sin\theta = l\,sin\theta$$

Equation of virtual work is

$$W\delta(AG) + T\delta(PQ) = 0$$

$$W\delta(lcos\theta) + T\delta(l\,sin\theta) = 0$$

$$-W\,lsin\theta\delta\theta + Tl\,cos\theta\delta\theta = 0$$

$$W\,sin\theta - T\,cos\theta = 0$$

$$T = W\,tan\theta = W\,\tan\left(\frac{A}{2}\right)$$

Example

Four equal uniform rods, each of weight w, are connected at one end of each by means of a smooth joint and the other ends rest on a smooth table and are connected by equal strings. A weight W is suspended form the joint. Show that the tension in each strings is

$$\frac{W + 2w}{4}\frac{a}{\sqrt{4l^2 - 2a^2}}$$

where l is the length of each rod and a is length of each string.

Solution

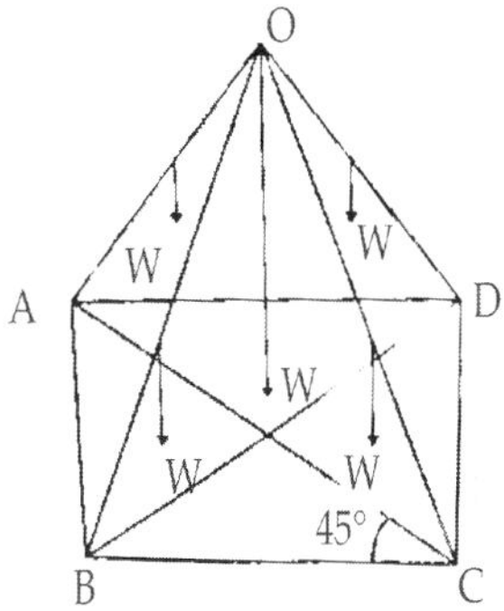

Here $OA = OB = OC = OD = l$

$$AB = BC = CD = AD = a$$

Let the suspended weight W makes angle θ with each rod. In triangle OCG

$$CG = OC\,sin\theta = l\,sin\theta$$

$$OG = OC\,cos\theta = l\cos\theta$$

Also $AC = 2(CG) = 2l\,sin\theta$

In triangle ABC

$$BC = AC\,cos45 = 2lsin\,\theta cos45 = \sqrt{2}\,l\,sin\theta$$

But, given that $BC = a,$ therefore

$$sin\theta = \frac{a}{\sqrt{2l}}$$

$$tan\theta = \frac{a}{\sqrt{2l^2 - a^2}}$$

Height of midpoint of each rod above the plane $ABCD$

$$\frac{1}{2}(OG) = \frac{1}{2}(lcos\theta)$$

Equation of virtual work is

$$-4w\delta\left(\frac{1}{2}cos\theta\right) - W\delta(lcos\theta) - 4T\delta\left(\sqrt{2}\,l\,sin\theta\right) = 0$$

$$2\,wlsin\theta\delta\theta + Wl\,sin\theta\delta\theta - 4\sqrt{2}\,lT\cos\theta\delta\theta = 0$$

$$2w + W - 4\sqrt{2}T\tan\theta = 0$$

$$T = \frac{W+2w}{4\sqrt{2}}tan\theta = \frac{W+2w}{4}\left(\frac{a}{\sqrt{4l^2 - 2a^2}}\right)$$

4.12. KINEMATICS

4.12.1. Dynamics

It is the branch of mechanics which deals with motion of objects. It is divided into two branches namely Kinematics and kinetics.

- **Kinematics**

  It deals with the geometry of motion without any reference to the forces acting on the body.

- **Kinetic**

It relates the motion of the body with the forces causing the motion.

4.13. CARTESIAN COMPONENT OF VELOCITY AND ACCELERATION

Consider the motion of the particle in thy xy-plane, as shown in figure below.

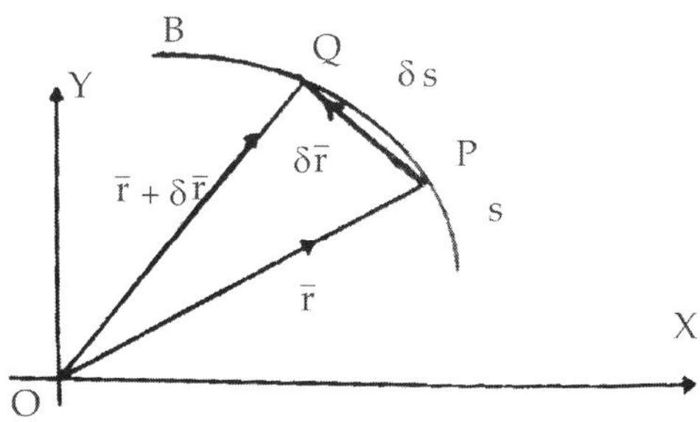

Let $P(\overline{r})\, and\, Q(\overline{r} + S\overline{r})$ be the position of the particle at time t and $t + \delta t$ respectively. Then $\overline{PQ} = \delta\overline{r}$ let S be the length of arc AB measured from some fixed point on the curve. Let are PQ = δs. Then $\frac{\delta\overline{r}}{\delta t}$ $Lim\frac{\delta\overline{r}}{\delta t}$ is called instantaneous velocity of the particle at P and denoted by $\overline{a}.Let\, P(x,y)$ be a point; and

$$\overline{r} = xi + yj.then$$

$$\overline{v} = \frac{d\overline{r}}{dt} = \frac{dx}{dt}i + \frac{dy}{dt}j$$

and $$\overline{a} = \frac{d^2\overline{r}}{dt^2} = \frac{d^2x}{dt^2}i + \frac{d^2y}{dt^2}j$$

$v_x = x -$ Component of velocity $= \dfrac{dx}{dt}$

$v_y = y -$ Component of velocity $= \dfrac{dy}{dt}$

$a_x = x -$ Component of acceleration $= \dfrac{d^{2x}}{dt^2}$

$a_y = x -$ Component of acceleration $= \dfrac{d^{2y}}{dt^2}$

$$v = |\overline{v}| = \sqrt{\left(\frac{dx}{dt}\right)^2 + \left(\frac{dy}{dt}\right)^2} = \frac{ds}{dt}$$

Example

Describe the path of a particle whose position vector is $\overline{r} = a\,cosh\,\omega t\overline{i} + b\,sinh\omega t\,\overline{j}\ (0 \le t < \infty)$, where a, b and are constants. Find its velocity and acceleration.

Solution

Here

$$\overline{r} = a\,cosh\omega\,\overline{i} + b\,sinh\omega\,\overline{j}, then$$

$$x = a\cosh \omega t, y = b\,sinh\omega t$$

$$\frac{x}{a} = cosh\omega t\,\frac{y}{b} = sinh\omega t$$

Thus $\frac{x^2}{a^2}-\frac{y^2}{b^2}=\cosh^2(\omega t)-\sinh^2(\omega t)=1$ is required path describe by the particle.

Now

$$\bar{v}=\frac{d\bar{r}}{dt}=a\omega\, sinh\omega\,\bar{i}+b\omega\cosh\omega t\,\bar{j}$$

$$\bar{a}=\frac{d^2\bar{r}}{dt^2}=a\omega^2 cosh\bar{i}+b\omega^2\, sinh\omega t\,\bar{j}$$

$$=\omega^2\left(a\, cosh\omega\bar{i}+bsinh\omega\bar{j}\right)=\omega^2\bar{r}$$

Thus

$$\bar{a}=\omega^2\bar{r}$$

Example

The position of the particle moving along an ellipse is given by $\bar{\mathrm{r}}=\mathrm{a}\cos\bar{\mathrm{i}}+\ \mathrm{b}\sin \mathrm{t}\ \bar{\mathrm{j}}$.if $\mathrm{a}>b$. Find the position of the particle where its velocity has a maximum or minimum magnitude

Solution

Given

$$\bar{r}=a\cos t\,\bar{i}+b\sin t\,\bar{j},$$ then

$$\bar{v}=\frac{d\bar{r}}{dt}=-a\, sint\,\bar{i}+b\cos t\,\bar{j}$$

$$v=|\bar{v}|=\sqrt{a^2\sin^2 t+b^2\cos^2 t}$$

$$=\sqrt{a^2\sin^2 t+b^2(1-\sin^2 t)}$$

$$=\sqrt{a^2\sin^2 t+b^2-b^2\sin^2 t)}$$

or

$$v=\sqrt{b^2+\left(a^2-b^2\right)\sin^2 t}$$

v is maximum when $\sin^2 t=1$, therefore

$$sint = \pm 1$$

when $\sin = 1, t = \pi /2$

when $\sin = -1, t = 3(\pi /2)$

Thus

$$t = \frac{\pi}{2}, 3\left(\frac{\pi}{2}\right)$$

For

$$t = \frac{\pi}{2}, {}_{3}\left(\frac{\pi}{2}\right)$$

$$\bar{r} = a\cos\left(\frac{\pi}{2}\right)\bar{i} + b\sin\left(\frac{\pi}{2}\right)\bar{j} = b\,\bar{j}$$

For

$$t = \frac{3\pi}{2}$$

$$r = a\cos\left(\frac{3\pi}{2}\right)\bar{i} + b\sin\left(\frac{3\pi}{2}\right)\bar{j} = -b\,\bar{j}$$

Therefore

$$\bar{r} = \pm b\bar{j}$$

v is minimum when $\sin^2 t = 0, then\, t = 0, \pi$

For

$$t = 0$$

$$\bar{r} = a\cos(0)\bar{i} + b\sin(0)\bar{j} = a\bar{i}$$

For

$$t = \pi$$

$$\bar{r} = a\cos(\pi)\bar{i} + b\sin(\pi)\bar{j} = a\,\bar{i}$$

Hence velocity has minimum magnitude when $\bar{r} = \pm a\,\bar{i}$

4.14. RADIAL AND TRANSVERSE COMPONENT OF VELOCITY AND ACCELERATION

Let P be the position of the particle at time t. Let $P(r,o)$ be its polar coordinates. The direction and that perpendicular to it in the direction of increasing θ is called the transverse direction.

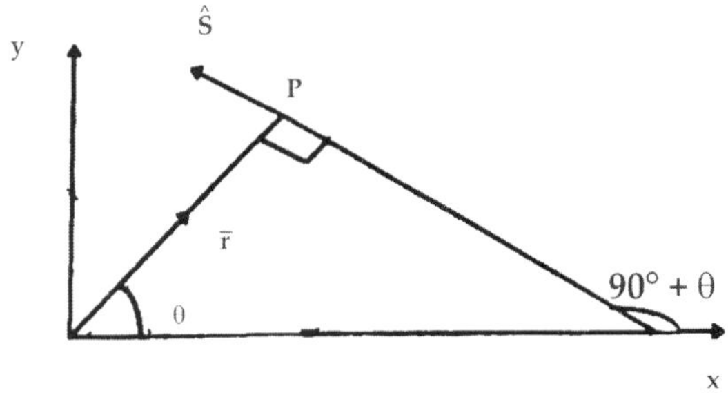

$$\overline{r} = cos\theta\,\overline{i} + sin\theta\,\overline{j}$$

$$\overline{s} = \cos(90+\theta)\overline{i} + \sin(90+\theta)\overline{j}$$

$$= -\sin\theta + \cos\theta\,\overline{j}$$

Now

$$\frac{d\overline{r}}{dt} = \frac{d\overline{r}}{d\theta}$$

$$= (-\sin\theta\overline{i} + cos\theta\overline{j})\frac{d\theta}{dt} = \overline{s}\theta = \overline{s}\theta; \frac{d\theta}{dt} = \theta$$

$$\frac{d\overline{s}}{dt} = \frac{d\overline{s}}{d\theta}\frac{d\theta}{dt}$$

$$= -(\cos\theta\overline{i} + sin\theta\overline{i})\frac{d\theta}{dt} = \overline{r}\theta = -\theta\overline{r}$$

Now

$$\overline{v} = \frac{d\overline{r}}{dt} = \frac{d}{dt}(r.\overline{r}) = \frac{dr}{dt}\overline{r} + \text{r}\frac{d\overline{r}}{dt} + r\frac{d\overline{r}}{dt} = r\overline{r} + r\theta\overline{s}$$

Thus

Radial component of the velocity = $v_r = r$

Transverse component of velocity = $v_\theta = r\theta.\overline{s}$

Also

$$\overline{a} = \frac{d\overline{v}}{dt} = \frac{d}{dt}\left(s\overline{r} + r\theta\overline{s}\right)$$

$$= \overset{ÿ}{r}\hat{r} + \dot{r}\dot{\theta}\hat{s} + \dot{r}\dot{\theta}\hat{s} + \overline{r}\,\ddot{\theta}\,\overline{s} + r\,\theta(\dot{\theta}\widehat{r})$$

$$= \left(\ddot{r} + r\,\theta^2)\hat{r} + (2\dot{r}\,\overset{ÿ}{\dot{\theta}} + r\,\ddot{\theta}\right)\hat{s}$$

Thus

Radial component of the acceleration = $a_r = \ddot{r} + r\theta^2$

Transverse component of acceleration = $a_0 = 2\overset{ÿ}{r}\,\overset{ÿ}{\theta} + r\,\ddot{\theta}$

Note that

$$a_0 = \frac{1}{r}\frac{d}{dt}\,(r^2\,\overset{ÿ}{\theta})$$

Example

Find the radial and transverse components of the velocity of a particle moving along the curve $ax^2 + by^2 = 1$ at any time t if the polar angle $\theta = ct^2$

Solution

Given that $\theta = ct^2, then\,\theta = 2ct$

Now changing the equation of the given curve into polar coordinates by,

$$x = r\cos\theta\; y = r\sin\theta$$

we get

$$ar^2\,cos^2\theta + br^2\sin^2\theta = 1$$

or

$$r = \left(a\,cos^2\theta + b\sin^2\theta\right)^{\frac{1}{2}}$$

Thus

$$r = \frac{dt}{dt} = \frac{dr}{d\theta}\frac{d\theta}{dt}$$

$$= -\frac{1}{2}\left(a\cos^2\theta + b\sin\theta\right)^{-\frac{3}{2}}(2)\left(-a+b\right)\sin\theta\, cps\,\theta\theta$$

$$= -\left(a\cos^2\theta + b\sin\theta\right)^{\frac{3}{2}}\left(-a+b\right)\left(2ct\right)\sin\theta\cos\theta$$

$$v_r = r = \frac{ct\left(a-b\right)sin2\theta}{\left(a\cos^2\theta + b\,sin^2\theta\right)^{\frac{3}{2}}}$$

$$v_\theta = r\theta = \frac{2ct}{\left(a\cos^2\theta + b\,sin^2\theta\right)^{\frac{1}{2}}}$$

Example

The velocities of particle along the perpendicular to the radius vector are λr^2 and $\mu\theta^2$, where $\lambda\, and\, \mu$ are constants. Find its radial and transverse components of acceleration in terms of r and θ

Solution

Here

$$v_r = r\,\lambda r^2\; and\; v_\theta = r\theta = \mu\theta^2$$

or

$$r = \lambda r^2\; and\; \theta = \frac{\mu\theta^2}{r}$$

Therefore, by differentiating, we get

$$r = 2\lambda rr = 2\lambda r\left(\lambda^2 r\right) = 2\lambda^2 r^3$$

Thus

$$a_r = r - r\theta^2 = 2\lambda^2 r^3 - r\left(\frac{\mu\theta^2}{r}\right)^2$$

$$= 2\lambda^2 r^3 - \frac{\mu^2\theta^4}{r}$$

$$a_\theta = \frac{1}{r}\frac{d}{dt}\left(r^2\theta\right) = \frac{1}{r}\frac{d}{dt}\left(r^2\theta\right) = \frac{1}{r}\frac{d}{dt}\left\{r^2\left(\frac{\mu\theta}{r}\right)\right\}$$

$$= \frac{1}{r}\frac{d}{dt}\left\{r\left(\mu\theta^2\right)\right\}$$

$$= \frac{\mu}{r}\frac{d}{dt}\left(r\mu^2\right)$$

$$= \frac{\mu}{r}\left(r\theta^2 + 2r\theta\right)$$

$$= \frac{\mu}{r}\left(\lambda r^2\right)\theta^2 + 2\frac{\mu}{r}r\theta\left(\frac{\mu\theta}{r}\right)$$

$$= \mu\lambda r\theta + 2\frac{\lambda^1\theta^3}{r}$$

4.15. TANGENTIAL AND NORMAL COMPONENTS OF VELOCITY AND ACCELERATION

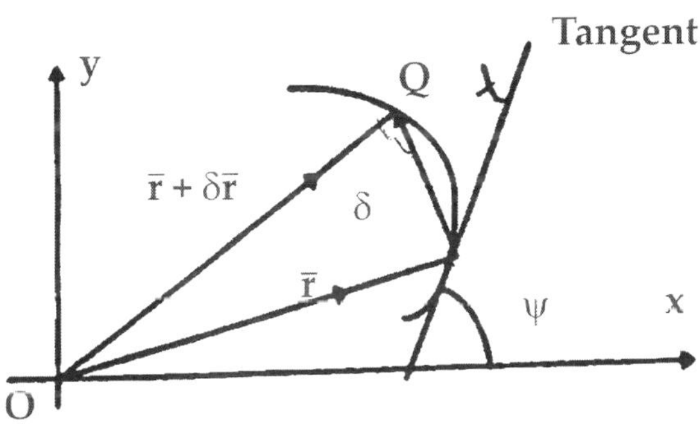

Let $P(\overline{r})$ and Q ($\overline{r}+\delta\overline{r}$) be the positions of the particles at time t and $t+\delta t$, respectively. Then $\overline{PQ} = \delta\overline{r}$ ...... $Let\, arc\, PQ = \delta s, then$

$$\overline{v} = \frac{d\overline{r}}{dt} = \frac{d\overline{r}}{ds}\frac{ds}{dt} = \frac{d\overline{r}}{dt} = \frac{d\overline{r}}{ds}v = v\frac{d\overline{r}}{ds}$$

Now

$$\left|\frac{d\overline{r}}{dt}\right| = Lim_{\delta\to 0}\frac{\left|\overline{\delta r}\right|}{\delta s} = Lim_{\delta\to 0}\frac{\left|\overline{PQ}\right|}{arc\,PQ} = 1$$

Also

$$\frac{d\overline{r}}{dt} = Lim_{\delta\to 0}\frac{\delta\overline{r}}{\delta s} = Lim_{\delta\to 0}\frac{\delta\overline{r}}{\delta s} = Lim_{\delta\to 0}\frac{1}{\delta s}(\delta\overline{r}),$$

which is a vector along the tangent at p. Therefore, $\frac{d\overline{r}}{ds}$ is a unit tangent at p. Let $\overline{t}$ be unit vector along the tangent at p. Therefore,

$$\frac{d\overline{r}}{ds} = \hat{t}$$

Put in eq. (1), then

$$\overline{v} = v\hat{t} = v\hat{t} + 0\hat{n}$$

where $\overline{n}$ is unit vector along normal at p. Thus

$v_t =$ Tangential component of the velocity = v, and

$v_n =$ Normal component of velocity = 0

Hence velocity is along the tangent to the path.

Let $\overline{a}$ be the acceleration, then

$$\overline{a} = \frac{d\overline{v}}{dt} = \frac{d}{dt}(v\hat{t}) = \frac{dv}{dt}\hat{t} + v\frac{d\hat{t}}{dt}$$

$$= \frac{dv}{dt}\hat{t} + v\frac{d\hat{t}}{d\psi}\frac{d\psi}{ds}\frac{ds}{dt}$$

$$= \frac{dv}{dt}\hat{t} + v\frac{d\hat{t}}{d\psi}Kv$$

where

$$K = \frac{d\psi}{ds} \text{ is curvature. Thus}$$

$$\overline{a} = \frac{dv}{dt}\hat{t} + \frac{v^2}{p}\frac{d\hat{t}}{d\psi}$$

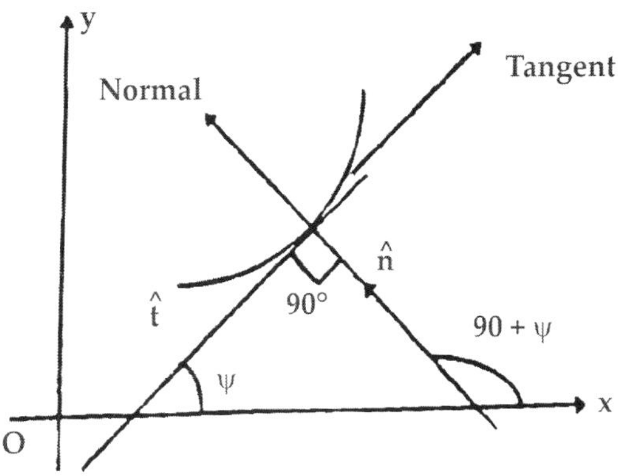

$$K = \frac{1}{\rho}$$

where ρ is radius of curvature. Let $\hat{t}$ and $\hat{n}$ be the unit vectors along tangent and normal at p, then

$$\hat{t} = cos\psi\hat{i} + sin\psi\hat{j}$$

$$\hat{n} = \cos(\psi + 90)\hat{i} + \sin(\psi + 90)\hat{j}$$

or

$$\hat{t} = cos\psi\hat{i} + sin\psi\hat{j}$$

$$\hat{n} = sin\psi\hat{i} + cos\psi\hat{j} = \hat{n}$$

Therefore

$$\frac{d\hat{t}}{d\psi} = -sin\psi\hat{i} + cos\psi\hat{j} = \hat{n}$$

Put in eq. (2), we get

$$\bar{a} = \frac{dv}{dt}\hat{t} + \frac{v^2}{p}\hat{n}$$

Hence

$a_t =$ tangential component of acceleration $= \frac{dv}{dt}$

$a_n =$ Normal component of acceleration $= \frac{v^2}{p}$

Example

Find the tangential and normal components of the acceleration of a particle describing the ellipse $\frac{x^2}{a^2}+\frac{y^2}{b^2}=1$ with uniform speed v when particle is at (0,b).

Solution

Given that

$$\frac{x^2}{a^2}+\frac{y^2}{b^2}=1$$

Differentiating $w.r.t.$ ‘xx’, we get

$$\frac{2x}{a^2}+\frac{2y}{b^2}y'=0\, or\, y'=-\frac{b^2x}{a^2y}$$

Again differentiating

$$y''=\frac{b^2}{a^2}\left(\frac{y-xy'}{y^2}\right)$$

Thus at $(0,b)$

$$y''=\frac{b}{a^2}$$

$$y'=0$$

Now from the differentiate calculus, we know that

$$p=\frac{\left[1+\left(\frac{dy}{dx}\right)^2\right]^{\frac{3}{2}}}{\frac{d^2y}{dx^2}}$$

$$so\, that\, p=\frac{\left[1+(0)^2\right]^{\frac{3}{2}}}{\frac{b}{a^2}}=\frac{a^2}{b}$$

Because p is always positive. Again, since velocity is uniform, therefore

$$\frac{dv}{dt} = 0,$$

hence,

$$a_t = \frac{dv}{dt} = 0, a_n = \frac{by^2}{a^2}$$

Example

A particle is moving with uniform speed along the curve $x^2 y = a(x^2 + \frac{a^2}{\sqrt{5}}$.) show that its acceleration has maximum value $\frac{10v^2}{9a}$.

Solution

Given that

$$x^2 y = a\left(x^2 + \frac{a^2}{\sqrt{5}}\right)$$

$$or\ y = a + \frac{a^3 x^{-2}}{\sqrt{5}}$$

Differentiating $w.r.t.'x', we\ get$

$$y' = \frac{2a^3 x^{-3}}{\sqrt{5}}$$

$$y'' = \frac{6a^3 x^{-4}}{\sqrt{5}}$$

Now from the ditterential calculus, we know that

$$\frac{\left[1 + \left(\frac{dy}{dx}\right)^2\right]^{\frac{3}{2}}}{\frac{d^2 y}{dx^2}}$$

so that

$$p = \frac{\left[1+\left(\frac{4a^6}{5x^6}\right)^2\right]^{\frac{3}{2}}}{\frac{6a^3}{\sqrt{5x^4}}}$$

Since speed is uniform therefore, $\frac{dv}{dt} = 0,$ thus,

$$\bar{a} = \frac{dv}{dt}\hat{t} + \frac{v^2}{p}\hat{n} = \frac{v^2}{p}\hat{n}$$

$$|\bar{a}| = \frac{v^2}{p}\hat{n}$$

$|\bar{a}|$ is maximum when p is minimum, $i.e. \frac{dp}{dx} = 0$

Differentiating (1), then

$$\frac{dp}{dx} = \frac{30a^3x^3\left(\frac{3}{2}\right)\left(5x^6+4a^6\right)^{\frac{1}{2}}\left(30x^5\right)-\left(5x^6+4a^6\right)^{\frac{3}{2}}\left(150a^3x^4\right)}{\left(30a^3x^5\right)^2}$$

$$= \frac{30a^3x^4\left(5x^6+4a^6\right)^{\frac{1}{2}}\left(45x^6-25x^6-20a^6\right)}{\left(30a^3x^5\right)^2}$$

$$= \frac{20\left(5x^6+4a^6\right)^{\frac{1}{2}}\left(x^6-a^6\right)}{30a^3x^6}$$

$$= \frac{20\left(5x^6+4a^6\right)^{\frac{1}{2}}}{30a^3x^6}\left(x+a\right)\left(x-a\right)\left(x^4+a^2x^2+a^4\right)$$

Thus $\frac{dp}{dx} = 0$ implies $x = a, x = -a.$ Hence

$$\frac{dp}{dx} < 0 \, before \, x = a.$$

$$\frac{dp}{dx} > 0 \, after \, x = a.$$

Therefore, p is minimum when $x = a$. Thus for $x = a \, eq.(1)$ gives

$$p = \frac{\left(5a^6 + 4a^6\right)^{\frac{3}{2}}}{30a^3a^5} = \frac{9a}{10}$$

Maximum value of magnitude of acceleration = $\frac{v^2}{p_{min}}$ or $a = \frac{10v^2}{9a}$

Example

A point describes a cycloid s = 4a $sin\psi$ with uniform speed v. Find the acceleration at any point.

Solution

Given that

$$S = 41 \, sin\psi$$

$$\frac{ds}{d\psi} = 4a \, cos\psi = p$$

Now

$$\bar{a} = \frac{dv}{dt}\hat{t} + \frac{v^2}{p} = \hat{n} = \frac{v^2}{p} = \hat{n}$$

Given that v is uniform $i.e \frac{dv}{dt} - 0$

By using (1), we have

$$\bar{a} = \frac{v^2}{4acos\psi}\hat{n} = \frac{v^2}{4a\sqrt{1 - \sin^2 \psi}}\hat{n}$$

or

$$\bar{a} = \frac{v^2}{4a\sqrt{1-\left(\frac{s^2}{16a^2}\right)}}\hat{n} = \frac{v^2}{4a\sqrt{16a^2-s^2}}\hat{n}$$

4.16. RELATIVE VELOCITY

Let $+(\bar{r}_p)$ and $Q(\bar{r}_Q)$ be the positions of two particles moving in $xy-$ plane.

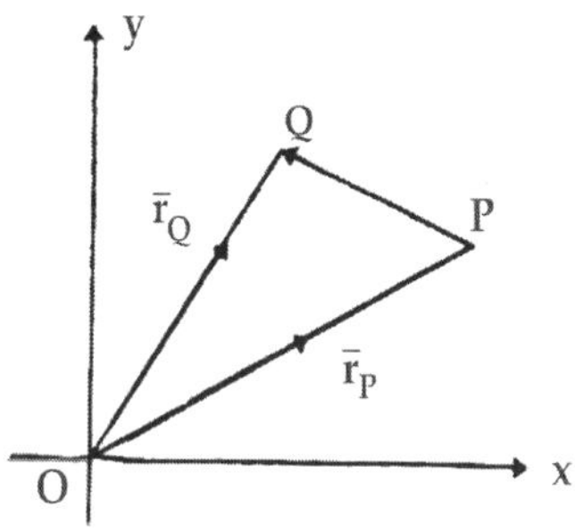

Then $(\overline{PQ}) = (\bar{r}_Q - \bar{r}_{p.})$

The rate of change of displacement vector $\overline{PQ}$ is called relative velocity of Q $w.r.t\ p.$ Thus relative velocity of $Q.w.r.t.p$ will be

$$\frac{d}{dt}(\overline{PQ})$$

or

$$v_{rel} = \frac{d}{dt}(\bar{r}_Q - \bar{r}_p) = \bar{r}_Q - \bar{r}_p$$

= (velocity of Q) – (velocity of P)

If the velocity of two particles is equal them relatives' velocity is zero.

4.17. ANGULAR VELOCITY

The angular velocity ω is the rate of change of angular displacement θ $w.r.t.$ time 't'. Let p be the position of the particles moving about O. Let

$$\angle XOP = \theta then\, \omega = \frac{d\theta}{dt}$$

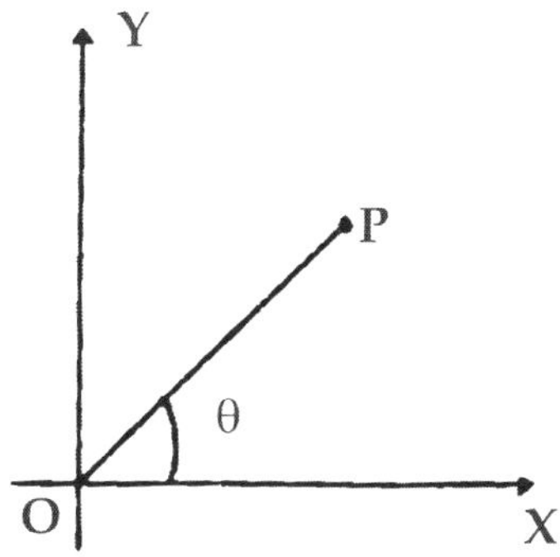

give the magnitude of angular velocity. The angular velocity is a vector quantity and its direction is perpendicular to the plane rotation of particles. Thus

$$\bar{\omega} = \frac{d\theta}{dt}\bar{k}$$

If the rotation is clockwise then the angular velocity is taken as positive otherwise it is negative.

Consider a particle is moving along a circle of radius r and center O. linear velocityvatpisalongthetangentatp.Let $\overline{OP} = \bar{r}, \angle POQ = \theta, arc\, PQ = s, then$

$$s = r\theta, or \frac{ds}{dt} = r\frac{d\theta}{dt}$$

Since r is constant. Thus

$$v = r\omega$$

In vector form this relation is written as

$$v = r\, x\, \omega$$

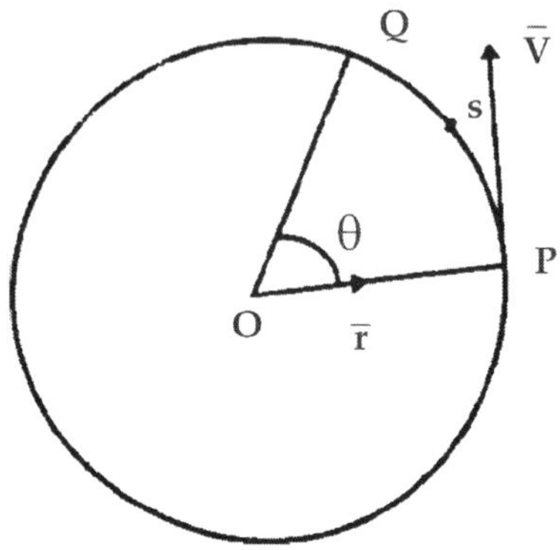

4.18. RECTILINEAR MOTION

4.18.1. Introduction

The motion of a particle along a straight line is called *rectilinear motion.* Let the particle start from point O along a line. We take the line as $x-axis$. Let after time t particle be at point P at a distance x from O. Let $\bar{r}$ be the position vector of point $p w.r.t.$ origin O. Then

$$\bar{r} = \overline{OP} = x\hat{i}$$

$$\bar{v} = \frac{d\bar{r}}{dt} = \frac{dx}{dt}\hat{i}$$

Let $v = |\bar{v}|$ and $a = |\bar{a}|$, then

$$v = \frac{dx}{dt}$$

$$a = \frac{d^2x}{dt^2}$$

Also

$$a = \frac{dv}{dt} = \frac{dv}{dx}\frac{dx}{dt} = \frac{dv}{dx}v = v\frac{dv}{dx}$$

4.18.2. Motion with Constant Acceleration

Let the particle start form O with velocity u at time t = 0 with constant acceleration a. Let the particle be at distance x from O after time t. Then

$a = \frac{dv}{dt}$, where a is constant.

Integration will yield

$$v = at + A$$

Since velocity $v = u$ at $t = 0$, therefore, $A = u$, thus

$$v = at + u = u + at$$

Now, since $v = \frac{dx}{dt}$, therefore eq. (1) gives

$$\frac{dx}{dt} = u + at$$

By integrating

$$x = ut + a\left(\frac{t^2}{2}\right) + B$$

Using $x = 0, at\, t = 0,$ we get, $B = 0,$ hence

$$x = ut + \frac{at^2}{2}$$

Also, $a = v\frac{dv}{dx}$, where a is constant, then

$$ax = v\,dv$$

Integrating

$$\frac{v^2}{2} = ac + C$$

Since v = u, at x = 0, therefore, C = $\frac{u^2}{2}$, and hence

$$v^2 - u^2 = 2ax$$

The equation $(1),(2)$ and (3) are called equation of motion. If the particle moves with constant retardation the acceleration a is taken as negative.

4.18.3. Vertical Motion under Gravity

For a failing body, the acceleration is constant. It is called acceleration due to gravity and is denoted by 'g'. In FPS system g = 32 ft/sec^2. In CGS system g =981 cm/sec^2 and in MKS system $g=9.81m/sec^2$. If the body is projected vertically upward then acceleration g is negative. For falling bodies equation of motion is

$$v=u+gt$$

$$x=ut+\frac{1}{2}gt^2$$

$$v^2-u^2=2gx$$

Example

A particle is projected vertically upward with velocity $\sqrt{2gh}$ and another is let fall from a height h at the same time. Find the height of the point where they meet each other.

Solution

Let both particles meet at point P at height x. Then for the first particle

$$x=ut-\frac{1}{2}gt^2$$

Put

$u=\sqrt{2gh}$, then

$$x=\sqrt{2gh}t-\frac{1}{2}gt^2 \qquad (1)$$

For the second particle

$$x=ut+\frac{1}{2}gt^2$$

$$x=ut+\frac{1}{2}gt^2$$

Put

$$u = 0, x = h - x, \text{ then}$$

$$h - x = 0 + \frac{1}{2}gt^2 \qquad (2)$$

Using $eqs.(1)$ and (2), we get

$$u = \sqrt{2gh}t$$

$$or t = \sqrt{2gh}t$$

Thus equation (1) gives

$$x = \sqrt{2gh}t\left(\frac{h}{\sqrt{2gh}}\right) - \frac{1}{2}g\left(\frac{h^2}{2gh}\right) = h - \frac{h}{4} = \frac{3h}{4}$$

Example

A particle is projected vertically upward. After time t another particle is sent up from the same point with the same velocity and meets the first at height h during the downward flight of the first. Find the velocity of the projection.

Solution

Let u be the velocity of the projection and $v\,be$ the velocity at height h, then

$$v^2 - u^2 = -2gh \text{ or } v = \sqrt{u^2 - 2gh}$$

Now, the time taken by the first particle from height h to the highest point and back to the height h is t, therefore, the time taken from the highest point is t /2.

$$Velocity\ at\ highest\ point = o$$

$$Velocity\ at\ height\ h = v$$

Here we use the formula

$$v = u - gt$$

Put

$$v = 0, u = v, t = \text{ t/2, then}$$

$$0 = v - \frac{gt}{2}$$

From (1) and (2)

$$\sqrt{u^2 - 2gh = \frac{gt}{2}}$$

$$u^2 - 2gh = \frac{g^2t^2}{4}$$

$$u^2 = 2gh + \frac{g^2t^2}{4}$$

$$u = \sqrt{\frac{8gh + g^2t^2}{4}}$$

4.18.4. Distance Traveled in n[th] Second

Let $x_1\ and\ x_2\ b$ e the distance travelled in the first n and n – 1 seconds, respectively, then from the equation of motion

$$x = ut + \frac{1}{2}at^2$$

Then

$$x_1 = un + \frac{1}{2}an^2$$

and

$$x_2 = u(n-1) + \frac{1}{2}a(n-a)^2$$

Distance travelled in the nth second = $x_1 - x_2$

$$x_1 - x_2 = un + \frac{a}{2}n^2 - \left\{u(n-a) + \frac{a}{2}(n-1)^2\right\}$$

$$= un + \frac{a}{2}n^2 - \left\{un - u + \frac{a}{2}\left(n^2 - 2n + 1\right)\right\}$$

$$= u + \frac{1}{2}a(2n-1)$$

4.18.5. Motion with Variable Acceleration

4.18.5.1. Time Dependent Acceleration

As the acceleration is time dependent, then it is function of time Therefore we can write

$$a = f(t)$$

or

$$\frac{dv}{dt} = f(t)$$

or

$$dv = f(t)\,dt$$

Integrating

$$\int dv = \int f(t)\,dt$$

or

$$v = g(t) + A$$

where g(t) = $\int f(t)\,dt,$ and A is constant of integration, which gives the velocity. Also since

$v = \frac{dx}{dt},$ Therefore

$$\frac{dx}{dt} = g(t) + A$$

or

$$dx = g(t)\,dt + A\,dt$$

By integrating, we get

$$x = \int g(t)\,dt + At + B$$

where B is another constant which gives the distance travelled by the particle· By using the initial conditions we can find the constants $A\, and\, B.$

Example

Discuss the motion of the particle moving in a straight line if it stars from rest t = 0 and its acceleration is equal to $accost + bsint$.

Solution

Give acceleration is

$$a = accost + b\,sint$$

By integrating, we have

$$v = asint - b\,cost + A$$

Since

$v = 0\;at\;t = 0$, therefore $A = b$, thus

$$C = asint - bcost + b$$

Also

v = $\frac{dx}{dt} = asint - bcost + b$

Again integrating

$$x = -acost - bsint + bt + B$$

where B is constant of integration.

Since x = 0, at t = 0, therefore B = a. Thus

$$x = -acost - bsint + bt + a$$

$$= a(1 - cost) + b(t - sint)$$

4.18.5.2. Velocity Dependent Acceleration

Let the acceleration a is function of velocity v, then

$a = f(v$

Since

$$a = v\frac{dv}{dx}, therefore$$

$$v\frac{dv}{f(v)} = dx$$

Integrating

$$\int v\frac{dv}{f(v)}=\int dx$$

or

$$x=g(v)+A$$

where $g(v)=\int v\frac{dv}{f(v)} and$ A is constant of integration which gives the distance travelled by the particles.

Again consider a = f(v), since

$a=\frac{dv}{dt}$, therefore

$$\frac{dv}{f(v)}=dt\,or$$

or

$$\int\frac{dv}{f(v)}=\int dt$$

or

$$t=h(v)+B$$

where $h(v)=\int\frac{dv}{f(v)} and\ B\ is$ constant of integration and gives the time.

Example

A particle moves in a straight line with an acceleration kv'. If its initial velocity is u, find the velocity and the time spend when the particle has travelled a distance of x.

Solution

Given that

$$a=kv^3$$

Since

$a = v\frac{dv}{dx}$, therefore

$$v\frac{dv}{dx} = kv^3$$

$$or v^{-2}\, dv = k\, dx$$

$$or \int v^{-2}\, dv = kdx$$

$$or\, v^{-1} = kx + A$$

Given $v = u, x = 0\, at\, t = 0,$ therefore $A = -\frac{1}{u}$. Thus

$$\frac{1}{v} = \frac{1}{u} - kx$$

or

$$c = \frac{u}{1 - kux}$$

Again since $v = \frac{dx}{dt},$ then

$$\frac{dx}{dt} = \frac{u}{1 - kux}$$

or $$(1 - kux)\, dx = u\, dt, thus$$

$$\int (1 - kux)\, dx = u \int dt$$

$$x \aleph ku\left(\frac{x^2}{2}\right) \quad ut \quad B$$

Using $x = 0, at\, t = 0, we\, get\, B = 0.$ Hence

$$x - ku\left(\frac{}{}\right) = ut$$ or $$t = \frac{}{}\left(2 - kux\right)$$

Example

A particle starts with a velocity u and move in a straight line. If the suffers a retardation equal to the square of velocity, find the distance travelled by the particle in time t.

Solution

Given retardation $=v^2$, therefore

$$a=-vt^2$$

We know that

$v\frac{dv}{dx}=a,$ therefore

$$\frac{dv}{dx}=-v$$

$$or\frac{dv}{v}=-dx$$

$$or\int\frac{dv}{v}=-\int dx$$

$$or\ In\,v=-x+A$$

where a is constant of integration.

Using $v=u,x=0\,at\,t=0,$ we have

$$A=In\,u,thus$$

$$\ln\left(\frac{u}{v}\right)=x\,or\frac{u}{v}=e^x\,or\,v=\frac{u}{e^x}$$

Since $v=\frac{dx}{dt},$ therefore

$$\frac{dx}{dt}=\frac{u}{e^x}$$

or

$e^x=u\,dt,$ then

$$e^x=ut+B$$

where B is constant of integration.

Using $x=0\, at\, t=0,$ we get $B=1$. Hence

$$e^{x}=ut+1 or x=in(ut+1)$$

4.18.5.3. Distance Dependent Acceleration

Let the acceleration be a function of distance x. Then a = f(x).

Since

$$v\frac{dv}{dx}=f(x)dx$$

$$\int vdv=\int f(x)dx$$

$$\frac{v^{2}}{2}=g(x)+A$$

where g(x) $\int f(x)dx\, and\, A$ is the constant of integration or $v=\pm\sqrt{2g(x)+2A}$.

We know that

$$\frac{dx}{dt}=v, therefore$$

$$\frac{dx}{dt}=\pm\sqrt{2g(x)+2A}$$

$$dt=\pm\frac{dx}{\sqrt{2g(x)+2A}}$$

Integrating, we get

$$t=\pm\int\frac{dx}{\sqrt{2g(x)\quad 2A}}+B$$

where B is constant of integration :

Example

Discuss the motion of the particle moving in a straight line if it starts at a distance ‘a’ from a point O and move with an acceleration equal to μ times its distance from the O.

Solution

Let x be the distance of particle from O, then

Acceleration $= a = \mu x$

Since $v\dfrac{dv}{dx} = a \, or \, v\dfrac{dv}{dx} = \mu \, x \, or \, vdv = \mu xdx$

or $$\int vdv = \mu \int xdx$$

$$\frac{v^2}{2} = \mu \frac{x^2}{2} + A$$

Given $v = 0, x = a \, at \, t = 0,$ so that

$A = -\mu \dfrac{a^2}{2},$ thus

$$\frac{v^2}{2} = \mu\left(\frac{x^2}{2} - \frac{a^2}{2}\right)$$ or

$$v = \pm\sqrt{\mu}\sqrt{x^2 - a^2}$$

Take $v = \sqrt{\mu}\sqrt{x^2 - a^2},$ since $v = \dfrac{dx}{dt}$ then

$$\int \frac{dx}{\sqrt{x^2 - a^2}} = \sqrt{\mu} \int dt$$ or

$$\cosh^{-1}\left(\frac{x}{a}\right) = \sqrt{\mu} \int dt$$

Using

$$x \aleph a \, at \, t \quad 0, B \quad 0,$$ thus

$$\cosh^{-1}\left(\frac{x}{a}\right) = \sqrt{\mu} \, t$$ or

$$x = a\cosh(\sqrt{\mu t})$$

If we take $v = -\sqrt{\mu}\sqrt{x^2 - a^2}, since \, v = \dfrac{dx}{dt},$ then

$$\int \frac{dx}{\sqrt{x^2 - a^2}} = -\sqrt{\mu} \int dt + C$$

$$\cosh^{-1}\left(\frac{x}{a}\right) = -\sqrt{\mu}\, t + C$$

Using

$x = a\, t = 0, C = 0,$ thus

$$\cosh^{-1}\left(\frac{x}{a}\right) = -\sqrt{\mu}\, t$$

or

$$a \cosh\left(-\sqrt{\mu}\, t\right) = a \cosh\left(\sqrt{\mu}\, t\right)$$

4.18.6. Graphical Solution of Rectilinear Problem

Consider the graph of the curve $y = f(x)$ in the $xy-$ plane, then

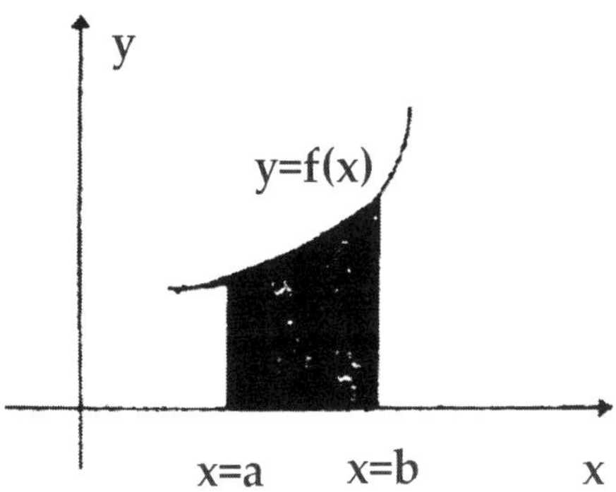

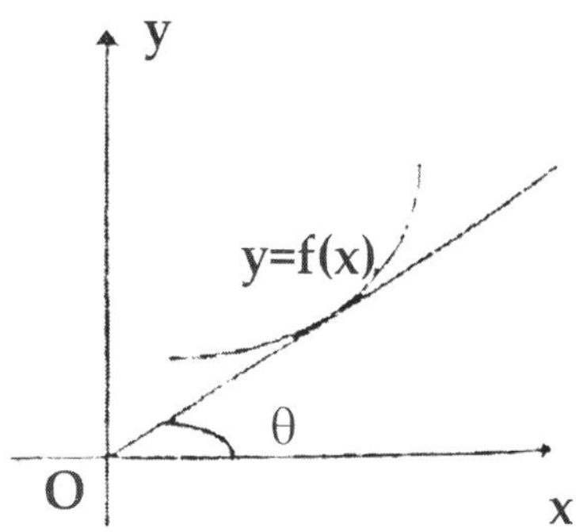

$(a) \frac{dy}{dx} =$ slope of the tangent at any point

(b) Area bounded by the curve $y = f(x), from\, x = a\, to\, x = b s\, is = \int_a^b y dx$

4.18.6.1. Velocity-Time Curve

Consider the graph of velocity time graph v = v(t), then acceleration = a = $\frac{dv}{dt}$ or $a =$ slope of velocity – time curve = $\tan\theta$

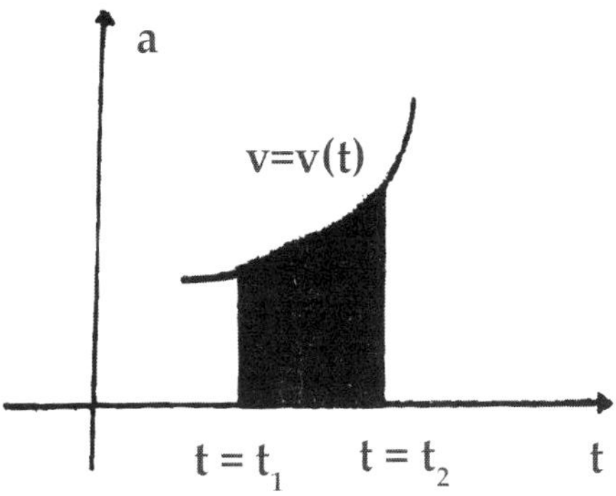

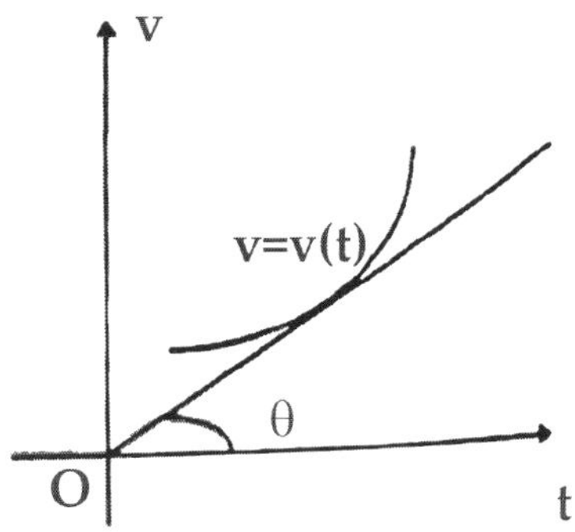

We know that $v = \frac{dx}{dt}$, thus

$dx = vdt$ or

$$\int_{t=t_1}^{t=t_2} dx = \int_{t=t_1}^{t=t_2} vdt$$ or

$$|x|_{t=t_1}^{t=t_2} = \int_{t=t_1}^{t=t_2} vdt$$ or

$$x(t_1) - x(t_2) = \int_{t=t_1}^{t=t_2} vdt$$

Change in distance $= area$ bounded by velocity time curve, t-axis $t = t_1 \, and \, t = t_2$

4.18.6.2. Space–Time Curve

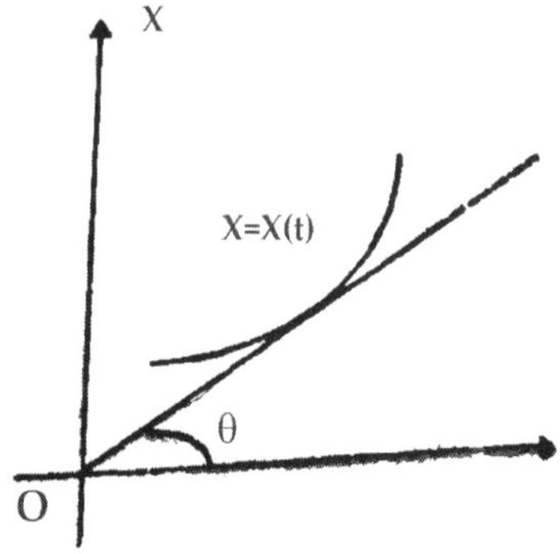

Consider the space- time curve $x = x(t)$, as

$$velocity = v = \frac{dx}{dt}$$

$velocity =$ slope of the space-time curve

$$= tan\theta$$

4.18.6.3. Acceleration–Time Graph

Consider the acceleration–time graph $a = a(t)$, then, since

$$a = \frac{dv}{dt} or dv = adt$$

$$\int_{t=t_1}^{t=t_2} dv = \int_{t=t_1}^{t=t_2} adt$$

$$|v|_{t=t_1}^{t=t_2} = \int_{t=t_1}^{t=t_2} adt$$

$$v(t_1) - v(t_2) = \int_{t=t_1}^{t=t_2} adt$$

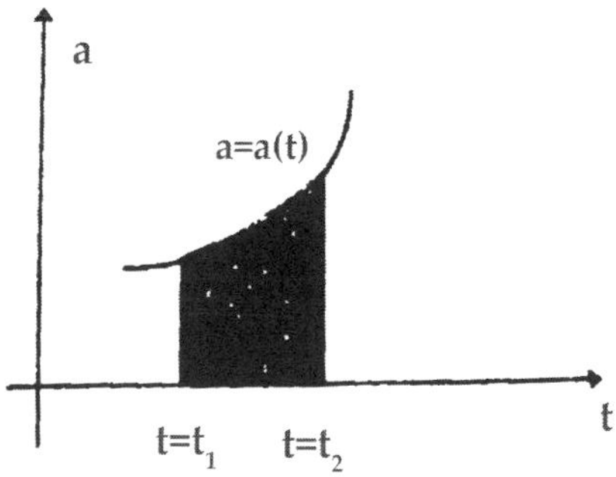

Change in velocity = area bounded be the acceleration- time curve, $t\ \aleph axis\ t\quad t_1\ and\ t\quad t_2$

4.18.6.4. Velocity–Space Curve

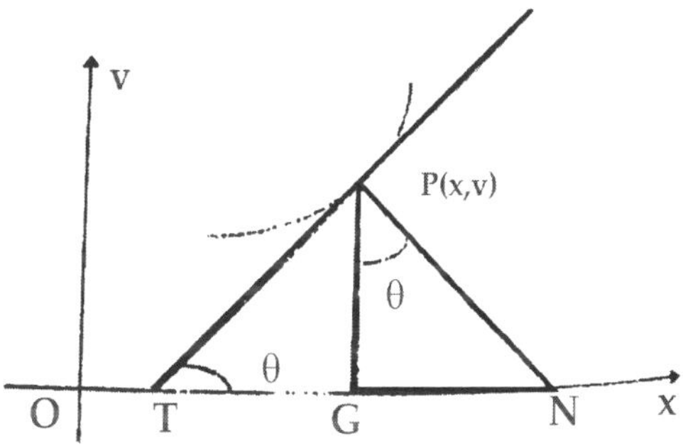

Consider the velocity–space curve $v = v(x)$. Now since

$$a = v\frac{dv}{dx}$$

where

Slope of tangent = $\frac{dv}{dx}$

Therefore

$$tan\theta = \frac{dv}{dx}$$

In triangle PNG

$$\tan\theta = \frac{GN}{GP} = \frac{GN}{v}$$

From (1) and (2), we have

$$\frac{dv}{dx} = \frac{GN}{v} \quad \text{or} \quad v\frac{dv}{dx} \quad GN$$

Therefore

Acceleration $= a = Lenght\ of\ subnormal\ at\ P$

Example

A particle starts from rest with a constant acceleration a. when its velocity acquires a certain value v, it moves uniformly and then its velocity start decreasing with constants retardation 2a till it comes to rest. Find the distance traveled by the particle. If the time taken from rest to rest is t.

Solution

Let $t_1, and\ t_2\ and\ t_3$ be the times for accelerated, uniform motion and retorted motion respectively, then

$$t = t_1 + t_2 + t_3$$

Acceleration = a = slope of OA $= \frac{v}{t_1}$ or

$$t_1 = \frac{v}{a}$$

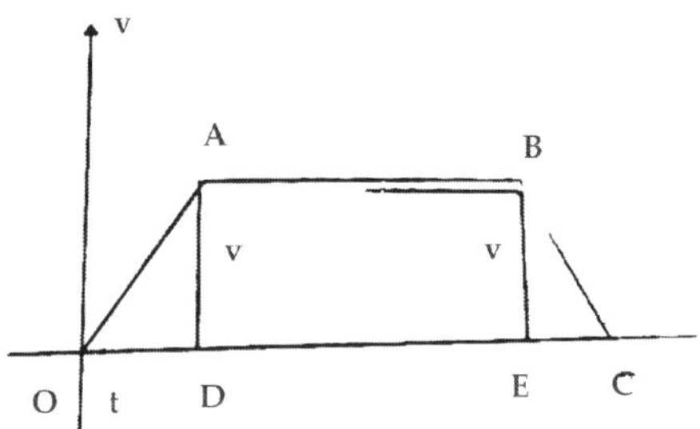

Similarly, retardation = $2a = \frac{v}{t^3}$ or $t_3 = \frac{v}{2a}$

Thus

$$t = t_1 + t_2 + t_3$$

$$= \frac{v}{a} + t_2 + \frac{v}{2a}$$

This implies that

$$t_2 = t - \frac{3v}{2a}$$

Distance = area under the velocity- time curve

= area of trapezium OAB

$$= \frac{1}{2}(OC + AB)(AD)$$

$$= \frac{1}{2}(t_1 + t_2 + t_3 + t_2)v$$

$$= -(t + t\ \ )v$$

$$= \frac{1}{2}\left(t + t - \frac{3v}{2a}\right)v = vt - \frac{3v^2}{4a}$$

Example

A particle moving along a straight line stars from rest and is accelerated uniformly till it attains the velocity v. The motion is the retorted and the particle come to rest after traversing a total distance x. if the acceleration is f, find the retardation and the total time take y the particle from rest to rest.

Solution

Let $t_1\ and\ t_2$ be the time for acceleration and retorted motion, respectively, then

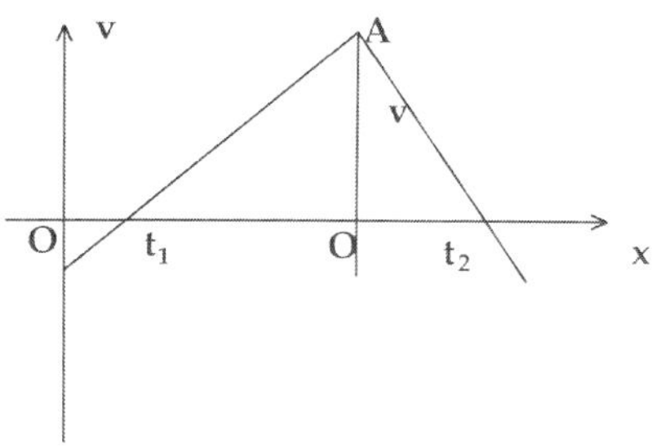

Acceleration = f $= slope\, of\, OA$

$$or\ f = \frac{v}{t_1} or\ t_1 = \frac{v}{f}$$

Let g be the retardation, then

g = slope of AB

$= \frac{v}{t^2}$ or

$$t^2 = \frac{v}{g}$$

Distance = x = area of triangle OAB $= \frac{1}{2}(t_1 + t_2)v$ or

$$t_1 + t_2 = \frac{2x}{v}$$

Thus, total time = $\frac{2x}{v}$

Put $t_1 = \frac{v}{f} and\ t_2 = \frac{v}{g} in(1), then$

$$x = \frac{1}{2}\left(\frac{v}{f} + \frac{v}{g}\right)v = \frac{1}{2}\left(\frac{1}{f} + \frac{1}{g}\right)v^2$$

$$or \frac{2x}{v^2} = \frac{1}{f} + \frac{1}{g}$$

$$or \frac{1}{g} = \frac{2x}{v^2} - \frac{1}{f} = \frac{2xf - V^2}{fv^2}$$

$$or\ g = \frac{fv^2}{2xf - v^2}$$

SUMMARY AND DISCUSSION

4.1. Work Done by a Force

If a constant force F acts on a particle and particle is displaced from point A to point B. Let the displacement AB = S, then work done W by the force F is defined as

$$W = F.S = FScos\theta$$

where θ is the angle between F and S

If $\theta = 0,$ then work done $= W = FScos\theta = FS$

If $\theta = \frac{\pi}{2},$ then work done $= W = FS\cos\frac{\pi}{2} = FS(0) = 0$

Therefore when the applied force is perpendicular to the displacement the work done will be zero.

If $\theta = \pi,$ then work done $= W = FS\cos\pi = FS(-1) = -FS$

If $0 \le \theta \le \frac{\pi}{2},$ then work done $= W$ is positive, because cos $\theta \ge 0.$

If $\frac{\pi}{2} \le \theta \le \pi,$ then work done = W is negative, because $cos\theta \le 0.$

4.2. Work Done by the Weight W

If body of weight W falls down vertically through a distance h, the work done by the gravity is positive and is Wh.

If the body rises vertically throughout the height h. the work done is negative and is equal to- Wh and we say that work is being done against gravity.

The total work done by all the forces is equal to the work done by their resultant.

4.6. Virtual Displacement and Virtual Work

If a set of particles or a body is in equilibrium under the action of forces then there is no motion and consequently there is no actual displacement.

Suppose that the set of particles or the body receives an imaginary displacement, the forces acting there on being regarded as constant during the displacement which is called *virtual displacement* and the work done by the forces during such displacement is called *virtual work.*

4.7. Constant

A constraint is a restriction imposed on a mechanical system to get a desired position or motion. In case of a particle moving down an inclined plane, the constraint on its motion is its motion is its contact with the plane. Any displacement along the plane is consistent with the constraint and the displacement away from the plane is not consistent with the constraint. Due to imposition of constraint we get unknown forces acting on the system. These forced are called reaction of constraint.

4.8. Ideal Constraint or Workless Constraints

If the forces of constraint do not virtual work, the constraints are said to be ideal or workless.

The tension on an inextensible string is workless.

4.8. Principle of Virtual Work for a Single Particle

A particle subject to workless constraints is in equilibrium if and only if zero virtual work is done by the applied forces in any arbitrary infinitesimal displacement consistent with the constraints.

4.9. Principle of Virtual Work for Set Particles

A set of particles, subject to workless constraints is in equilibrium if and only if zero virtual works is one by the applied forces in an arbitrary infinitesimal displacement consistent with constraints.

4.10. Principle of Virtual Work of a Single Rigid Body

A rigid body, subject to workless constraints is equilibrium if and only if zero virtual work is done by the applied forces and applied torques in any arbitrary infinitesimal displacement consistent with constraints.

4.11. Principle of Virtual Work for a Set of Rigid Bodies

A set of rigid bodies. Subject to workless constraints, is in equilibrium if and only if zero virtual work is done by the applied forces and applied torques in any arbitrary infinitesimal displacement consistent.

4.12.1. Dynamics

It is the branch of mechanics which deals with motion of objects. It is divided into two branches namely Kinematics and kinetics.

- **Kinematics**

It deals with the geometry of motion without any reference to the forces acting on the body.

- **Kinetic**

It relates the motion of the body with the forces causing the motion.

4.13. Cartesian Component of Velocity and Acceleration

If $\overline{r} = xi + yj$, then

$$\overline{v} = \frac{d\overline{r}}{dt} = \frac{dx}{dt}i + \frac{dy}{dt}j$$

and

$$\overline{a} = \frac{d^2\overline{r}}{dt^2} = \frac{d^2x}{dt^2}i + \frac{d^2y}{dt^2}j$$

$v_x = x -$ Component of velocity $= \frac{dx}{dt}$

$v_y = y -$ Component of velocity $= \frac{dy}{dt}$

$a_x = x -$ Component of acceleration $= \frac{d^{2x}}{dt^2}$

$a_y = x -$ Component of acceleration $= \frac{d^{2y}}{dt^2}$

$$v = |\overline{v}| = \sqrt{\left(\frac{dx}{dt}\right)^2 + \left(\frac{dy}{dt}\right)^2} = \frac{ds}{dt}$$

4.14. Radial and Transverse Component of Velocity and Acceleration

Radial component of the velocity = $v_r = r$

Transverse component of velocity = $v_\theta = r\theta\overline{s}$

Radial component of the acceleration = $a_r = \ddot{r} + r\theta^2$

Transverse component of acceleration = $a_0 = 2\dot{r}\,\dot{\theta} + r\ddot{\theta}$

$$a_0 = \frac{1}{r}\frac{d}{dt}(r^2\dot{\theta})$$

4.15. Tangential and Normal Components of Velocity and Acceleration

$v_t =$ Tangential component of the velocity = v, and

$v_n =$ Normal component of velocity = 0

Hence velocity is along the tangent to the path.

Let $\overline{a}$ be the acceleration, then

$$\overline{a} = \frac{d\overline{v}}{dt} = \frac{d}{dt}\left(v\hat{t}\right) = \frac{dv}{dt}\hat{t} + v\frac{d\hat{t}}{dt}$$

$$= \frac{dv}{dt}\hat{t} + v\frac{d\hat{t}}{d\psi}\frac{d\psi}{ds}\frac{ds}{dt}$$

$$= \frac{dv}{dt}\hat{t} + v\frac{d\hat{t}}{d\psi}Kv$$

where K = $\frac{d\psi}{ds}$ *is curvature.* Thus

$$\overline{a} = \frac{dv}{dt}\hat{t} + \frac{v^2}{p}\frac{d\hat{t}}{d\psi}$$

Hence $a_t =$ tangential component of acceleration $= \frac{dv}{dt}$

$a_n =$ Normal component of acceleration $= \frac{v^2}{p}$

4.17. Angular Velocity

The angular velocity ω is the rate of change of angular displacement θ *w.r.t.* time 't'.

$$\omega = \frac{d\theta}{dt}$$

$$\bar{\omega} = \frac{d\theta}{dt}\bar{k}$$

If the rotation is clockwise then the angular velocity is taken as positive otherwise it is negative.

$$v = r\omega$$

In vector form this relation is written as

$$v = r \, x \, \omega$$

4.18. Rectilinear Motion

The motion of a particle along a straight line is called *rectilinear motion.*

4.18.3. Vertical Motion Under Gravity

For a failing body, the acceleration is constant. It is called acceleration due to gravity and is denoted by 'g'. In FPS system g = $32\,ft/sec^2$.*In* CGS system g =981 cm/sec^2 and in MKS system $g = 9.81m/sec^2$. If the body is projected vertically upward then acceleration g is negative. For falling bodies equation of motion is

$$v = u + gt$$

$$x = ut + \frac{1}{2}gt^2$$

$$v^2 - u^2 = 2gx$$

4.18.4. Distance Traveled in n^th^ Second

Let x_1 *and* x_2 *b* e the distance travelled in the first n and n – 1 seconds, respectively, then from the equation of motion

$$x = ut + \frac{1}{2}at^2$$

then $$x_1 = un + \frac{1}{2}an^2$$

and $$x_2 = u(n-1) + \frac{1}{2}a(n-a)^2$$

Distance travelled in the nth second = $x_1 - x_2$

$$x_1 - x_2 = un + \frac{a}{2}n^2 - \left\{u(n-a) + \frac{a}{2}(n-1)^2\right\}$$

$$= un + \frac{a}{2}n^2 - \left\{un - u + \frac{a}{2}\left(n^2 - 2n + 1\right)\right\}$$

$$= u + \frac{1}{2}a(2n-1)$$

Motion with variable acceleration:

(i) time dependent acceleration;

(ii) velocity dependent acceleration; and

(iii) distance dependent acceleration.

CHAPTER 5

PROJECTILE MOTION

CONTENTS

5.1. PROJECTILE

An object thrown into the space with certain velocity from a gun or dropped from a moving plane under the action of gravity is called a projectile. Thus, a projectile moves with a constant horizontal velocity and at the same falls freely under the action of gravity. The path of projectile is called the *trajectory*.

5.2. EQUATION OF THE TRAJECTORY OF A PROJECTILE

Let the particle m be projected from point O with an initial velocity v_o making an angle α with the horizontal. Take O as the origin and the horizontal and vertical lines through O as x-axis and y-axis, respectively.

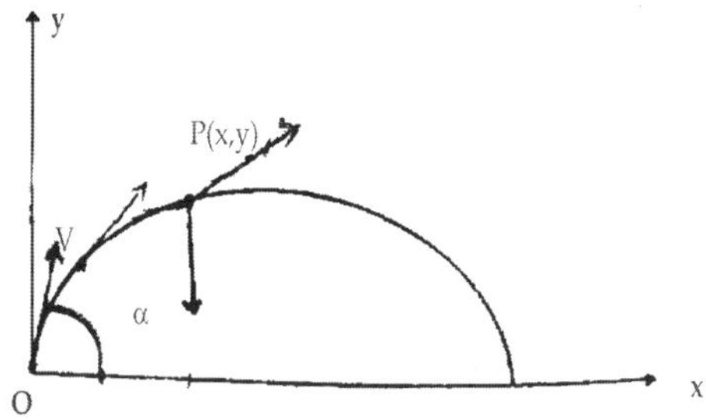

Suppose that after time t, the particle is at point $p(x,y)$ whose position vector is $\bar{r}$. i.e.,

$$\bar{r} = x\hat{i} + y\hat{j}$$

$$\dot{\bar{r}} = \dot{x}\hat{i} + \dot{y}\hat{j}$$

$$\ddot{\bar{r}} = \ddot{x}\hat{i} + \ddot{y}\hat{j}$$

The gravitational force $\overline{F}$ acting on the particle at P is

$$\overline{F} = mg\hat{j}$$

$$m\ddot{\bar{r}} = -mg\hat{j}$$

$$or\,\ddot{\bar{r}} = g\hat{j}$$

$$\ddot{x}\hat{i}+\ddot{y}\hat{j}=-g\hat{j}, then$$

$$\ddot{x}=0 and\ \ddot{y}=-g$$

Integrating, we get

$$\dot{x}=A\, and\ \dot{y}=-gt+B$$

At $t=0,$ the particle is at O, therefore $tt=0,\dot{x}=v_0\, cos\alpha,\dot{y}=v_o\, sin\alpha$

Therefore

$$A=v_o\, cos\alpha, B=v_o\, sin\alpha$$

Hence

$$\dot{x}=v_0\cos\alpha,\dot{y}=-gt+v_o\, sin\alpha$$

which gives the horizontal and vertical component of velocity at any time t. again integrate

$$x=(v_0\cos\alpha)t, and\ y=-g\frac{t^2}{2}+t+D$$ at $t=0,x=0,y=0.$

Therefore,

$$C=0,D=0,$$

Hence

$$x=(v_0\cos\alpha)t$$

$$y=v_0\, sin\alpha-\frac{1}{2}gt^2$$

The eqs (1) and (2) are parametric equation of trajectory. To find the Cartesian equation, eliminate t from the $eqs.$ (1) and (2), we get

which is the Cartesian equation of trajectory· From $(3),$ we have

$$\frac{gx^2}{2v_o^2\cos^2\alpha}-x\frac{sin\alpha}{cos\alpha}=-y$$

$$or x^2 - x\frac{sin\alpha}{cos\alpha}\left(\frac{2v_o^2\cos^2\alpha}{g}\right) = -y\frac{2v_o^2\cos^2\alpha}{g}$$

$$or\, x^2 - \frac{2x v_0^2\, sin\alpha\cos\alpha}{g} = -2y\frac{v_o^2\cos^2\alpha}{g}$$

$$or\, x^2 - 2x\frac{v_0^2 sin\alpha\cos\alpha}{g} + \left(\frac{v_0^2 sin\alpha\cos\alpha}{g}\right)^2$$

$$= -2y\frac{v_o^2\cos^2\alpha}{g} + \left(\frac{v_0^2 sin\alpha\cos\alpha}{g}\right)^2$$

$$or\left(x - \frac{v_0^2 sin\alpha\cos\alpha}{g}\right)^2 = \frac{2v_o^2\cos^2\alpha}{g}\left(y - \frac{v_o^2\sin^2\alpha}{2g}\right)$$

Now

5.2.1. Vertex

$\left(\frac{v_0^2 sin\alpha\cos\alpha}{g}, \frac{v_o^2\sin^2\alpha}{2g}\right)$ is the vertex of the *parabola.*

5.2.2. Latus Rectum

Latus Rectum = $|4a| = \frac{2v_o^2\cos^2\alpha}{g}$

5.2.3. Maximum Height

Maximum height attained above x-axis

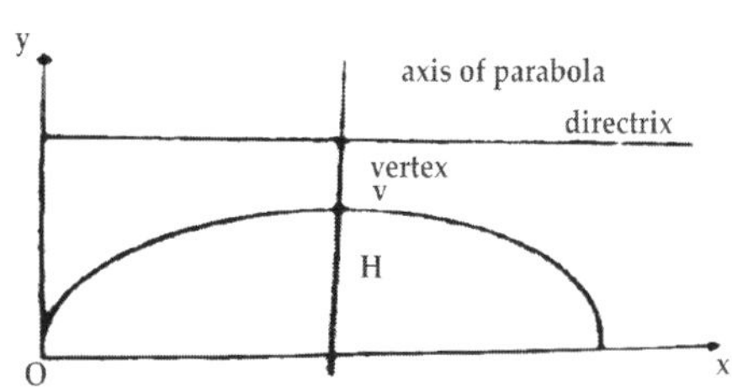

H = Maximum height attained above x-axis

$= y -$ Coordinate of vertex

$$= \frac{v_o^2 \sin^2 \alpha}{2g}$$

This equation can also be obtained by putting $\frac{dy}{dx} = 0$ in the equation of trajectory.

5.2.4. Equation of Directrix

Height of Directrix above $x - axis$

$$= H + \frac{1}{4}(latus\,rectum)$$

$$= \frac{v_o^2 \sin^2 \alpha}{2g} + \frac{v_o^2 \cos^2 \alpha}{2g}$$

$$= \frac{v_o^2}{2g}\left(\sin^2 \alpha + \cos^2 \alpha\right) = \frac{v_o^2}{2g}$$

Equation of Directrix is

$$y = \frac{v_o^2}{2g}$$

5.2.5. Focus

X-coordinate of focus = x-coordinate of vertex

$$= \frac{v_0^2 sin\alpha \cos\alpha}{g}$$

y-coordinate of focus = H - $\frac{1}{4}(latus\,rectum)$

$$= \frac{v_o^2 \sin^2 \alpha}{2g} + \frac{v_o^2 \cos^2 \alpha}{2g}$$

$$=\frac{v_o^2}{2g}\left(\sin^2\alpha+\cos^2\alpha\right)$$

$$=-\frac{v_o^2}{2g}cos2\alpha$$

Thus, the focus

$$=\left(\frac{v_0^2 sin\alpha\cos\alpha}{g},-\frac{v_o^2}{2g}cos2\alpha\right)=\left(\frac{v_o^2}{2g}sin2\alpha-\frac{v_o^2}{2g}cos2\alpha\right)$$

5.2.6. Time of Flight

The time taken by the particle to reach the horizontal plane through the point of projection is called its time of flight. In this case put y = 0, in the equation

$$y=g\frac{t^2}{2}+(v_o\sin\alpha)t$$

Then

$$0=-g\frac{t^2}{2}+(v_o\sin\alpha)t$$

$$since\, t\neq 0,\text{therefore}$$

$$v_o\sin\alpha-\frac{1}{2}gt=0$$

$$or\, t=\frac{2v_o sin\alpha}{g}$$ is the time of flight.

5.2.7. Range

5.2.7.1. Horizontal Range

The horizontal range of the projectile is the horizontal distance described by the particle during the time of flight. Therefore,

Range R = (horizontal velocity) (time of flight)

$$v_o\cos\alpha\left(\frac{2v_o sin\alpha}{g}\right)$$

$$= \frac{v_o^2 2(\sin\alpha \cos\alpha)}{g} = \frac{2v_o^2 \sin 2\alpha}{g}$$

This range can also be obtained by putting y = 0 in the Cartesian equation of trajectory, i.e., putting y = 0 in the equation

$$y = x\tan\alpha - \frac{gx^2}{2v_o^2}\sec^2\alpha$$

Also, it is notable that this range will remain same if we replace $\alpha by \alpha = \frac{\pi}{2} - \alpha$

5.2.7.2. Maximum Range

Range R = $\frac{v_o^2 \sin 2\alpha}{g}$

For maximum range $\frac{dr}{d\alpha} = 0$

$$\Delta \frac{2v_o^2}{g}\cos 2\alpha = 0$$

$$\Delta \cos 2\alpha = 0$$

$$\Delta 2\alpha = \frac{\pi}{2}$$

$$\Delta \alpha = \frac{\pi}{4}$$

Therefore $R_{max} = \frac{v_o^2}{g}$

5.3. RANGE IN INCLINED PLANE

Let β the inclination of the inclined plane. Let a particle is projected from point O with velocity v_o by making an angle α to the horizontal where

$\beta < \alpha$. Let the projectile meet the inclined plane at $P(x, y)$ then OP = r is called the range of the inclined plane.

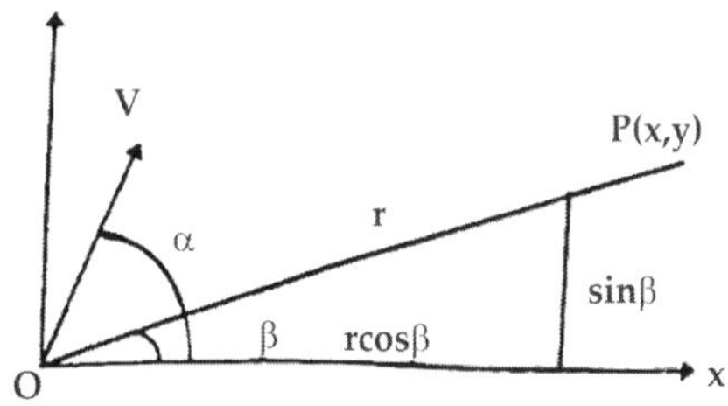

$$x = rcos\,\beta, y = rsin\beta$$

$$P(x, y) = (rcos\beta, rsin\beta)$$

Equation of trajectory is

$$y = \tan\alpha - \frac{1}{2} g \frac{x^2}{v_o^2} \sec^2\alpha$$

The point $p(rcos\beta, rsin\beta)$ lies on it, then

$$r\,sin\beta = r\beta tan\alpha - \frac{1}{2} g \frac{r\cos^2\beta}{v_o^2\cos^2\alpha}$$

$$\frac{grcos^2\beta}{2v_o^2\cos^2\alpha} = \cos\beta \frac{sin\alpha}{cos\alpha} - sin\beta$$

$$= \frac{sin\alpha\, cos\beta - cos\alpha\, sin\beta}{cos\alpha}$$

$$= \frac{\sin(\alpha - \beta)}{cos\alpha}$$

Thus

$$r = \frac{2v_o^2\cos^2\alpha}{g\cos^2\beta} \times \frac{\sin(\alpha - \beta)}{cos\alpha}$$

$$= \frac{2v_o^2\cos^2\alpha\sin(\alpha - \beta)}{g\cos^2\beta}$$

For range down the inclined plane, change $\beta\, to -\beta.$

5.3.1. Maximum Range

Equation (1) can be written as, by using

$$\sin(\alpha+\beta)-\sin(\alpha-\beta)=2\cos\alpha sin\beta$$

$$r=\frac{v_o^2}{gcos^2\beta}(\sin(2\alpha-\beta)=1, so\, that$$

$$2\alpha-\beta=\frac{\pi}{2} or\, \alpha=\frac{\beta}{2}+\frac{\pi}{4}$$

Thus

$$r_{max}=\frac{v_o^2}{gcos^2\beta}(1-\sin\beta)$$

$$\frac{v_o^2(1-sin\beta)}{g(1-\sin^2\beta)}=\frac{v_o^2}{g(1-\sin\beta)}$$

For maximum range put $\frac{dr}{dx}=0$

$\cos(2\alpha-\beta)=0$

$2\alpha-\beta=\frac{\pi}{2}$

5.3.2. Time of Flight

Time of flight = $\frac{Horizontol\, distance}{Horizontol\, velocity}$

$$=\frac{x}{v_o\cos\alpha}=\frac{rcos\beta}{v_o\cos\alpha}$$

Using (1), we get

Time of flight = $\frac{2v_o^2\cos^2\alpha\sin(\alpha-\beta)}{g\cos^2\beta}\frac{cos\beta}{v_o cos\alpha}$

$$=\frac{2v_o^2\sin(\alpha-\beta)}{g\cos\beta}$$

For time of flight down the plane, change $\beta\, to -\beta.$

Show that the speed at a point P of the projectile is the same as attained by a particle of same mass at P falling freely from the point vertically above P on the Directrix of the trajectory.

$Let\, v_o$ beThe Velocity of projection and v be the velocity at point $p(x,y)$ of trajectory, then

$$\overline{v}=\dot{x}\hat{i}+\dot{y}\hat{j}$$

$$\overline{v}=(v_0\, cos\alpha)\hat{i}+(v_o sin\alpha-gt)\hat{j}$$

$$v=|\overline{v}|=\sqrt{(v_0\cos^2\alpha)+(v_o\sin\alpha-gt)^2}$$

$$v=\sqrt{v_o^2\cos^2\alpha+v_o^2\sin^2\alpha-2gtv_o\, sin\alpha+g^2t^2}$$

$$v=\sqrt{v^2\, \_\, 2g(v_o t\, sin\alpha-\frac{1}{2}gt^2}$$

Since

$$y=v_o t\sin\alpha-\frac{1}{2}gt^2$$

Therefore

$$v=\sqrt{v_0^2-2gy}$$

Let N be the point on the Directrix vertically above $P(x,y)$. From fig.

$$MN=MP+PN$$

$$\frac{V_0^2}{2g}=y+PN$$

Let v' be the velocity of the particle in failing freely a distance NP. Using the equation of motion $v^2-U^2=2gx$ where u is initial velocity and v is final velocity.

Here $u=0, x=PN, v=v'$, then

$$v'^2-0=2g(PN)$$

$$v'^2=2g\left(\frac{v_o^2}{2g}-y\right)$$

$$v'^2=v_o^2-2gy$$

$$v'=\sqrt{v_o^2-2gy}$$

From $(1)\,and\,(2)$, we have

$$v'=v$$

5.4. PARABOLA IF SAFETY

As the equation of the projectile is

$$y=x\,tan\alpha-\frac{1}{2}g\frac{x^2}{v^2}\sec^2\alpha$$

which shows that for different values of α we have different trajectories. So, we define that the locus of all the points on different trajectories which lie at maximum distance from the point of projection O is called the parabola of safety. All the trajectories from O must touch the parabola of safety at some point. Thus, it is envelope of the family of all the trajectories from with same velocity of protestation v_o, α being regarded as parameter. Equation of trajectory is

$$y=x\tan\alpha-\frac{1}{2}g\frac{x^2}{v_o^2}\sec^2\alpha$$

$$y=x\tan\alpha-\frac{1}{2}g\frac{x^2}{v^2}$$

$$\frac{gx^2}{2v_o^2}\tan^2\alpha-x\tan\alpha+\left(\alpha+\frac{gx^2}{2v_o^2}\right)=0$$

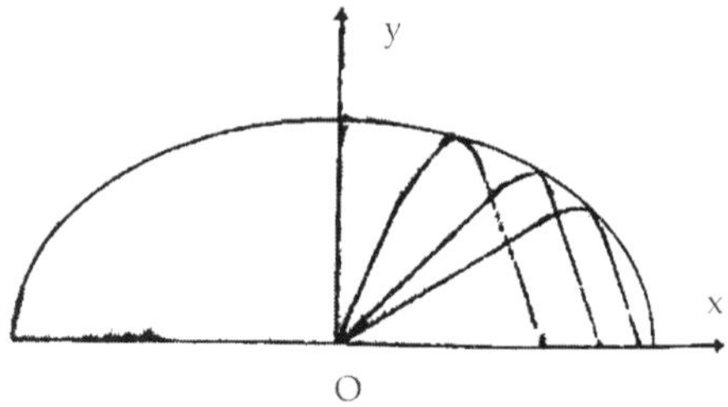

It is quadratic equation in $\tan\alpha$. For the equation of envelope, put discriminate = 0, *i.e*

$$x^2 - 4\left(\frac{gx^2}{v_o^2}\right)\left(y + \frac{gx^2}{v_o^2}\right) = 0$$

Since $x \neq 0$, therefore dividing by x^2, we get

$$-2\left(\frac{g}{v_o^2}\right)\left(y + \frac{gx^2}{v_o^2}\right) = 0$$

$$-\frac{2gy}{v_o^2} - \frac{g^2x^2}{v_o^2} = 0$$

$$\frac{g^2x^2}{v_o^2} = \frac{2gy}{v_o^2} + 1$$

$$x^2 = -2\frac{v_o^2}{g}\left(y - \frac{v_o^2}{2g}\right)$$

which is the equation of parabola of safety having vertex $\left(0, \frac{v_o^2}{2g}\right)$

Latus Rectum = $\frac{2v_o^2}{g}$

Equation of Directrix is

$$y = \frac{v_o^2}{2g} + \frac{1}{4}(\textit{latus rectum})$$

$$y=\frac{v_o^2}{2g}+\frac{1}{4}\left(\frac{2v_o^2}{g}\right)$$

$$y=\frac{v_o^2}{2g}+\frac{v_o^2}{2g}=\frac{v_o^2}{g}$$

$$x-coordinate\,of\ focus=x-coordinate\,of\ vertex=0$$

$$y-coordinateoffocus\ =y-\ coordinateofvertex\ =0$$

$$-\frac{1}{4}(latus\,rectum)$$

$$=\frac{v_o^2}{2g}-\frac{1}{4}\frac{2v_o^2}{g}=0$$

Therefore, focus is $(0,0)$.

5.5. PROJECTION WITH AIR RESISTANCE

Let a particle of mass m be projected with an initial velocity v_o making an angle α with the horizontal as shown in the figure.

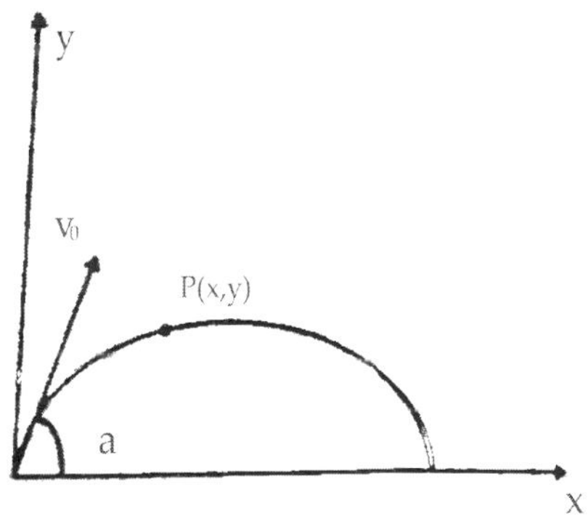

Here we restrict ourselves to the simple case of air resistance proportional to the first power of velocity. We suppose that after some time t the particle is at P(x, Y). The equations of motion are, with mass canceled,

$$\ddot{x}=-k\dot{x}$$

$$\ddot{y}=-g-k\dot{y}$$

$$From(1)$$

$$\frac{dx}{dt}=-kx$$

$$\frac{dx}{x}=-kdt$$

Integrate

$$In(x)=-kt+c_1$$

at $t=0\,x=v_o\cos\alpha, therefore\,c_1=in(v_o cos\alpha)$

In(x) $=-kt+in(v_o cos\alpha)$

$$x=v_o\cos\alpha\, e^{-kt} \qquad (3)$$

$$x=-\frac{v_o}{k}cos\alpha\, e^{-kt}+c_2$$

at $t=0, x=0,$ therefore

$$c_2==\frac{v_o}{k}cos\alpha, thus$$

$$x=\frac{v_o}{k}\cos\alpha\, e^{-kt}+\frac{v_o}{k}cos\alpha$$ or

$$x=-\frac{v_o}{k}\left(1-e^{-kt}\right)cos\alpha \qquad (4)$$

Now, consider equation (2)

$$\frac{dy}{dt}=-g-ky$$ or

$$\frac{dy}{g+ky}=-dt$$

Integrate

$$in(g+ky)=-kt+c_3$$

at $t = 0, y = v_o \sin\alpha,$ this implies that

$$c_3 = in\ (g + kv_o \sin\alpha)$$

In $\left(g + ky\right) = -kt + In\left(g + kv_o sin\alpha\right)$

In $In\dfrac{g + ky}{g + kv_o \sin\alpha} = -kt$

$$g + ky = \left(g + kv_o \sin\alpha\right)e^{-kt}$$

$$y = -\frac{g}{k} + \left(\frac{g}{k} + kv_o\ sin\alpha\right)e^{-kt} \qquad (5)$$

Again integrating

$$y = -\frac{g}{k} - \frac{1}{k}\left(\frac{g}{k} + v_o sin\alpha\right)e^{-ky} + c_4$$

at $t = 0, y = 0,$ this implies that

$$c_4 = \frac{1}{k}\left(\frac{g}{k} + v_o\ sin\alpha\right)$$

Thus

$$y = -\frac{g}{k} - \frac{1}{k}\left(\frac{g}{k} + v_o\ sin\alpha\right)e^{-kt} + \frac{1}{k}\left(\frac{g}{k} + v_o\ sin\alpha\right)$$

$$y = -\frac{g}{k}t - \frac{1}{k}\left(\frac{g}{k} + v_o\ sin\alpha\right)\left(1 - e^{-kt}\right) \qquad (6)$$

The equation of the path is obtained by climinating t from equations (4) and (6) as

$$y = -\frac{xcos\alpha}{v_o}\left(\frac{g}{k} + v_o\ sin\alpha\right) + \frac{g}{K^2}In\left(1 - \frac{xcos\alpha}{v_o}\right)$$

Example

A shell burst on contact with the ground and pieces from it fly in all directions with all speed upto 80 ft./sec. Prove that a man 100 ft. is in danger for $\dfrac{5}{\sqrt{2}}$ seconds.

Solution

We know that the range of the particle is

$$R = \frac{v_o^2 sin2\alpha}{g}$$

Here $R=100\ ft., v_o = 80\ ft\ /\ sec., g = 32\ ft\ /\ sec^2.,$ therefore

$$100 = \frac{6400 sin2\alpha}{32}$$

$$or\ \sin 2\alpha = \frac{1}{2} or\ \alpha = 15^o, 75^o$$

For the range $100\ ft.$ there are two directions of projection. Let $t_1\ and\ t_2$ be the times of flight, respectively, then

$$t_1 = \frac{2v_o sin15^o}{g} \quad \text{and} \quad t_2 = \frac{2v_o sin75^o}{g}$$

The man is in danger for time = $t_2 - t_1$

$$= \frac{2v_o sin15^o}{g} - \frac{2v_o sin75^o}{g}$$

$$= \frac{2v_o}{g}\left(sin15^o - sin75^o\right)$$

$$= \frac{160}{32}\left(2\cos\left(\frac{75+15}{2}\right)\sin\left(\frac{75-15}{2}\right)\right)$$

$$=5\left(2cos45^o\ sin30^o\right) = \frac{5}{\sqrt{2}} sec.$$

Example

Find the range of a rifle bullet when α is the elevation of projection and v_o is speed. Show that, if the rifle is field with the same elevation and speed from a car travelling with v towards the target, the range will be increase by $\frac{2vv_o sin\alpha}{g}$.

Solution

We know that the range is given by

$$R = (horizontal\,velocity)(time\,of\;flight)$$

$$R = v_o cos\alpha\left(\frac{2v_o sin\alpha}{g}\right) = \frac{2v_0^2 \sin\alpha cos\alpha}{g}$$

When the shell is fired from a car moving with speed v towards the target the horizontal velocity is increased by v.

Horizontal velocity = $v_o\; cos\alpha + v$

New $range + R'$

$$= (new\,horizontal\,time\,of\;flight)$$

$$= (v_o cos\alpha + v)\left(\frac{2v_o sin\alpha}{g}\right)$$

$$R' = \frac{2v_0^2 \sin \aleph cos}{g} + \left(\frac{2v_o v\,sin}{g}\right)$$

Increase in range = $r' - R = \left(\frac{2v_o v\,sin\alpha}{g}\right)$

Example

A battleship is steaming ahead with speed v and a gun is mounted on a battleship so as to point straight backward and is set at an angle of elevation α if v_o is the speed of projection (relative to gun), show that the range is

$$\left(\frac{2v_o sin\alpha}{g}\right)(v_o\; cos\alpha - v)$$

Prove also that the angle of elevation for maximum rang is

$$\cos^{-1}\left(\frac{v + \sqrt{v^2 + 8v_o^2}}{4v_o}\right)$$

Solution

The ship is moving with speed v away from the target, so the horizontal velocity of the projectile will be decreased by v.

Actual horizontal velocity = $v_o \cos\alpha - v$

Actual range = R = (horizontal velocity) (time of flight)

$$(v_o cos\alpha - v)\left(\frac{2v_o sin\alpha}{g}\right)$$

$$\left(\frac{2v_o sin\alpha}{g}\right)(v_o cos\alpha - v)$$

Since

$$R = \left(\frac{2v_o sin\alpha}{g}\right)(v_o cos\alpha - v)$$

$$\frac{v_o}{g}(v_o \sin 2\alpha - 2v\, sin\alpha)$$

Therefore, for maximum range

$$\frac{dR}{d\alpha} = \frac{2v_o}{g}(v_o\, cos2\alpha - v\, cos\alpha) = 0$$

$$v_o \cos 2\alpha - vcos\alpha = 0$$

$$v_o\ (2\cos^2\alpha - 1) - vcos\alpha = 0$$

$$v_o(2\cos^2\alpha - 1) - v\, cos\alpha = 0$$

$$2v_0 \cos^2\alpha - vcos\alpha - v_0 = 0$$

$$cos\alpha = \frac{v \pm \sqrt{v^2 + 8v_o^2}}{4V_o}$$

$$\frac{d^2R}{d\alpha^2} = \frac{2v_o}{g}(-2v_o\, sin2\alpha + vsin\alpha)$$

$$= -\left(\frac{2v_o sin\alpha}{g}\right) 4v_o \, cos\alpha -$$

$$\cos\alpha = \frac{v + \sqrt{v^2 + 8v_o^2}}{4V_o}$$

when

$$\frac{dR}{d\alpha} = \frac{2v_o sin\alpha}{g} \left(\sqrt{v^2 + 8v_0^2}\right) = -ve$$

For 0 < $\alpha < 90$, sina is positive

Thus, R is maximum when

$$cos\alpha = \frac{v + \sqrt{v^2 + 8v_o^2}}{4V_o}$$ or

$$\alpha = \cos^{-1}\left(\frac{v + \sqrt{v^2 + 8v_o^2}}{4V_o}\right)$$ or

$$\alpha = \cos^{-1}\left(\frac{v + \sqrt{v^2 + 8v_o^2}}{4V_o}\right)$$

Example

Prove that the speed required to project a particle from a height h to fall a horizontal distance a from the point of projection is at least

$$\sqrt{g(\sqrt{a^2 + h^2 - h)}}$$

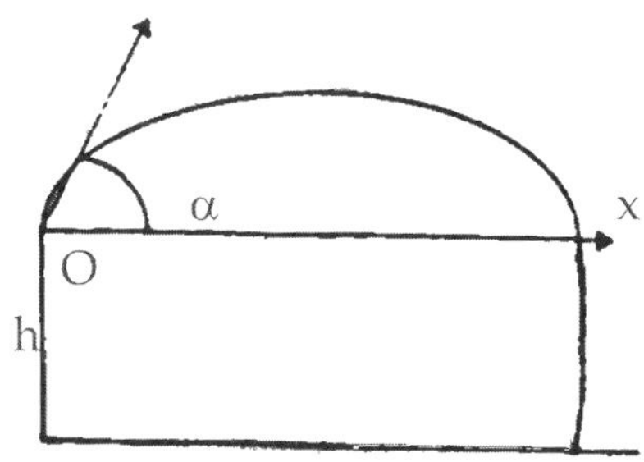

Solution

Equation of trajectory is

$$y = x\tan\alpha - \frac{gx^2}{2v_o^2}\sec^2\alpha$$

Point $(a,-h)$ lies on it, then

$$-h = a\tan\alpha - \frac{ga^2}{2v_0^2}\sec^2\alpha$$

$$-h = a\tan\alpha - \frac{ga^2}{2v_0^2}(1+\tan^2\alpha)$$

$$ga^2\tan^2\alpha - 2av_o\tan\alpha + ga^2 - 2hv_o^2) \geq 0$$

Since $\tan\alpha$ is real therefore, Discriminant ≥ 0, *i.e.*

$$4a^2v_o^4 - 4ga^2\left(ga^2 - 2hv_o^2\right) \geq 0$$

$$\left(v_0^2 + gh\right)^2 \geq g^2\left(a^2 + h^2\right)$$

$$v_0^2 + gh \geq g\sqrt{a^2 + h^2}$$

$$v_0^2 \geq g\left(\sqrt{a^2 + h^2} - h\right)$$

$$v_o \geq \sqrt{g\left(\sqrt{a^2 + h^2} - h\right)}$$

Therefore, least value of v_o is $\sqrt{g\left(\sqrt{a^2 + h^2} - h\right)}$

Example

A particle is launched at an angle a from a cliff of height H above sea level. If it falls into the sea at a distance D from the base of the *cliff,* prove that the maximum height above the sea level is

$$H + \frac{D^2\tan^2\alpha}{4(H + D\tan\alpha)}$$

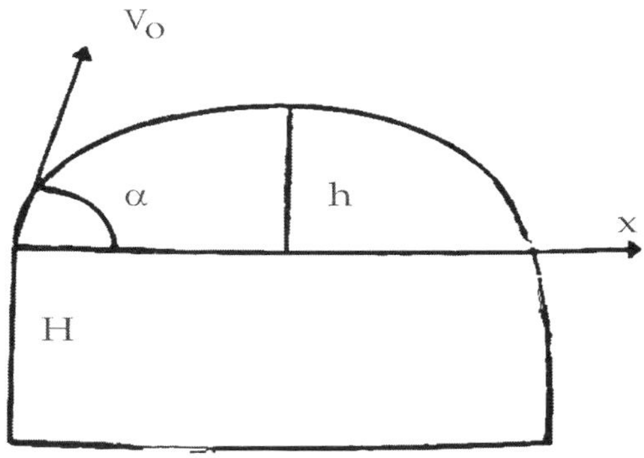

Solution

Equation of the trajectory is

$$y = xtan - \frac{gx^2}{2v_0^2}\sec^2\alpha$$

Point (D, -H) lies on it, therefore

$$-H = D\tan\alpha - \frac{gD^2}{2v_o^2}\sec^2\alpha$$

$$H + D\tan\alpha - \frac{gD^2}{2v_o^2\cos^2\alpha}$$

$$v_o^2 = \frac{gD^2}{2\cos^2\alpha(H + D\tan\alpha)}$$

Now, $h = height\,above\,x - axis$, so that

$$h = \frac{v_o^2\sin^2\alpha}{2g}$$

Putting the value of v_o^2, then

$$h = \frac{gD^2}{2\cos^2\alpha\left(H + D\,tan\alpha\right)}\frac{\sin^2\alpha}{2g}$$

$$= \frac{D^2\tan^2\alpha}{4\left(H + D\,tan\alpha\right)}$$

Height above sea level $= H + h$

$$= H + \frac{D^2 \tan^2 \alpha}{4(H + D tan\alpha)}$$

Example

A front and a slip are both armed with guns which give their projectile a muzzle velocity $\sqrt{2gk}$ *and* guns in the fort are at height h above the guns in the ship. If d_1 and d_2 are the greatest horizontal ranges at which the fort and ship, respectively can engage, prove that

$$\frac{d_1}{d_2} = \sqrt{\frac{k+h}{k-h}}$$

Solution

Let S be the ship and F be the fort.

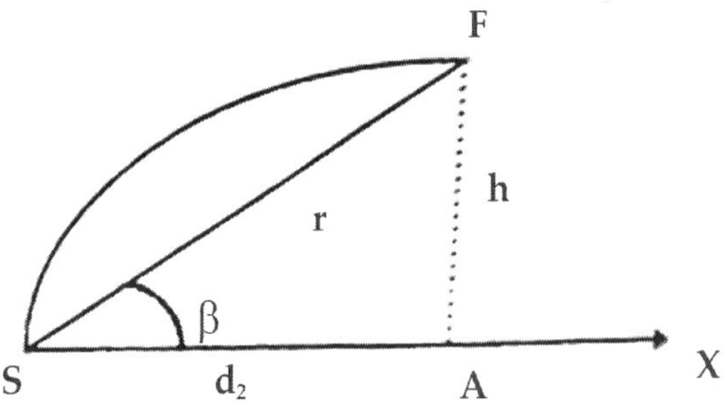

Let $\angle XSF = \beta$ and $SF = r$

d_2 *is* greatest horizontal range for gun in ship, so r is maximum range on inclined plane with inclination β. Now

$$r = \frac{v_o^2}{g(1 + sin\beta)}$$

Put $v_o = \sqrt{2gk}$ in the above equation, then

$$r = \frac{2gk}{g(1 + sin\beta)} = \frac{2k}{1 + \sin\beta}$$

$$r + r\sin\beta = 2k$$

Since form the figure $h = r sin\beta$, therefore

$$r + h = 2k$$

Also from right angle triangle SAF, we have

$$r^2 = h^2 + d_2^2$$

So that

$$h^2 + d_2^2 = (2k - h)^2 = 4k^2 - 4kh + h^2$$

$$d_2^2 = 4k^2 - 4kh = 4k(k - h) \qquad (1)$$

For gun in fort, change h to – h, then

$$d_1^2 = 4k(K + h) \qquad (2)$$

Dividing (2) by (1), then

$$\frac{d_1}{d_2} = \sqrt{\frac{k+h}{k-h}}$$

Example

The horizontal rang of a ball for two different projection is same and equal to R. if $h_1 \; and \; h_2$ are the greatest height in the two parts for which this is possible prove that

$$R^2 = 16h_1h_2$$

Solution

Let R be the range, when α is angle of the projection and v_o is velocity of projection, then

$$R = \frac{2v_o^2 \, sin\alpha cos\alpha}{g}$$

Let R' be the range when angle of projectile is $\frac{\pi}{2} - \alpha,$ then

$$R' \frac{2v_o^2 \sin\left(\frac{\pi}{2}-\alpha\right) cos\left(\frac{\pi}{2}-\alpha\right)}{g}$$

$$R' = \frac{2v_o^2 \, sin\alpha cos\alpha}{g}$$

which show that R is the same when angle of projection are α and $\frac{\pi}{2}-\alpha$

Let $h_1 and\, h_2$ be the maximum height in these cases, respectively, therefore

$$h_1 = \frac{v_o^2 \sin^2 \alpha}{2g}$$

$$h_2 = \frac{v_o^2 \sin^2\left(\frac{\pi}{2}-\alpha\right)}{2g} = \frac{v_o^2 \cos^2 \alpha}{2g}$$

Hence

$$16h_1h_2 = 16\frac{v_o^2 \sin^2 \alpha}{2g}\left(\frac{v_o^2 \cos^2 \alpha}{2g}\right)$$

$$= 4\frac{v_o^4 \sin^2 \alpha \cos^2 \alpha}{g} = R^2$$

$$i.e\, R^2 = 16\, h_1 h_2$$

Example

Prove that if time of flight of a bullet over a horizontal range R is T seconds then the inclination of direction to the horizontal is

$$tan^{-1}\left(\frac{gt^2}{2r}\right)$$

Solution

We know that

$$T = \frac{2_o \sin \alpha}{g}$$

$$R = \frac{2v_o^2 \, sin\alpha cos\alpha}{g}$$

$$\frac{gT^2}{2R} = \frac{4v_o^2 \sin^2 \alpha}{g} \frac{g}{4v_o^2 \, sin\alpha cos\alpha} = tan\alpha$$

Hence

$$\alpha = \tan^{-1}\left(\frac{gT^2}{2R}\right)$$

Example

A cricket ball thrown from height of 6 feet at an angle of 30° with the horizontal speed $of\ 60\ ft./sec.$ is caught by another player at height 2 feet from the ground. How far apart were the two men?

Solution

Let two men are x feet apart. As the ball is caught at a height of 2 feet from the ground, so the distance it falls vertically is 6–2 = 4 ft. The point $(x,-4)$ lies on the trajectory. Now

$$y = (v_o \, sin\alpha)t - \frac{1}{2}gt^2$$

Here $y = -4, v_0 = 6 - ft\ /sec., \alpha = 3 =^0, t = ?$

$$4 == (60 sin 30)t - \frac{1}{2}(32)t^2$$

$8t^2 - 15t - 2 = 0$ or (t–2) (8t + 1) = 0

$$t = 2\, or\, t = -\frac{1}{8}$$

Since $t \neq -\frac{1}{8}$, therefore $t = 2$.

$$x = Horzontal\ distance$$

$$= (Horizontal\ velocity)(time)$$

$$= (v_o cos\alpha)t = (60 cos 30^0)(2)$$

$$x = 60\sqrt{3}\ ft.$$

SUMMARY AND DISCUSSION

5.1. Projectile

Definition

An object thrown into the space with certain velocity from a gun or dropped from a moving plane under the action of gravity is called a projectile. Thus, a projectile move with a constant horizontal velocity and at the same falls freely under the action of gravity. The path of projectile is called the *trajectory.*

Horizontal and vertical component of velocity at any time t are $\dot{x} = v_0 \cos\alpha, \dot{y} = -gt + v_o\, sin\alpha$,

$x = (v_0 \cos\alpha)t$, $y = v_0\, sin\alpha - \frac{1}{2}gt^2$ are the parametric equation of trajectory.

$$\left(x - \frac{v_0^2 sin\alpha \cos\alpha}{g}\right)^2 = \frac{2v_o^2 \cos^2\alpha}{g}\left(y - \frac{v_o^2 \sin^2\alpha}{2g}\right)$$

5.2.1. Vertex

$$\left(\frac{v_0^2 sin\alpha \cos\alpha}{g}, \frac{v_o^2 \sin^2\alpha}{2g}\right)$$

5.2.2. Latus Rectum

Latus Rectum = $|4a| = \frac{2v_o^2 \cos^2\alpha}{g}$

5.2.3. Maximum Height

$$= \frac{v_o^2 \sin^2\alpha}{2g}$$

5.2.4. Equation of Directrix

$$y = \frac{v_o^2}{2g}$$

5.2.5. Focus

$$= \left(\frac{v^2 sin\alpha \cos\alpha}{g}, -\frac{v_o^2}{2g} cos2\alpha \right) = \left(\frac{v_o^2}{2g} sin2\alpha - \frac{v_o^2}{2g} cos2\alpha \right)$$

5.2.6. Time of Flight

The time taken by the particle to reach the horizontal plane through the point of projection is called its time of flight.

$t = \dfrac{2v_o\, sin\alpha}{g}$ is the time of flight.

5.2.7. Range

5.2.7.1. Horizontal Range

The horizontal range of the projectile is the horizontal distance described by the particle during the time of flight. Therefore

Range R = (horizontal velocity) (time of flight)

$$v_o \cos\alpha \left(\frac{2v_o\, sin\alpha}{g} \right)$$

$$= \frac{v_o^2\, 2(\sin\alpha\, cos\alpha)}{g} = \frac{2v_o^2 sin2\alpha}{g}$$

This range can also be obtained by putting y = 0 in the Cartesian equation of trajectory, i.e., putting y = 0 in the equation

$$y = xtan\alpha - \frac{gx^2}{2v_o^2} \sec^2\alpha$$

Also, it is notable that this range will remain same if we replace

$\alpha by \alpha = — - \alpha$

5.2.7.2. Maximum Range

Range R = $\dfrac{v_o^2 sin2\alpha}{g}$

For maximum range $\dfrac{dr}{d\alpha} = 0$

$$\Delta \frac{2v_o^2}{g} \cos 2\alpha = 0$$

$$\Delta cos 2\alpha = 0$$

$$\Delta 2\alpha = \frac{\pi}{2}$$

$$\Delta \alpha = \frac{\pi}{4}$$

Therefore $R_{max} = \dfrac{v_o^2}{g}$.

5.3.2. Time of Flight

Time of flight = $\dfrac{Horizontol\ distance}{Horizontol\ velocity}$

Time of flight $= \dfrac{x}{v_o \cos \alpha} = \dfrac{rcos\beta}{v_o \cos \alpha}$

Using (1), we get

Time of flight = $\dfrac{2v_o^2 \cos^2 \alpha \sin(\alpha - \beta)}{g \cos^2 \beta} \dfrac{cos\beta}{v_o cos\alpha}$

$$\text{Time of flight} = \frac{2v_o^2 \sin(\alpha - \beta)}{g \cos \beta}$$

For time of flight down the plane, change $\beta\ to - \beta$.

5.4. Parabola of Safety

As the equation of the projectile is

$$y = x\tan\alpha - \frac{1}{2}g\frac{x^2}{v^2}\sec^2\alpha$$

which shows that for different values of α we have different trajectories. So we define that the locus of all the points on different trajectories which lie at maximum distance from the point of projection O is called the parabola of safety.

CHAPTER 6

HARMONIC OSCILLATION

CONTENTS

6.1. SIMPLE HARMONIC MOTION

It is a motion of a particle moving in a straight line with an acceleration which is always directed toward a fixed point in the line and is proportional to the distance of the particle from that point.

$$\underline{AopB}$$

$$x=-a\,x=0\,x=x\,x=a$$

Let O be the fixed point on the line along the particle is moving. Let the particle be at point P. Let OP= x. Then equation of motion is

$$accleration=-\lambda x$$

where λ is constant of proportional. The negative sign shows that acceleration is directed in opposite directed in which x increase.

Since acceleration $=a=v\dfrac{dv}{dx}$, therefore

$$v\frac{dv}{dx}=-\lambda x$$

$$vdv=-\lambda xdx$$

$$\int vdv=-\lambda\int xdx$$

$$\frac{v^2}{2}=-\lambda\left(\frac{x^2}{2}\right)+B$$

where B is constant of integration.

Let the velocity of particle be zero at point A whose distance from O is a. Thus, when $x=a,v=0$ so that

$$0=-\lambda\frac{a^2}{2}+B$$

$$B=\lambda\frac{a^2}{2}$$

Thus

$$\frac{v^2}{2}=-\lambda\left(\frac{x^2}{2}-\frac{a^2}{2}\right)$$

$$\lambda\left(a^2-x^2\right)$$

$$v=\pm\sqrt{\lambda}\sqrt{a^2-x^2}$$

which gives the velocity for a given displacement. If the particle is moving to the right then as t increase, x also increase and so the velocity is positive,

$$v=\sqrt{\lambda}\sqrt{a^2-x^2}$$

Since $v=\frac{dx}{dt}$, therefore

$$\frac{dx}{dt}=\sqrt{\lambda}\sqrt{a^2-x^2}$$

$$\frac{dx}{\sqrt{a^2-x^2}}=\sqrt{\lambda}\,dt$$

$$\int\frac{dx}{\sqrt{a^2-x^2}}=\sqrt{\lambda}\int dt$$

$$\sin^{-1}\left(\frac{x}{a}\right)=\sqrt{\lambda}\,t+C \qquad (1)$$

where C is a constant of integration.

Case (i)

Let the particle Be at point A at t = 0, then x = a at t = a, so that (1) gives

$$\sin^{-1}\left(\frac{a}{a}\right)=\sqrt{\lambda}\,(0)+C$$

$$C=\frac{\pi}{2}$$

Then $\frac{x}{a}=\sin\left(\sqrt{\lambda}\,t+\frac{\pi}{2}\right)$

$$x=a\sin\left(\sqrt{\lambda}\mathrm{t}+\frac{\eth}{2}\right)$$

$$x = a\cos(\sqrt{\lambda}\mathrm{t})$$

which gives the displacement at time t.

Case (ii)

Let the particle be at O at t = 0, then x = 0 at t = 0, so that form (1)

$$\sin^{-1}\left(\frac{O}{a}\right) = \sqrt{\lambda}(0) + C$$ or $C = 0$

then

$$\sin^{-1}\left(\frac{x}{a}\right) = \sqrt{\lambda}\mathrm{t}$$

$$x = a\sin\left(\sqrt{\lambda}\mathrm{t}\right)$$

6.1.1. General solution

The differential equation of simple harmonic motion is

$$a = -\sqrt{\lambda}\,x$$

Since $a = \dfrac{d^2x^2}{dt}$

$$\frac{d^2x}{dt^2} = -\sqrt{\lambda}x$$

The characteristics equation is $D^2 + \lambda = 0$, whose roots are

$$D = \pm i\sqrt{\lambda}$$

Thus, general solution is

$$x = c_1\cos\sqrt{\lambda}\,t + c_2\sin\sqrt{\lambda}\,t$$

6.2. NATURE OF SIMPLE HARMONIC MOTION

Consider the equation:

$$x = a\cos\sqrt{\lambda}\,t$$

Since $-1 \le \cos\left(\sqrt{\lambda t}\right) \le 1,$ therefore $-a \le x \le a,$ thus the maximum displacement from the fixed-point O is called *center of the motion*. The maximum displacement from the center is called *amplitude of the motion*.

$\underline{AopB}$

$x = -a\, x = 0\, x = x\, x = a$

Let

$x = a\cos\sqrt{\lambda t}$ (1)

Then

$x = a\cos\left(\sqrt{\lambda}\, t + 2\pi\right) = a\cos\sqrt{\lambda}\ (t + \frac{2\pi}{\sqrt{\lambda}})$ (2)

From (1) and (2), the distance at time t and $t + \frac{2\pi}{\sqrt{\lambda}}$ is same.

Differentiate (1), then

$\frac{dx}{dt} = -a\sqrt{\lambda}\sin\left(\sqrt{\lambda t}\right)$ (3)

$$\frac{dx}{dt} = -a\sqrt{\lambda}\sin\left(\sqrt{\lambda t} + 2\pi\right)$$

$\frac{dx}{dt} = -a\sqrt{\lambda}\sin\sqrt{\lambda}\left(t + \frac{2\pi}{\sqrt{\lambda}}\right)$ (4)

From (3) and (4), the velocity at time $t + \frac{2\pi}{\sqrt{\lambda}}$ is also same.

Therefore, the motion is repeated after time $t = \frac{2\pi}{\sqrt{\lambda}}$ and the particle oscillates between x = -a and x = a.

$$velocity = v = \sqrt{\lambda}\sqrt{a^2 - x^2} = \sqrt{\lambda}\sqrt{a^2 - x^2} = 0$$

and acceleration = a = - $\sqrt{\lambda x} = -\sqrt{\lambda}\, a.$ which has maximum magnitude. Thus, at point B. x = a, its velocity is zero but its acceleration is maximum and directed towards O. Due to acceleration the particle moves toward O,

At point O, At point O, x = 0, so its acceleration is zero but its velocity is maximum and due to velocity, the particle moves toward point A.

At point A. x = -a velocity is zero and acceleration is maximum and due to acceleration. The particle moves toward O and finally comes to rest at A. So the particle complete *oscillation* or vibration. The motion is then repeated after time $2\pi / \lambda$ the time taken to complete one oscillation is called time period and denoted by T, therefore, $T = \dfrac{2\pi}{\sqrt{\lambda}}$.

The number of oscillation complete in unit time is called frequency and is denoted by υ thus

$$\upsilon = \frac{1}{T} = \frac{\sqrt{\lambda}}{2\pi}$$

If a particle described a cycle with a constant angular velocity, prove that the foot of the perpendicular from the particle on a diameter moves with simple harmonic motion.

Proof: Let the particle moves along a circle with center O and radius 'a' let ω be the angular velocity. Let center O be origin and diameter AA' on x-axis. Let $Q(x, y)$ be position of the particle at time t let $p(x, 0)$ be foot of the perpendicular from Q to diameter AA' let $\angle AOQ = \theta$, then

$$\frac{d\theta}{dt} = \omega, also$$

$$x = a\, cos\theta$$

Differentiate $w.r.t.t$, then

$$x = -a \sin \frac{d\theta}{dt} = -\omega sin\theta$$

Differentiating again $w.r.t.'t'$, we get

$$x = -a\omega \cos \frac{d\theta}{dt} = -a\omega^2 \cos\theta = -a\theta^2 x$$

The acceleration is proportional to the distance. Hence foot of perpendicular P describes simple harmonic motion. Here

$\lambda = \omega^2$ or $\sqrt{\lambda} = \text{ù}$

Time period = T = $\frac{2\pi}{\sqrt{\lambda}} = \frac{2\pi}{\omega}$

Example

If a P moves with a velocity v given by $v^2 = -n^2\left(ax^2 + 2bx + c\right)$ show that P describes a simple harmonic motion. Find the center, the amplitude and time period of motion.

Solution

Given $v^2 = -n^2\left(ax^2 + 2bx + c\right)$.

Differentiate $w.r.t$ 'x' we get

$$2v\frac{dv}{dx} = n^2\left(2ax + 2b\right)$$

$$v\frac{dv}{dx} = -n^2\left(2ax + 2b\right)$$

$$v\frac{dv}{dx} = -n^2\left(ax + b\right)$$

$$v\frac{dv}{dx} = -an^2\left(x + \frac{b}{a}\right)$$

Let X = x + $\frac{b}{a}$, then above equation gives

$$v\frac{dv}{dx} = -an^2X$$

$$i.e\,v\frac{dv}{dx} \propto X$$

Therefore, P excusive simple harmonic oscillation.

To find the center put X = 0, then $x + \frac{b}{a} = 0,$ so that $x = -\frac{b}{a}$ is center of motion.

Also, here

$$\lambda = an^2 .so\sqrt{\lambda} = \sqrt{a}\,n$$

Time period = $\dfrac{2\pi}{\sqrt{\lambda}} = \dfrac{2\pi}{\sqrt{an}}$

To find amplitude put v = 0, then

0 = $-n^2\left(ax^2 + 2bx + c\right)$

$$ax^2 + 2bx + c = 0$$

$$x = \frac{-b \pm \sqrt{b^2 - ac}}{a}$$

Let O be the origin, then

$$OA = \frac{-b + \sqrt{b^2 - ac}}{a}$$

and

OA' = $\dfrac{-b - \sqrt{b^2 - ac}}{a}$

Let C be the center, then OC = $-\dfrac{b}{a}$ and therefore, the amplitude = CA. Thus

$$Amplitude = CA == \frac{-b + \sqrt{b^2 - ac}}{a} - \left(\frac{\text{b}}{\text{a}}\right) = \frac{\sqrt{b^2 - ac}}{a}$$

$\underline{A'C\,O\,A}$

$$x = \frac{-b - \sqrt{b^2 - ac}}{r}\; x = -\frac{b}{a}\; x = \frac{-b + \sqrt{b^2 - ac}}{a}$$

Example

A particle describing simple harmonic motion has velocities 5 ft /sec and 4 ft/sec when its distance from center are 12 m and 13 m, respectively. Find the time period of motion.

Solution

We know that $v^2 = \lambda\left(a^2 - x^2\right)$, therefore' when $x = 12\,ft, v = 5\,ft\,/sec$, thus

$$25 = \lambda\left(a^2 - 144\right)$$

when $x = 13\ ft, v = 4\ ft\ /sec,$ so that

$$16 = \lambda\left(a^2 - 169\right) \qquad (2)$$

and

$$g = -\lambda x_2$$

Now

(i) Let $x_2 > x_1, therefore\, u > v.$ From (1) and (3), we have

$$U^2 - v^2 = \lambda\left(a^2 - x_1^2\right) - \lambda(a^2 - x_2^2)$$

$$= -\lambda\left(x_1^2 - x_2^2\right) = -\lambda\left(x_1 + x_2\right)\left(x_1 - x_2\right)$$

From (2) and (4), we have and

$$\text{f} + \text{g} = -\lambda\left(x_1 + x_2\right) \qquad (6)$$

Combining (5) and (6), we have

$$x_1 - x_2 = \frac{u^2 - v^2}{f + g}$$

(ii) $if\ x_2 > x_1,$ then g > f. From (2), (4) and (7)

$$g - f = \lambda\left(x_1 - x_2\right) = \lambda\left(\frac{u^2 - v^2}{f + g}\right)$$

$$\lambda = \frac{g^2 - f^2}{u^2 - v^2}$$

Thus, time period == $\dfrac{2\pi}{\sqrt{\lambda}} = 2\pi\sqrt{\dfrac{u^2 - v^2}{g^2 - f^2}}$

(iii) from (2), $= -\dfrac{f}{\lambda},$ therefore (1) becomes

$$u^2 = \lambda\left(a^2 - \frac{f^2}{\lambda^2}\right)$$

$$\frac{u^2}{\lambda} = a^2 - \frac{f^2}{\lambda^2}$$

$$a^2 = \frac{u^2}{\lambda} + \frac{f^2}{\lambda^2} = \frac{1}{\lambda}\left(u^2 + \frac{f^2}{\lambda}\right)$$

Subtracting (2) from (1), we get

$\lambda = \frac{9}{25}$ or $\sqrt{\lambda} = \frac{3}{5}$

Hence time period = T = $\frac{2\pi}{\sqrt{\lambda}} = \frac{10\pi}{3} sec.$

Example

A particle described simple harmonic motion in such a way that velocity and acceleration at a point p are u and f, respectively and the corresponding quantities at another point Q are b and g, show that

(i) distance PQ = $\frac{u^2 - v^2}{f + g}$

(ii) Time period $= 2\pi\sqrt{\frac{u^2 - v^2}{g^2 - f^2}}$

(iii) amplitude = $\sqrt{\frac{(u^2 - v^2)(u^2 g^2 - v^2 f^2)}{g^2 - f^2}}$

Solution

Let O be the origin and OP $= x_1$ *and* $PQ = x_2.$ we know that

$v^2 = \lambda(a^2 - x^2)$ and $a = -\lambda x$

$$x = x_1 x + x_2$$

$\overrightarrow{OPQ}$

Acceleration = f acceleration = g

Velocity = u velocity = v

Therefore, at point P $u^2 - \lambda\left(a^2 - x_1^2\right)$ and $f = -\lambda x_1$ and at point Q

$$v^2 = \lambda\left(a^2 - x_2^2\right)$$

u sin (8),

we get

$$a^2 = \left(\frac{u^2 - v^2}{g^2 - f^2}\right)\left\{u^2 + f^2\left(\frac{u^2 - v^2}{g^2 - f^2}\right)\right\}$$

$$= \left(\frac{u^2 - v^2}{g^2 - f^2}\right)\left\{\frac{u^2 g^2 - u^2 f^2 + f^2 u^2 - f^2 v^2}{g^2 - f^2}\right\}$$

Thus $a = \sqrt{\dfrac{(u^2 - v^2)\left(u^2 g^2 - v^2 f^2\right)}{g^2 - f^2}}$ is amplitude.

Example

(a) Show that if the displacement of the particle at any time is given by

$$x = a\cos(nt) + b\sin(nt),$$

The particle executes simple harmonic motion of amplitude $\sqrt{a^2 + b^2}$ *and period* $\dfrac{2\pi}{n}$

(b) if a − 3, b = 4, n = 4. Find the time period, amplitude, maximum velocity and maximum acceleration

Solution

Given

$$x = a\cos(nt) + b\sin(nt)$$

Therefore

$$\frac{dx}{dt} = = -an\,sn(nt) + bn\cos(nt)$$

$$\frac{d^2x}{dt^2} = -an^2\cos(nt) + bn^2\sin(nt) = n^2x$$

which shows that acceleration is proportional to the distance. Thus, the motion is a simple harmonic motion

Now $\lambda = n^2$ or $\sqrt{\lambda} = n$ then

Time period $= \dfrac{2\pi}{\sqrt{\lambda}} = \dfrac{2\pi}{n}$

$x = a\cos(nt) + b\sin(nt)$, therefore

$$x = \sqrt{a^2+b^2}\left\{\frac{a}{\sqrt{a^2+b^2}}\cos(nt) + \frac{b}{\sqrt{a^2+b^2}}\sin(nt)\right\}$$

Let $sin\theta = \dfrac{a}{\sqrt{a^2+b^2}}$ and $cos\theta = \dfrac{b}{\sqrt{a^2+b^2}}$

Then $tan\theta = \dfrac{b}{a}$ or $\theta = \tan^{-1}\left(\dfrac{b}{a}\right)$. Thus

$$x = \sqrt{a^2+b^2}\{sin\theta\cos(\mathrm{nt}) + \cos\theta\sin(\mathrm{nt})\}$$

$$x = \sqrt{a^2+b^2}\sin(nt+\theta)$$

$$x = \sqrt{a^2+b^2}\sin(nt + \tan^{-1}\frac{b}{a})$$

The maximum value of sin $\left(ny + tan^{-1}\dfrac{b}{a}\right) = 1, thus$

Maximum displacement $= x = \sqrt{a^2+b^2}$ or amplitude $= \sqrt{a^2+b^2}$

(b) when a =3, b = 4, n = 2, then

Time period $T = \dfrac{2\pi}{n} = \pi$

Amplitude $= \sqrt{a^2+b^2} = 5$

Since $v = \sqrt{\lambda}\sqrt{a^2-x^2}, v$ is maximum when $x = 0$, thus

$$Maximum\,velocity = \sqrt{\lambda}\sqrt{5^2 - 0} = 2(5) = 10$$

Acceleration = $-\lambda x$

Acceleration is maximum when x = a, thus maximum value of magnitude of acceleration = $\lambda a = (4)(5) = 20$.

6.3. DAMPED HARMONIC OSCILLATOR

A harmonic oscillator whose amplitude decrease gradually with time and vanishes ultimately is known as Damped Harmonic Oscillator.

Consider a free (no driving force) motion as a mass less spring whose one end is attached to an immovable wall and other end to a particle of mass m as shown in figure.

The whole system being immersed in a fluid which affords a retarding force on m, proportional to the velocity of m. Let x be the displacement of m from the mean position, then equation of motion can be written as

$$mx = -c_1\, x - c_2 x$$

where c_1 is the force constant of the spring and c_2 is constant of proportionality for retarding force, ivied above equation by m, then

$$x + 2kx + \omega^2\, x = 0$$

where 2k $= \frac{c^2}{m}\, and\; \omega^2 = \frac{c_1}{m}.\omega$ is know as the angular frequency of undamped.

Eq.(1) can be written as

$$\frac{d^2x}{dt^2} + 2k\frac{dx}{dt} + \omega^2 x = 0$$

$$\left(D^2 + 2kD + \omega^2\right)x = 0; d = \frac{d}{dt}$$

The auxiliary equation is

$$D^2 + 2kD + \omega^2 = 0$$

Here are three cases as given in the following subsections.

6.3.1. Under Damped Case (Light Damping)

In this case $k^2 < \omega^2$, therefore $k^2 - \omega^2$ is negative. Put $k^2 - \omega^2 = -n^2$, then

$$D = K \pm in$$

Thus, the solution of (1) is

$$x = e^{-kt}\{a\sin(nt) + b\cos(nt)\}$$

where a and b are constants.

By putting a = d $\sin\alpha, b = d\ cos\alpha x = de^{-ki}\cos(nt-\alpha)$ shows an oscillatory motion where $d = \sqrt{a^2 + b^2}$ and $\alpha = \tan^{-1}\left(\frac{a}{b}\right)$

If K is small quantity, then amplitude de^{-kt} decrease gradually with time and tends to zero as $t \to \infty$

6.3.2. Critically Damped Case

In this case $k^2 = \omega^2$, therefore $k^2 - \omega^2$ is zero. Put $k^2 - \omega^2 = 0, then$,

$$D = -\omega \pm 0$$

Thus the solution of (1) is

$$x = e^{-\omega t}\ (\mathrm{A + Bt})$$

where a and b are constants.

For limiting case when $t \to \infty$.

$$x = LIM_{t\to\infty}\left(\frac{A+Bt}{e^{\omega t}}\right)$$

This $\frac{\infty}{\infty}$ from, so using L. Hospital Rule.

$$x = LIM_{t\to\infty}\frac{B}{\omega e^{\omega t}} \to 0$$

6.3.3. Over Damped Case (Heavy Damping)

In this case $k^2 > \omega^2 .therefore\, k^2 - \omega^2$ is positive. Put $k^2 - \omega^2 = \lambda^2 (say)$, then

$$D = -K \pm \lambda$$

Thus, the solution of (1) is

$$x = Ce^{(-k+\lambda)t} + De^{(-k-\lambda)t}$$

$$x = Ce^{(-k-\lambda)t} + De^{(-k+\lambda)t}$$

which does not represent an oscillatory motion. Also, as \ $t \to \infty, t \to 0$, which is dead beat motion.

6.3.4. Damped Force Oscillator

Let a particle of mass m executes simple harmonic motion can be written as

$$mx = -m\omega^2 x$$

$$x = -\omega^2 x$$

If there is a damping force proportionality for damping force. Also, we consider a driving force (applied force) F cos(pt) upon it, then equation (2) becomes

$$x = -\omega^2 x - 2kx + F \cos pt$$

$$x = 2kx + \omega^2 x = F \cos pt$$

For complimentary function put

$$x = 2kx + \omega^2 x = 0$$

whose auxiliary equation is

$$D^2 + 2kd \mid \omega^2 = 0$$

This implies that

$$D = -K \pm \sqrt{k^2 - \omega^2}$$

For $k^2 < \omega^2, put\, k^2 - \omega^2 = -n^2$, then

$$D = k \pm in$$

And therefore,

$$x_c = Ce^{-kt}\left\{A\sin(nt) + A\cos(nt)\right\}$$

$$x_c = Ce^{-ki}\sin(nt+\alpha)$$

where C $\sqrt{A^2+B^2}$ $and\ \alpha = \tan^{-1}\left(\frac{A}{B}\right)$

For particular integral

$$x_p = \frac{F}{D^2+2KD+\omega^2}\cos pt$$

$$x_p = \frac{F}{-P^2+2KD+\omega^2}\cos pt$$

$$\pm_p = \frac{F}{2\pm \quad +\left(\omega^2 - \quad ^2\right)}\cos$$

$$x_p = \frac{F\left[2kd-\left(\omega^2-p^2\right)\right]}{4k^2D^2-\left(\omega^2-p^2\right)^2}\cos pt$$

$$x_p = \frac{F\left[2kd-\left(\omega^2-p^2\right)\right]}{-4k^2D^2-\left(\omega^2-p^2\right)^2}\cos pt$$

$$x_p = \frac{-F\left[2kd-\left(\omega^2-p^2\right)\right]}{4k^2D^2+\left(\omega^2-p^2\right)^2}\cos pt$$

$$x_p = \frac{-F\left[2kd\cos pt-\left(\omega^2-p^2\right)\cos pt\right]}{4k^2D^2+\left(\omega^2-p^2\right)^2}$$

$$x_p = \frac{-F\left[2kp\sin pt-\left(\omega^2-p^2\right)\cos pt\right]}{4k^2D^2+\left(\omega^2-p^2\right)^2}$$

$$x_p = \frac{F\left[2kp\sin pt + \left(\omega^2 - p^2\right)\cos pt\right]}{4k^2D^2 + \left(\omega^2 - p^2\right)^2}$$

$$x_p = \frac{FR\sin\left(pt + \alpha\right)}{R^2} = \frac{FR\sin\left(pt + \alpha\right)}{R}$$

where

$$2kp = R\,cos\alpha.\omega^2 - p^2 = R\sin\alpha$$

$$R^2 = 4k^2p^2 + \left(\omega^2 - P^2\right)^2.\alpha = \tan^{-1}\left(\frac{\omega^2 - p^2}{2kp}\right)$$

Thus complete solution is

$$x = x_e + x_p$$

$$= Ce^{-ki}\sin\left(nt + \alpha\right) + \frac{F\sin\left(pt + \alpha\right)}{R}$$

Hence x_e (complementary function) $\to 0$ as t $\to\infty$ Therefore

$$x = \frac{F\sin\left(pt + \alpha\right)}{R}$$

The amplitude of particle integral is F/R. therefore oscillate between F/R and has no tendency of becoming permanently zero. For this reason complimentary function x_c is called transient (temporary) and the particular integral x_p is called the steady state solution.

Example

A particular of mass m moving under the action of forces

$$F_1 = -m\omega^2 x, F_2 = mF_o t, F_3 = -2m\mu x$$

Assuming that damping is small, set up and solve the equation of otion

Solution

The equation of motion for $F_1\, can be$ written as

$$mx = m\omega^2 x$$

there is damping

6.3.5. Vertical Motion with Air Resistance

When a body moves in the atmosphere, air retards motion. For ordinary velocities, neither very small not very large, this retardation is very nearly proportional to the square of the velocity. Anyhow, for academic interest, any other law of resistance may be assumed

6.3.6. Downward Motion

A particle of mass m is related from rest at an initial height above the ground. The resistance of the air is assumed to be proportional to the square of the velocity.

SUMMARY AND DISCUSSION

6.1. Simple Harmonic Motion

It is a motion of a particle moving in a straight line with an acceleration which is always directed toward a fixed point in the line and is proportional to the distance of the particle from that point.

$\underline{AopB}$

$x = -a\, x = 0\, x = x\, x = a$

i. $x = a\cos(\sqrt{\lambda}\mathrm{t})$ Which gives the displacement at time t.

6.2. Nature of simple Harmonic Motion

Consider the equation

$x = a\cos\sqrt{\lambda}\, t$

Since $-1 \leq \cos\left(\sqrt{\lambda t}\right) \leq 1$, therefore $-a \leq x \leq a$, thus the maximum displacement from the fixed-point O is called *center of the motion*. The maximum displacement from the center is called *amplitude of the motion.*

$\underline{AopB}$

$x = -a\, x = 0\, x = x\, x = a$

Therefore, the motion is repeated after time $t = \dfrac{2\pi}{\sqrt{\lambda}}$ and the particle oscillates between x = -a and x = a.

$$velocity = v = \sqrt{\lambda}\sqrt{a^2 - x^2} = \sqrt{\lambda}\sqrt{a^2 - x^2} = 0$$

and acceleration = a = – $\sqrt{\lambda}x = -\sqrt{\lambda}\,a.$ which has maximum magnitude. Thus, at point B. x = a, its velocity is zero but its acceleration is maximum and directed towards O. Due to acceleration the particle moves toward O, At point O, At point O, x = 0, so its acceleration is zero but its velocity is maximum and due to velocity, the particle moves toward point A.

At point A. x = -a velocity is zero and acceleration is maximum and due to acceleration. The particle moves toward O and finally comes to rest at A. So the particle complete *oscillation* or vibration. The motion is then repeated after time $2\pi / \lambda$ the time taken to complete one oscillation is called time period and denoted by T, therefore, $T = \frac{2\pi}{\sqrt{\lambda}}$

The number of oscillation complete in unit time is called frequency and is denoted by υ thus

$$\upsilon = \frac{1}{T} = \frac{\sqrt{\lambda}}{2\pi}$$

If a particle described a cycle with a constant angular velocity, prove that the foot of the perpendicular from the particle on a diameter moves with simple harmonic motion.

6.3. Damped Harmonic Oscillator

A harmonic oscillator whose amplitude decrease gradually with time and vanishes ultimately is known as Damped harmonic Oscillator.

6.3.1. Under Damped Case (Light Damping)

$$x = e^{-kt}\{a\sin(nt) + b\cos(nt)\}$$

where a and b are constants.

By putting a = d $\sin\alpha, b = d\,cos\alpha x = de^{-ki}\cos(nt - \alpha)$. shows an oscillatory motion where $d = \sqrt{a^2 + b^2}$ and $\alpha = ü^{-1}\left(\frac{a}{b}\right)$

If K is small quantity. then amplitude de^{-kt} decrease gradually with time and tends to zero as $t \to \infty$.

6.3.2. Critically Damped Case

$x = e^{-\omega t}$ (A + Bt) where A and B are constants. For limiting case when $t \to \infty. x = LIM_{t\to\infty}\left(\frac{A+Bt}{e^{\omega t}}\right)$. This $\frac{\infty}{\infty}$ from, so using L. Hospital Rule.

$x = LIM_{t\to\infty}\frac{B}{\omega e^{\omega t}} \to$ 0.

6.3.3. Over Damped Case (Heavy Damping)

$$x = Ce^{(-k+\lambda)t} + De^{(-k-\lambda)t}$$

$$= Ce^{(-k-\lambda)t} + De^{(-k+\lambda)t}$$

which does not represent an oscillatory motion. Also, as \ $t \to \infty, t \to 0$, which is dead beat motion.

CHAPTER 7

CENTRAL FORCE AND APPLICATIONS

CONTENTS

7.1. CENTRAL FORCE

A force acting on a P particle which always passes through a fixed point is called a Central force. The fixed point is known as center of force, usually it is taken as origin. The central force is either attractive (toward the center of force) or repulsive (away from the center of force).

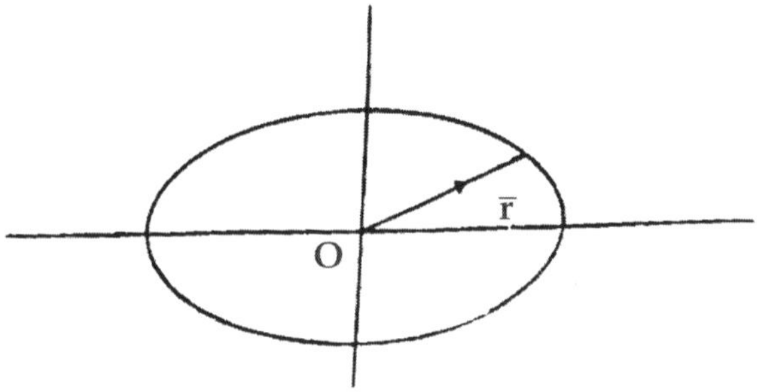

The central force at a point is a function of distance of that point from the origin. This relation is called the *law of force*. The path described by the particle moving under a central is called the central orbit thus $\bar{F}$ is a central force if

$$\bar{F} = f(r)\hat{r}$$

where $\hat{r}$ is a unit vector in the direction of $\hat{r}\, and\, r\, is\, the\, magnitude$ of the position vector $\hat{r}.Also$

$$\bar{r}\, x\bar{F} = f(r)\bar{r}\, x\bar{r} = f(r)r\hat{r}\, x\hat{r} = 0$$

$$\text{i.e}\bar{r}\, x\bar{F} = 0$$

7.2. PROPERTIES

(i) Show that the orbit of a particle moving under a central force must be a plane must be a plane curve.

Proof:

Let $\bar{r}$ be the position vector of a particle which is moving under the action of central force $\bar{F}$. Then

$$\bar{r}\, x\bar{F} = 0$$

$$\bar{r}\, x \frac{d\bar{v}}{dt} = 0$$

$$\frac{d}{dt}(\bar{r}\, x \bar{v}) = 0$$

By integrating

$$\bar{r}\, x \bar{v} = \bar{c};\bar{c}\ is\, a\, constant\, vector$$

This show that normal to the plane containing $\bar{r}\,(\,posiyion\, vector\,)$ and $\bar{v}$ (tangent vector) has a constant direction, which is possible in the case of a plane curve i.e path of the particle lie on a plane.

(ii) Show that angular momentum is constant and find its magnitude.

Proof.

The angular momentum of a particle of mass m, moving with a velocity $\bar{v}$ is defined as

$$\bar{J} = \bar{r}\, x\, m\bar{v}$$

$$\bar{j} = m\bar{r}\, x \bar{v}$$

as m is constants calar and $\bar{r}\, x \bar{v}\, is\, a$ constant vector by property (i), therefore $\bar{j}$ ia a constant vector. Hence its magnitude is constant.

Take the velocity vector $\bar{v}$ in radial and transverse components as

$$\bar{v} = \dot{r}\hat{r} + r\theta\,\hat{s}$$

$$\bar{j} = m\bar{r}\, x\left(\overset{\ddot{y}}{r}\hat{r} + r\,\theta\hat{s}\right)$$

$$- mr\,\hat{r}\, x\,(r\,\hat{r} + r\,\theta\hat{s}$$

$$= 0 + mr^2\,\theta(\hat{r}\, x\,\hat{s})$$

$$j = |\bar{j}| = mr^2\,\theta\,|\hat{r}x\,\hat{s}| = mr^2\theta$$

$$\frac{j}{m} = r^2\theta$$

$$h = r^2\theta$$

where h = $\frac{j}{m}$.

7.3. DIFFERENTIAL EQUATION OF THE ORBIT IN POLAR FORM

Let a particle of mass m moving under the action of an attractive central force of magnitude $f(r)$ per unit mass then equation of motion can be written as

$$m\ddot{\bar{r}} = mf(r)\hat{r}$$

$$\ddot{\bar{r}} = f(r)\hat{r}$$

By using the radial and transverse components

$$\left(\ddot{\bar{r}} - r\theta^2\right)\hat{r} + \frac{1}{r}\frac{d}{dt}\left(r^2\theta\right)\hat{s} = -f(r)\hat{r}$$

Equating radial and transverse components, we get

$$\left(\ddot{\bar{r}} - r\theta^2\right) = -f(r)\hat{r} \qquad (1)$$

$$\frac{1}{r}\frac{d}{dt}\left(r^2\theta\right) = 0 \qquad (2)$$

Consider eq. (2),

$\frac{d}{dt}\left(r^2\theta\right) = 0,$ this implies that

$$r^2\theta = h \qquad (3)$$

where h is constant of integration and known as angular momentum of a particular of unit mass. Now consider equation (1),

$\left(\ddot{\bar{r}} - r\frac{h^2}{r^4}\right) = -f(r);$ using (3)

$$\left(\ddot{r}-\frac{h^2}{r^3}\right)=-f(r)$$

Let $r=\frac{1}{u}$, then $\dot{r}=\frac{1}{u^2}\frac{du}{d\theta}=-\frac{1}{u^2}\frac{du}{d\theta}\dot{\theta}$

$\dot{r}=-h\frac{du}{d\theta}$; using (3)

$$\ddot{r}=-h\frac{d^2u}{d\theta}\dot{\theta}$$

$\ddot{r}=-h^2u^2\frac{d^2u}{d\theta^2}$; using (3)

Put value of r and r into equation, we get

$$-h^2u^2\frac{d^2u}{d\theta^2}-h^2u^3=-f(u)$$

$$h^2u^2\left(\left(\frac{d^2u}{d\theta^2}\right)+u\right)=f(u)$$

which is the differential equation of the orbit

7.4. DIFFERENTIAL EQUATION OF THE ORBIT IN PEDAL FORM

From calculus we know that

$$\frac{1}{p^2}=\frac{1}{r^2}+\frac{1}{r^4}\left(\frac{dr}{d\theta}\right)^2$$

Put $r=\frac{1}{u}, then\frac{dr}{d\theta}=-\frac{1}{u^2}\frac{du}{d\theta}$

Now eq. (1) becomes

$$\frac{1}{p^2}=u^2+\left(\frac{du}{d\theta}\right)^2$$

Differentiate $w.r.t.$ 'r', we get

$$-\frac{2}{p^3}\frac{dp}{dr}=2u\frac{du}{d\theta}\frac{d\theta}{dr}+2\left(\frac{du}{d\theta}\right)\frac{d^2u}{d\theta^2}\frac{d\theta}{dr}$$

$$-\frac{2}{p^3}\frac{dp}{dr}=\frac{du}{d\theta}\frac{d\theta}{dr}\left(u\frac{d^2u}{d\theta^2}\right)$$

$$\frac{d^2u}{d\theta^2}+u=\frac{f(u)}{h^2u^2},$$ therefore

$$-\frac{1}{p^3}\frac{dp}{dr}=\frac{f(u)}{h^2u^2}\frac{du}{dr}$$

Since

$$u=\frac{1}{r},therefore,\frac{du}{dr}=-\frac{1}{r^2}=-u^2$$

Thus

$$\frac{1}{p^3}\frac{dp}{dr}=\frac{f(r)}{h^2}$$

$$f(r)=\frac{h^2}{p^3}\frac{dp}{dr}$$

which is the pedal equation of the orbit

7.5. APSE AND APSIDAL DISTANCE

An apse is a point on an orbit at which the radius vector drawn from the center of force has extreme (maximum or minimum) magnitude. Therefore, at an apse

$$\frac{dr}{dt}=0 \quad (1)$$

i.e., the radial component of velocity is zero. From (1)

$$\frac{dr}{dt}=\frac{dr}{d\theta}\frac{d\theta}{dt}=0$$

Since $\frac{d\theta}{dt}\neq 0$, therefore

$$\frac{dr}{d\theta} = 0$$

Let r = 1 /u, then by differentiating

$$\frac{dr}{d\theta} = \frac{1}{u^2}\frac{du}{d\theta}$$

Therefore, by using (2), we have

$\frac{dr}{d\theta} = 0$ or $\frac{du}{d\theta} = 0$

The magnitude of the radius vector at an apse is known as apsidal distance. The line joining an apse to the center of the force is an apse line. The angle between two consecutive apse line is an apse angle.

N.B.: The radius vector is perpendicular to tangent at an apse.

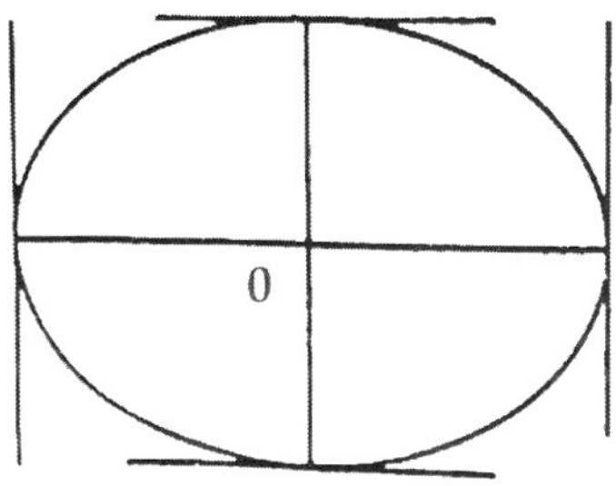

Let $p(r,\theta)$ be a point on orbit whose position vector is $\bar{r}$. If $\varnothing$ is angle between radius vector and tangent at p then from calculus

$$\tan\varnothing = r\frac{d\theta}{dr}$$

where $\varnothing$ is the angle between radius vector and tangent.

$$\cot\varnothing = \frac{1}{r}\frac{dr}{d\theta}$$

But at an apse

$\frac{dr}{d\theta} = 0,$ then

$\cot \varnothing = 0$ or $\varnothing = \frac{\pi}{2}, \frac{3\pi}{2}$

Hence radius vector is perpendicular to the tangent at an apse.

Example

Find the apse of an ellipse

Solution

The equation of the ellipse is

$$\frac{x^2}{a^2} + \frac{y^2}{b^2} = 1$$

where 'a' is semi major axis and b is semi minor axis. To change in polar coordinates put

$$x = r\cos\theta, y = r\,sin\theta$$

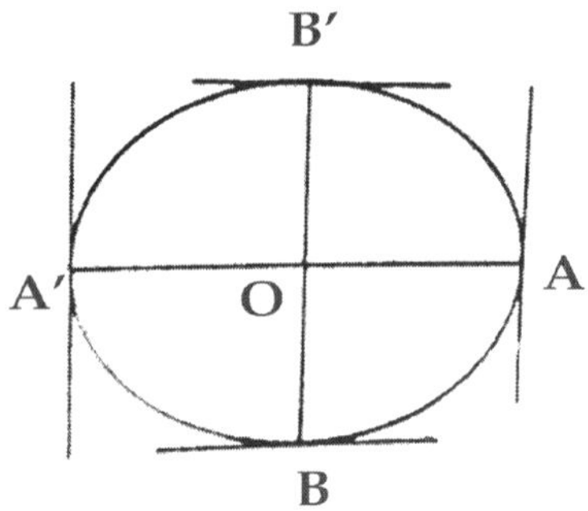

Then equation of ellipse become

$$r^2\left(\frac{\cos^2\theta}{a^2} + \frac{\sin^2\theta}{b^2}\right) = 1$$

Differentiate w.r.tθ, we get

$$2r\frac{dr}{d\theta}\left(\frac{\cos^2\theta}{a^2} + \frac{\sin^2\theta}{b^2}\right) + 2r^2\sin\theta\cos\theta\left(-\frac{1}{a^2} + \frac{1}{b^2}\right) = 0$$

For apse put $\frac{dr}{d\theta} = 0,$ then

$$0 + 2r^2 \sin\theta cos\theta\left(-\frac{1}{a^2}+\frac{1}{b^2}\right)=0$$

$$\sin\theta\, cos\theta = 0$$

This implies that either $sin\theta = 0$,i.e., $\theta = 0, \pi or\ \cos\theta = 0 i.e\, \theta = \frac{\pi}{2}, \frac{3\pi}{2}$.

Hence the apse of an ellipse are A, A', B and B'.

7.6. AREAL VELOCITY

Let $\bar{r}$ *be* the position vector of a point P on the orbit and Q is the neighboring point whose position vector is $\bar{r} + d\bar{r}$ and the angle between $\bar{r}$ and $\bar{r} + d\bar{r}$ is $d\theta$, then from the figure

$$\left|\bar{r} + d\bar{r}\right| \sin(d\theta) = \left|d\bar{r}\right|$$

where $\left|d\bar{r}\right| = rd\theta.$

Area swept by the position vector $\bar{r}$ is

$$dA = \frac{1}{2}\left|\left\{\bar{r} x\left(\bar{r} + d\bar{r}\right)\right\}\right| = \frac{1}{3}\left|\bar{r}\right|\left|\bar{r} + d\bar{r}\right| \sin(d\theta)$$

But

$$\left|\bar{r} + d\bar{r}\right| \sin(d\theta) = \left|d\underline{r}\right|$$

$$\frac{1}{2}\left|\bar{r}\right|\left|d\bar{r}\right| = \frac{1}{2} r\, rd\theta = \frac{1}{2} r^2\, d\theta$$

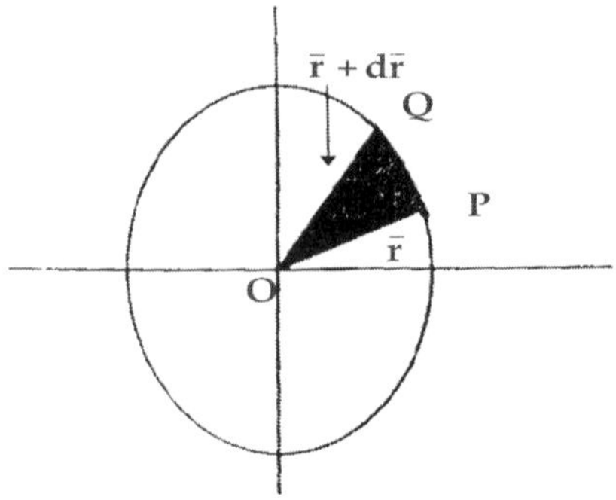

Areal Velocity = $\frac{dA}{dt} = \frac{1}{2} r^2 \theta$

For circular orbit

$r^2 \theta = h$, therefore

Area velocity = $\frac{dA}{dt} = \frac{h}{2}$

Area described in any time t_1 from a fixed point where $t = 0$ can be obtained by integrating the real velocity, therefore

$$A = \frac{h}{2}\int_0^{t_1} dt = \frac{h}{2} t_1$$

7.7. PERIODIC TIME

The time required to complete on rotation is known as periodic time. Since h/**2** is the area described in unit time, it follows that if T is the time that the particle takes to described in unit time, it follows that if T is the time that the particle takes to described the whole area of the ellipse, then

Area of the ellipse = $\frac{h}{2} T$

$$\pi ab = \frac{h}{2} T$$

$$T = \frac{2\pi ab}{h}$$

7.8. KEPLER'S LAWS

In 1609 Johannes Kepler stated the first two laws of planetary motion, which are:

7.8.1. First Law

Every planet described a ellipse with the sun at one focus.

7.8.2. Second Law

The radius vector drawn from the sum to a planet described equal areas intervals of times, ten years later, in 1619, Kepler announced the third law.

7.8.3. Third Law

The squares of the periods of the different planets are proportional to the cube s of their respective semi axes.

After half a century Newton derived gravitational law from Kepler's Law

7.9. DERIVATION OF NEWTON'S LAW OF GRAVITATION FROM KEPLER'S LAWS

A particle moves in an ellipse under a central force directed towards its focus, whose polar equation is

$$\frac{l}{r} = 1 + e\,cos\theta$$

Put $\frac{l}{r} = u,$ then

$$lu = 1 + e\,cos\theta \qquad (1)$$

Therefore, by differentiating two times w.r.t. θ

$$l\frac{d^2u}{d\theta} = -ecos\theta$$

By using (1)

$$l\frac{d^2u}{d\theta} = 1 - lu$$

$$\frac{d\ \ u}{d} = \frac{-}{l} \qquad (2)$$

We know that the differential equation of the orbit is

$$h^2u^2\left(\frac{d^2u}{d\theta^2} + u\right) = f(u)$$

Using (2), we have

$$h^2u^2\left(\frac{1}{l} - u + u\right) = f(u)$$

$$f(u) = \frac{h^2}{l}u^2$$

$$f(r) = \frac{h^2}{lr^2} = \frac{\mu}{r^2} \text{ where } \mu = \frac{h^2}{l}$$

which is Newton's Law of Gravitation.

7.10. DERIVATION OF KEPLER'S LAWS

7.10.1. First Law

Differential equation of circular orbit is

$$h^2u^2\left(\frac{d^2u}{d\theta^2} + u\right) = f(u)$$

By Newton's Law of gravitation

$$f(u) \quad —u$$

Therefore (1) becomes

$$\frac{d^2u}{d\theta^2} + u = \frac{1}{l}$$

For complimentary function

$$\frac{d^2u}{d\theta^2}+u=0$$

whose solution is $u_c = A\,cos\theta + B\sin\theta = C\cos(\theta-\alpha)$ where $C=\sqrt{A^2+B^2}, A = C\,cos\alpha, B = C\sin\alpha$

For particular integral

$$u_p = \frac{1}{D^2+1}\left(\frac{1}{1}\right)=\frac{1}{l}$$

Thus

$$u = u_c + u_p = C\cos(\theta-\alpha)+\frac{1}{l}$$

Rotating initial line through an angle α

$$u = C\cos(\theta)+\frac{1}{l}$$

$$u = lC\cos(\theta)+1$$

$$\frac{l}{r}=1+e\,cos\theta$$

This equation represents an ellipse where $e = lC\,and\,l$ is the semi latus rectum of an ellipse which is equal to $\frac{b^2}{a}, b$ is semi minor axis and a is semi major axis.

7.10.2. Second Law

We know that Areal velocity of a particle $=\frac{dA}{dt}=\frac{h}{2}$.

Suppose that $A_1 and\ A_2$ are the areas sweeps out by the particle in time $t_1\ and\ t_2$, respectively, then $A_1 = \frac{h}{2}\int_0^{t_1} dt = \frac{h}{2}t_1$

$$A_2 = \frac{h}{2}\int_0^{t_2} dt = \frac{h}{2}t_2$$

By subtracting $A_2 - A_1 = \frac{h}{2}(t_2 - t_1)$ which shows that radius vector drawn from the sun to the planet sweeps out equal areas in equal times.

7.10.3. Third Law

We know that periodic time of a particle is $T = \frac{2\pi ab}{h}$ where a is semi major axis and b is semi minor axis. Also

$$h = \sqrt{\mu l}$$

where l is semi latus rectum and equal to $\frac{b^2}{a}$, therefore

$$h = \sqrt{\mu l} = \frac{\sqrt{\mu}}{\sqrt{a}}b$$

Put this value of h in periodic time, then

$$T^2 = \frac{4\pi^2}{\mu}a^3$$

Hence, the squares of the periods of the planets are proportional to the cubes of the major axis of their orbits.

7.11. COLLISIONS: THEORY OF COLLISION

A collision is said to be taken place when two bodies physically collide with each other.

7.12. LINEAR MOMENTUM OF A PARTICLE

If a particle of mass m moving with velocity v, then linear momentum of the particle is the product of mass and velocity it is usually denoted by "L". Therefore

$l = mv$

7.13. ANGULAR MOMENTUM

If l Is the linear momentum of the particle. Then angular momentum $\underline{\grave{U}}$ is define as

$\underline{\Omega} = \underline{r} + \underline{l}$

If particle moving in $xy - plane$ then

$$\underline{\Omega} = (x\underline{i} + y\underline{j})x(L\underline{i} + L\underline{j})$$

$$\underline{\Omega} = (x\underline{i} + y\underline{j})x(mv\ \underline{i} + mv\ \underline{j})$$

where $\underline{L}\ m\ v = m(v\underline{i} + v\underline{j})$

$$\underline{\Omega} = m(xv - yv)\underline{K}$$

$$\underline{\Omega} = m(xy - yx)\underline{k}$$

7.14. PRINCIPAL OF LINEAR MOMENTUM

We know prove the principle. The sudden change in linear momentum of a system is equal to the total external impulsive force.

Proof. *Let* m, be mass of the i-th particle and X, Y, the components of external force on it in the direction of the axes. Then

$$\sum_i \frac{d}{dt}(m,x) = \Sigma X, \sum_I \frac{d}{dt}(m_I, y_I) = \sum_i y_I$$

Integrating this equation with respect to t over the interval (t_1, t_2) *and* taking the limit as t_2 approaches t_1, we have

$$Lt_{t_2 \to t_1} \left| \sum_I m_I x_I \right|_{t_1}^{t_2} = Lt_{t_2 \to t_1} \int_{t_1}^{t_2} \Sigma X_I\, dt.$$

$$Lt_{t_2 \to t_1} \left| \sum_I m_I y_I \right|_{t_1}^{t_2} = Lt_{t_2 \to t_1} \int_{t_1}^{t_2} \Sigma Y_I\, dt.$$

The left hand members of these equation denotes the instantaneous change in the momentum and the right hand members represent the total impulsive forces in the directions of the coordinates axes; hence the principal. The above equations may be written in the vector from as

$$\underline{M} = \hat{I}$$

where $\hat{I} = \hat{X}\underline{i} + \hat{Y}\underline{j}$

$$M\bar{v} = \hat{I} \left(\because \underline{M} = M\bar{v} \right)$$

where M is mass of the system and $\bar{v}$,

The sudden change in the velocity of the mass center. If these are no external impulsive forces, we have

$$\underline{\hat{I}} = 0$$

and hence

$$\underline{\hat{v}} = 0$$

Thus when the impulsive forces are purely internal, there is no sudden change in the velocity of the mass center. This result is important for solving problems of collision and explosions.

Note. It should be remembered that a blow changes only the velocity or momentum of a particle, and it does not change its instantaneous position.

7.15. THE PRINCIPLE OF ANGULAR MOMENTUM

Now we prove that the sudden change in angular momentum of a system about a fixed is equal to the moment of the external impulsive forces about the axis. We prove the principal for a single particle. The proof can be extended to a system quite easily.

Taking the fixed axis as the z-axis we have, by the principal of angular momentum,

$$\frac{d}{dt}\left[m\left(x\dot{y} - y\dot{x} \right) \right] = xY - yX.$$

where the external force on the particle is $\underline{F} = X\underline{i} + Y\underline{j}$

Integrating the last equation over the interval $\left(t_1, t_2 \right)$ and proceeding to the limit t_2 approaches t_1 we have

$$Lt_{t_2 \to t_1} \left| m\,(xu - yx \right|_{t_1}^{t_2} = Lt_{t_2 \to t_1} \int_{t_1}^{t_2} (xY - yX)\, dt$$

$$x\left\{ Lt_{t_2 \to t_1} \int_{t_1}^{t_2} (Y)\, dt \right\} - y\left\{ Lt_{t_2 \to t_1} \int_{t_1}^{t_2} (X)\, dt \right\}$$

($\because x, y\, are$ remain unchanged ins $tan\,taneously$)

$x\hat{Y} - y\hat{X} = \hat{N}$ say

Thus left hand member represent h, the instantaneous change in the angular momentum and the right hand member of denotes the moment of the external force about z- axis. This complete the proof.

Similarly, considering the motion relative to the center of mass, we can prove that the sudden change in angular momentum relative to the mass center is equal to the moment of the external impulsive forces about the mass center.

7.16. IMPULSE

If a particle of mass m is moving under the action of a constant force $\underline{F}$ producing an acceleration $\underline{a}$ then by second law of Newton $\underline{F} = m\underline{a}$ and impulse of the force during an interval of time t is defined as the product of the force and the time t. Thus, impulse $\underline{i}$ is given by

$$\underline{i} = \underline{F}t$$

By using second law of Newton's

$$\underline{I} = m\underline{a}t$$

Also using first law of motion we have

$$v_f - v_t = \underline{a}t$$

$$\underline{I} = m\left(v_f - v_t\right)$$

where v_f is the final velocity and v_t is the initial velocity of the particle. Hence impulse of the constant force is equal to change in momentum of the particle. When the force is variable in magnitude but constant in direction, then impulse $\underline{I}$ during the interval from $t = t_1$ to $t = t_2$ is defined as

$$\underline{I} = \int_{t_2}^{t_1} dt$$

$$\underline{I} = \int_{t_2}^{t_1} m\frac{dv}{dt} : F = m\underline{a} = m\frac{dv}{dt}$$

$$\underline{I} = \int_{t_2}^{t_1} m\underline{dv}$$

where $v_1 \, and \, v_2$ are terminal velocity $t = t_1 \, and \, t = t_2$, respectively. Hence whether the force is constant or variable momentum in the interval.

7.17. IMPULSE FORCE

When bat strikes a cricket ball the beginning and end of the collision can be determined fairly precisely. The bat is in contact with the ball for an interval that is quite short compared with time during which we are watching the ball. During the collision the bat exert large force on the ball. This force varies with time in a example manner that we can measure only with difficulty.

Both the bat and ball are deformed during collision. "Force that act for a time that is short compared with the time of observation of the system are called impulsive force."

We reconsider integral in a manner that the force $\underline{F} = X\underline{i} + Y\underline{j}$ increase without limit and time interval $t_2 - t_1 = t$ tend to zero in such way that the integral remains constant or approaches a finite limit. In such a case, the value $\underline{I}$ of the integral is known as impulsive force or below and its components $\hat{X}, \hat{Y}$ are given

$$\hat{X} = Lt_{t_2 \to t_1} \int_{t_1}^{t_2} X dt \; and \; \hat{Y} = Lt_{t_2 \to t_1} \int_{t_1}^{t_2} Y dt$$

Force which remains finite during the interval (t_1, t_2) contribute nothing to the impulsive force, for if $\underline{F}$ is the magnitude of such a force, the corresponding impulsive force

$$= Lt_{t_2 \to t_1} \int_{t_1}^{t_2} \underline{F}\, dt = \underline{F}.0 = 0$$

7.18. COLLISION OF TWO SPHERE AND COEFFICIENT OF RESTITUTION

Newton's law of restitution states that when two bodies impinge, their relative velocity along the common normal after impact, bears a constant ratio- e to that before impact, bear a constant depending upon the nature of the material of where e is a constant depending upon the nature of the material of which the bodies are composed and $0 \leq e \leq 1$ This constant e is called the coefficient of restitution, For an elastic impact, e = 1 and for a perfectly inelastic impact, e = 0. If 0< e < 1, the collision is said to be inelastic.

Now we consider the problem of collision of two sphere which are moving along the line joining their centers. We take ox along the line of center, and use the following notation:

Masses of center = m_1, m_2

Velocities o center before collision u_1, u_2

Velocities of center after collision = v_1, v_2

Let $\hat{P}$ be the magnitude of the impulsive reaction. Then we have

$$m_1(v_1 - u_1) = -\hat{P}$$

$$m_2(v_2 - u_2) = \hat{P}$$

$$m_1(v_1 - u_1) = --m_2(v_2 - u_2)$$

$$m_1 v_1 + m_2 v_2 = m_1 u_1 + m_2 u_2 + m_2 u_2$$

The latest equation is evident otherwise also, since impulsive action and reaction are only internal to system and therefore the total linear momentum of the system should remain unaffected by the collision By Newton's law of restitution.

$$v_2 - v_1 = -e(u_2 - u_1)$$

$$= e(u_1 - u_2)$$

Solving equation (1) and (2) for $v_1 and\, v_2.$ we get

$$v_1 = \frac{m_1 - em_2}{m_1 + m_2} u_1 + \frac{m_2}{m_1 + m_2} u_2$$

$$v_2 = (1+e)\frac{m_1}{m_1 + m_2} u_1 + \frac{m_2 - em_1}{m_1 + m_2} u_2$$

For perfectly inelastic sphere, e = o and we obtained

There is thus no rebound and after the collision the spheres move with the same velocity. It can be proved that if spheres are elastic (e = 1) and have the same mass, they exchange velocities after impact.

7.19. POISSON HYPOTHESIS

This is an alternative to Newton's law of restitution. According ` to this hypothesis, when two bodies come into contact, there is a short interval in which they undergoes compression during this interval the centers of the two bodies get closer and closer to each them as a result of the deformation of the two bodies due to impact This is the followed by another short interval in which the original shape of the bodies move with common velocity and the centers if two bodies are nearest to each other. During the interval of restitution (the interval in which bodies regain their shape), the impulsive pressure between the bodies.

SUMMARY AND DISCUSSION

7.1. Central Force

A force acting on a P particle which always passes through a fixed point is called a Central force. The fixed point is known as center of force, usually it is taken as origin. The central force is either attractive (toward the center of force) or repulsive (away from the center of force).

The central force at a point is a function of distance of that point from the origin. This relation is called the *law of force*. The path described by the particle moving under a central is called the central orbit thus $\overline{F}$ is a central force if

$$\bar{F} = f(r)\hat{r}$$

where $\hat{r}$ is a unit vector in the direction of $\hat{r}$ *and r is the magnitude* of the position vector $\hat{r}$.*Also*

$$\bar{r}\,x\,\bar{F} = f(r)\bar{r}\,x\bar{r} = f(r)r\hat{r}\,x\,\hat{r} = 0$$

$$\text{i.e}\bar{r}\;x\,\bar{F} = 0$$

7.2 Properties

i. The orbit of a particle moving under a central force must be a plane must be a plane curve.

ii. The angular momentum is constant.

7.3. Differential Equation of the Orbit in Polar Form

$$h^2u^2\left(\left(\frac{d^2u}{d\theta^2}\right)+u\right) = f(u)$$

7.4. Differential Equation of the Orbit in Pedal Form

$$f(r) = \frac{h^2}{p^3}\frac{dp}{dr}$$

7.5. Apse and Apsidal Distance

An apse is a point on an orbit at which the radius vector drawn from the center of force has extremum (maximum or minimum) magnitude. Therefore, at an apse $\frac{dr}{dt} = 0$, i.e., the radial component of velocity is zero.

i. The magnitude of the radius vector at an apse is known as apsidal distance.

ii. The line joining an apse to the center of the force is an apse line

iii. The angle between two consecutive apse line is an apse angle.

iv. The radius vector is perpendicular to tangent at an apse.

7.6. Areal Velocity

Let $\bar{r}$ *be* the position vector of a point P on the orbit and Q is the neighboring point whose position vector is $\bar{r} + d\bar{r}$ and the angle between $\bar{r}$ and $\bar{r} + d\bar{r}$ is $d\theta,$ then

Areal Velocity = $\dfrac{dA}{dt} = \dfrac{1}{2} r^2 \theta$

For circular orbit

$$r^2 \theta = h, therefore$$

Area velocity = $\dfrac{dA}{dt} = \dfrac{h}{2}$

Area described in any time t_1 from a fixed point where $t = 0$ can be obtained by integrating the real velocity, therefore

$$A = \frac{h}{2} \int_0^{t_1} dt = \frac{h}{2} t_1$$

7.7. Periodic Time

The time required to complete on rotation is known as periodic time. Since h/2 is the area described in unit time, it follows that if T is the time that the particle takes to described in unit time, it follows that if T is the time that the particle takes to described the whole area of the ellipse, then

Area of the ellipse = $\dfrac{h}{2} T$

$$\pi ab = \frac{h}{2} T$$

$$T = \frac{2\pi ab}{h}$$

7.8. Kepler's laws

In 1609, Johannes Kepler stated the first two laws of planetary motion, which are given in the following subsections.

7.8.1. First Law

Every planet described an ellipse with the sun at one focus.

7.8.2. Second Law

The radius vector drawn from the sum to a planet described equal areas intervals of times, ten years later, in 1619, Kepler announced the third law.

7.8.3. Third Law

The squares of the periods of the different planets are proportional to the cube s of their respective semi-axes. After half a century Newton derived gravitational law from Kepler's Law.

7.9. Derivation of Newton's Law of Gravitation from Kepler's laws.

A particle moves in an ellipse under a central force directed towards its focus, whose polar equation is

$$\frac{l}{r} = 1 + e\cos\theta$$

$$f(r) = \frac{h^2}{lr^2} = \frac{\mu}{r^2} \text{ where } \mu = \frac{h^2}{l}$$

which is Newton's Law of Gravitation.

7.11. Collision

A collision is said to be taken place when two bodies physically collide with each other.

7.12. Linear Momentum of a particle

If a particle of mass m moving with velocity v, then linear momentum of the particle is the product of mass and velocity it is usually denoted by "L". Therefore, $l = mv$

7.13. Angular momentum

If l Is the linear momentum of the particle. Then angular momentum $\grave{\underline{U}}$ is define as

$\grave{\underline{U}} = \underline{r} + \underline{1}$

7.14. Principal of Linear Momentum

The sudden change in linear momentum of a system is equal to the total external impulsive force.

7.15. The Principle of Angular Momentum

The sudden change in angular momentum of a system about a fixed is equal to the moment of the external impulsive forces about the axis.

7.16. Impulse

IF a particle of mass m is moving under the action of a constant force $\underline{F}$ producing an acceleration $\underline{a}$ then by second law of Newton $\underline{F} = m\underline{a}$ and impulse of the force during an interval of time t is defined as the product of the force and the time t. Thus, impulse $\underline{i}$ is given by

$$\underline{i} = \underline{F}t$$

7.17. Impulse Force

When bat strikes a cricket ball. The beginning and end of the collision can be determined fairly precisely. The bat is in contact with the ball for an interval that is quite short compared with time during which we are watching the ball. During the collision the bat exert large force on the ball. This force varies with time in a example manner that we can measure only with difficulty.

Both the bat and ball are deformed during collision. "Force that act for a time that is short compared with the time of observation of the system are called impulsive force".

7.18. Collision of Two Sphere and Coefficient of Restitution

Newton's law of restitution states that when two bodies impinge, their relative velocity along the common normal after impact, bears a constant ratio- e to that before impact, bear a constant depending upon the nature of the material of where e is a constant depending upon the nature of the material of which the bodies are composed and $0 \le e \le 1$ This constant e is called the coefficient of restitution, For an elastic impact, e = 1 and for a perfectly inelastic impact, e = 0. If 0< e < 1, the collision is said to be inelastic.

7.19. Poisson Hypothesis

This is an alternative to Newton's law of restitution. According ` to this hypothesis, when two bodies come into contact, there is a short interval in which they undergoes compression during this interval the centers of the two bodies get closer and closer to each them as a result of the deformation of the two bodies due to impact This is the followed by another short interval in which the original shape of the bodies move with common velocity and the centers if two bodies are nearest to each other. During the interval of restitution (the interval in which bodies regain their shape), the impulsive pressure between the bodies.

CHAPTER 8

BASIC CONCEPTS OF DIFFERENTIAL EQUATIONS

CONTENTS

8.1. INTRODUCTION

Mathematical models for real world phenomena often take the form of equations relating various quantities and their rates of change (derivatives), For example, the motion of a particle involves the distances covered in time t and velocity v and/or acceleration a. Now the rate of change $\frac{ds}{dt}$ of s with respect to t is the velocity v and rate of change $\frac{dv}{dt}$ of velocity with respect to t is the acceleration a. A particle moving in a straight line has an equation of motion as $s = f(t)$, where t is in seconds and s is in meters. Its velocity satisfies the equation

$$v = \frac{ds}{dt} = 4t^2 + 5t - 3$$

This leads us to the definition of a differential equation (D.E).

A differential equation is a mathematical equation that relates some function with its derivatives. In applications, the functions typically represent physical quantities, the by-product represents their rates of change, and the equation defines an association between the two. Because such relations are exceptionally common, differential equations amusement a prominent role in many disciplines including engineering, physics, economics, and biology.

In pure mathematics, differential equations are studied from several different perspectives, mostly concerned with their solutions-the set of functions that satisfy the equation. Only the simplest differential equations are solvable by explicit formulas; however, some ownership of solutions of a given differential equation may be determined without discovery their exact form.

8.2. DIFFERENTIAL EQUATION AND THEIR CLASSIFICATION

An equation involving one dependent variable and its derivatives with respect to one or more independent variables is called differential equation. For example

i. $$\frac{dy}{dx} + y\cos x = \sin x$$

ii. $$\frac{d^2y}{dx^2}+xy\left(\frac{dy}{dx}\right)^2=0$$

iii. $$\left[1+\left(\frac{dy}{dx}\right)^2\right]^{\frac{3}{2}}=\frac{d^2y}{dx^2}$$

8.2.1. Ordinary Differential Equation (O. D. E.)

A differential equation, in which ordinary derivatives of the dependent variable with respect to a single independent variable occur, is called an ordinary differential equation (O. D. E.). Above equations (i), (ii), and (iii) are examples of Ordinary Differential Equation.

8.2.2. Partial Differential Equation

A differential equation involving partial derivatives of the dependent variable with respect to more than one independent variable is called a partial differential equation.

iv. $$x\frac{\partial z}{\partial x}+y\frac{\partial z}{\partial y}=nx$$

v. $$\frac{\partial^2u}{\partial x^2}+\frac{\partial^2u}{\partial y^2}+\frac{\partial^2u}{\partial z^2}=0$$

8.3. ORDER OF DIFFERENTIAL EQUATION

The order of a differential is the order of the highest derivative that occurs in the equation.

8.4. DEGREE OF A DIFFERENTIAL EQUATION

The degree of a differential equation is the greatest exponent of the highest order derivative that appears in the equation. (The dependent variable and its derivatives should be expressed in a form free from radicals and fractions).

i. $$\frac{dy}{dx}+y\cos x=\sin x$$ (order 1, degree 1)

ii. $$\frac{d^2y}{dx^2}+xy\left(\frac{dy}{dx}\right)^2=0$$ (order 2, degree 1)

iii. $\left[1+\left(\frac{dy}{dx}\right)^2\right]^{\frac{3}{2}}=\frac{d^2y}{dx^2}$ (order 2, degree 2)

exponent of $\frac{d^2y}{dx^2}$ is 2 after removing the radical by squaring both sides of the equation

iv. $x\frac{\partial z}{\partial x}+y\frac{\partial z}{\partial y}=nx$ (order 1, degree 1)

v. $\frac{\partial^2 u}{\partial x^2}+\frac{\partial^2 u}{\partial y^2}+\frac{\partial^2 u}{\partial z^2}=0$ (order 2, degree 1)

In this chapter, we shall study ordinary differential equation only.

Recall that a function $T:U\Delta V$, where U,V, are vector spaces over the same field F, is called linear if, for $\alpha,\beta\in F,x,y\in U$,

$$T(\alpha x+\beta y)=\alpha T(x)+\beta T(y).$$

8.5. LINEAR DIFFERENTIAL EQUATION

An ordinary differential equation $F\left(x,y,\frac{dy}{dx},\frac{d^2y}{dx^2},\ldots,\frac{d^ny}{dx^n}\right)=0$ is said to be linear if F is a linear function of the variables $x,\frac{dy}{dx},\frac{d^2y}{dx^2},\ldots,\frac{d^ny}{dx^n}$

A similar definition applies to partial differential equations.

Thus, the general linear ordinary differential equation of order n is

$$a_0(x)\frac{d^ny}{dx^n}+a_1(x)\frac{d^{n-1}y}{dx^{n-1}}+\ldots+a_{n-1}(x)\frac{dy}{dx}+a_n(x)y=F(x)$$

where $a_0(x)$ is not identically zero.

The equation $\frac{d^3y}{dx^3}+2e^x\frac{d^2y}{dx^2}+y\frac{dy}{dx}=x^3$ is not linear because of the term $y\frac{dy}{dx}$.

It should be carefully noted that in a linear ordinary differential equation

i. The dependent variable y and its derivative are all of degree one.

ii. No product of y or any of its derivative appear.

iii. No transcendental function of y and/or its derivative occur.

A differential equation that is not linear is called a nonlinear differential equation. Differential equations occur in the mathematical formulation of many problems in science and engineering. Some such problems are

a. Determining the motion of projectile, rocket, satellite or planet.

b. Finding the charge or current in an electric circuit.

c. Study of chemical reactions.

d. Determination of curves with given geometrical properties.

8.6. SOLUTION OF A DIFFERENTIAL EQUATION

A solution (or integral) of a differential equation is a relation between the variables, not containing derivatives, such that this relation and the derivatives obtained from it satisfy the given differential equation identically.

For example, the equation $\frac{dy}{dx} = -\mu y$ has a solution $y = c.e^{-\mu x}$, where c is an arbitrary constant. The differential equation $\frac{d^2y}{dx^2} + y = 0$ has solution $y = A\cos x,\ y = B\sin x\, and \quad y = A\cos x + B\sin x,$ where $A\, and\, B$ are arbitrary constants.

A solution of a differential equation which contains as many arbitrary constants as the order of the equation is called a general solution (or integral) of the differential equation. A solution obtained from the general solution by giving particular values to the constants is called a particular solution. The graph of a particular integral is called an integral curve of the differential equation.

8.7. FORMATION OF A DIFFERENTIAL EQUATION

Given a relation $f\left(x, y, c_1, c_2, \ldots, c_n\right) = 0$ (1)

between variables x, y and containing n constants $c_1, c_2, \ldots, c_n$ it is always possible to form a differential equation of order n such that the given relation (1) is the general solution of the equation. This is done by differentiating (1) n times thereby obtaining equations and then eliminating the n constants from the original relation and n derived equations. The method is illustrated by means of examples.

Example: The equation $y = x + a$ (1)

Represents a family of parallel straight lines for different values of a. Elimination of one constant 'a'

Requires two equations. The second equation is obtained by differentiating (1). Thus $\frac{dy}{dx} = 1$ is the differential equation of the relation (1), with a eliminated.

Example: From the differential equation by eliminating the two constants A and B from the relation

$$y = A\sin x + B\cos x. \quad (1)$$

Solution: It is clear that three equations are required to eliminate two unknowns A and B. We obtain two other needed equations by successive differentiation of (1). Thus, from (1), we have

$$\frac{dy}{dx} = A\cos x - B\sin x \quad (2)$$

$$\frac{d^2y}{dx^2} = -A\sin x - B\cos x = -y \text{, using (1)} \quad (3)$$

So

$$\frac{d^2y}{dx^2} + y = 0 \quad (4)$$

is the required differential equation and (1) is its general solution.

Example: Find the differential equation of all parabolas whose axes are parallel to the $y-axis$.

Solution: General equation of a parabola whose axis is parallel to the $y-axis$ is

$$y = ax^2 + bx + c. \quad (1)$$

In order to obtain its differential equation, we have to eliminate a, b, c from (1). For that, we need three more equations. Differentiating (1) successively, we have

$$\frac{dy}{dx} = 2ax + b$$

$$\frac{d^2y}{dx^2} = 2a,$$

$$\frac{d^3y}{dx^3} = 0.$$

The last equation does not contain any of the constants a, b and c. Thus $\frac{d^3y}{dx^3} = 0$ is the differential equation of all parabolas whose axes are parallel to the $y-axis.$

8.8. INITIAL AND BOUNDARY CONDITION

We have observed that general solution of a differential equation contains the same number of arbitrary constants, as is the order of the equation. Sometimes we need to find the solutions of differential equations subject to supplementary conditions. Two types of conditions will be often encountered.

8.8.1. Initial Condition

It is often required to find the solution of a differential equation subject to certain conditions. If the conditions relate to one value of the independent variable such as $y - y_0$ at $x = x_0$ (written as $y(x_0) = y_0$) and $\frac{dy}{dx} = y'(x_0)$ at $x = x_0$, where x_0 belongs to some interval $]\alpha, \beta[$ then they are called initial conditions (or one- point boundary conditions) and x_0 is called the initial point. An initial value problem consists of a differential equation (of any order) together with a collection of initial conditions that must be satisfied by the solution of the differential equation and derivatives at the initial point.

8.8.2. Boundary Conditions

The problem of finding the solution of a differential equation such that all the associated constraints relate to two different values of the independent variable is called a two-point boundary value problem (or simply a boundary value problem). The associated supplementary boundary conditions are called two-point boundary conditions.

Example: Solve $\frac{dy}{dx} = 2x$ (1)

Such that $y(1)=4.$

Solution: This is an initial value problem. We note that $y = x^2 + c,$ c being arbitrary constant, is the general solution of (1). Since $y(1) = 4,$ we have

$$y(1) = 1^2 + c.$$

Therefore,

$$4 = 1 + c or c = 3.$$

Thus $y = x^2 + 3$ is the solution of the initial value problem (1). Note that the general solution represents a family of parabolas for different values of c. The solution $y = x^2 + 3$ is a particular member of the family that passes through $(1,4)$.

Example: Solve $\frac{d^2y}{dx^2} + y = 0$ (1)

Subject to the conditions $y\left(\frac{\pi}{4}\right) = \sqrt{2},\ y'\left(\frac{\pi}{4}\right) = \frac{1}{\sqrt{2}}.$

Solution: Since both the conditions relate to one value of $x,$ namely $x = \frac{\pi}{4},$ this is an initial value problem. We have already noted in previous example that

$$y = A\sin x + B\cos x \quad (2)$$

is the general solution of (1). Differentiating (2) w. r. t $x,$ we get

$$y' = A\cos x - B\sin x. \quad (3)$$

By the given conditions, we have, respectively from (2) and (3)

$$y\left(\frac{\pi}{4}\right)=\sqrt{2}=A\sin\frac{\pi}{4}+B\cos\frac{\pi}{4}=\frac{A}{\sqrt{2}}+\frac{B}{\sqrt{2}}$$

$$y'\left(\frac{\pi}{4}\right)=\frac{1}{\sqrt{2}}=A\cos\frac{\pi}{4}-B\sin\frac{\pi}{4}=\frac{A}{\sqrt{2}}-\frac{B}{\sqrt{2}}.$$

Hence $A+B=2$ and $A-B=1$

Solving for $A\,and\,B,$ we find that

$$A=\frac{3}{2},B=\frac{1}{2}$$

With these values of $A\,and\,B,$ the particular solution of (1) is

$$y=\frac{3}{2}\sin x+\frac{1}{2}\cos x$$

Example: Verify that $y=c_1\cos x$ and $y=c_2\sin x$ are solutions of $\frac{d^2y}{dx^2}+y=0.$ Find a particular solution of the equation satisfying the boundary conditions

$$y(0)=1,y\left(\frac{\pi}{2}\right)=2.$$

Solution: We have $y=c_1\cos x.$

Differentiating twice, we get

$$\frac{dy}{dx}=-c_1\sin x$$

$$\frac{d^2y}{dx^2}=c_1\cos x=-y$$

$$\frac{d^2y}{dx^2}+y=0.$$

Thus $y=c_1\cos x$ is a solution of $\frac{d^2y}{dx^2}+y=0$ (1)

Similarly, it can be checked that $y = c_2 \sin x$ is also a solution of (1).

General solution of (1) is

$$y = c_1 \cos x + c_2 \sin x \qquad (2)$$

Applying the boundary conditions, we obtain from (2) $y(0) = 1 = c_1 + 0$ and $y\left(\frac{\pi}{2}\right) = 2 = 0 + c_2$.

Thus $c_1 = 1$ and $c_2 = 2.$ Hence the particular solution of (1) satisfying the given conditions is

$$y = \cos x + 2\sin x$$

Example: Solve the boundary value problem $\frac{d^2y}{dx^2} + y = 0, y(0) = 1, y(\pi) = 5.$

Solution: Applying the boundary conditions to the general solution $y = A\sin x + B\cos x$ of the given equation, we have

$$y(0) = 1 = B$$

$$y(\pi) = 5 = -B$$

Thus, we obtain two values of B and we are unable to determine any definite value of $A.$ Hence the boundary value problem has no solution. It follows that a boundary value problem need not always have a solution.

In above section we classified the differential equations and saw how a differential equation can be formed by eliminating constants from a given functional equation. The problem of finding general solution (integral) of a given differential may or may not exist. Even if the integral of a given equation exists, it may not be easy to find. We shall only discuss methods of solutions of special types of differential equations.

8.9. SEPARABLE EQUATIONS

Definition. A differential equation of the type

$$F(x)G(x)dx + f(x)g(y)dy = 0 \qquad (1)$$

is called an equation with separable variables or simply a separable equation.

Equation (1) may be written as

$$\frac{F(x)}{f(x)}dx+\frac{g(y)}{G(y)}dy=0$$

which can be easily integrated.

Example: Solve $\frac{dy}{dx}=\frac{x^2}{y}$ (1)

Solution: Equation (1) can be written as

$$y\,dy=x^2dx.$$

Integrating both the sides, we get

$$\frac{y^2}{2}=\frac{x^3}{3}+c_1$$

$$3y^2=2x^2+c$$

which is the required solution of (1).

Example: Solve $\frac{dy}{dx}=\frac{1}{x\tan y}$.

Solution: The given equation is separable and can be written as

$$\frac{dx}{x}=\tan y\,dy$$

Integrating, we have

$$\ln|x|=-\ln|\cos y|+c$$

$$\ln|x|+\ln|\cos y|=c$$

$$\ln|x\cos y|=c$$

$$|x\cos y|=e^c=a(say)$$

$x\cos y=a$ because $a>0$.

Example: Solve $x\sin y\,dx+\left(x^2+1\right)\cos y\,dy=0$ (1)

Solution: Dividing (1) by $\left(x^2+1\right)\sin y$, we have

$$\frac{x}{x^2+1}dx + \cot y\,dy = 0$$

which is a separable equation. Therefore,

$$\int \frac{x\,dx}{x^2+1} + \int \cot y\,dy = c_1 = \ln|c_2|$$

$$\frac{1}{2}\ln\left(x^2+1\right) + \ln|\sin y| = \ln|c_2|$$

$$\ln\left(x^2+1\right) + 2\ln|\sin y| = 2\ln|c_2|.$$

Now, we can write $2\ln|\sin y| = \ln|\sin y|^2 = \ln(\sin y)^2 = \ln\sin^2 y$ and $2\ln|c_2| = \ln|c_2|^2 = \ln c_2^2 = \ln c, where\ c_2^2 = c > 0$. Hence $\ln(x^2+1)\sin^2 y = c$ is the required solution.

Example: Solve the initial value problem

$$\frac{dy}{dx} = \frac{2x}{y+x^2y}, y(0) = -2.$$

Solution: We have $\frac{dy}{dx} = \frac{2x}{y\left(1+x^2\right)}$ or $y\,dy = \frac{2x}{1+x^2}dx$. Integrating, we obtain $\frac{y^2}{2} = \ln(1+x^2) + \ln c,$ c being a constant.

$$= \ln\left(c\left(1+x^2\right)\right)$$

$$y^2 = \ln(c^2\left(1+x^2\right)^2) \qquad \text{(i)}$$

Now setting $y(0) = -2$ into (1), we have

$$4 = \ln c^2\ or\ c^2 = e^4.$$

So (i) becomes

$$y^2 = \ln\left(e^4\left(1+x^2\right)^2\right)$$

which is the required solution.

8.10. HOMOGENOUS EQUATIONS

Definition. A function $f(x,y)$ is called homogeneous of degree n if

$$f(tx,ty)=t^n f(x,y),$$

where t is a non-zero real number. Thus, $\sqrt{xy},\frac{x^{10}+y^{10}}{x^2+y^2}$ *and* $\sin\left(\frac{x}{y}\right)$ are homogeneous function of degree 1.8 and 0, respectively.

A first order differential equation

$$\frac{dy}{dx}=f(x,y) \qquad (1)$$

is said to be homogeneous function of any degree. If (1) is written in the form

$$M(x,y)dx+N(x,y)dy=0$$

Then it is called homogeneous functions of the same degree.

The equation $\frac{dy}{dx}=\ln x-\ln y+\frac{x+y}{x-y}=\ln\frac{1}{\frac{y}{x}}+\frac{1+\frac{y}{x}}{1-\frac{y}{x}}$ is homogenous, but $\frac{dy}{dx}=\frac{y^3+2xy}{x^2}=y\left(\frac{y}{x}\right)^2+2\left(\frac{y}{x}\right)$ is not homogeneous.

Theorem. A homogeneous equation $\frac{dy}{dx}=g\left(\frac{y}{x}\right)$ (1)

can be transformed into a separable equation (in the variables v and x) by the substitution $y-vx$.

Proof: Put $y=vx$ into (1). Then $\frac{dy}{dx}=v+x\frac{dv}{dx}$ and (1) becomes

$$v+\frac{xdv}{dx}=g(v)$$

$$v-g(v)+x\frac{dv}{dx}=0$$

$$\left[v-g(v)\right]dx+x\,dv=0$$

This equation is separable and can be solved as in the previous section.

Example: Solve $\dfrac{dy}{dx}=\dfrac{x^3+y^3}{x^2y+xy^2}$

Solution: We have $\dfrac{dy}{dx}=\dfrac{1+\left(\dfrac{y}{x}\right)^3}{\left(\dfrac{y}{x}\right)+\left(\dfrac{y}{x}\right)^2}$

So that the equation is homogeneous. Setting $y=vx$ into (1), we obtain

$$v+x\frac{dv}{dx}=\frac{1+v^3}{v+v^2}=\frac{1+v^3}{v(1+v)}=\frac{v^2-v+1}{v}$$

$$x\frac{dv}{dx}=\frac{v^2-v+1}{v}-v=\frac{1-v}{v}$$

$$\frac{v}{1-v}dv=\frac{dx}{x}$$

$$\left(-1+\frac{1}{1-v}\right)dv=\frac{dx}{x}$$

Integrating, we get

$$-v-\ln|1-v|=\ln|x|+\ln|c|$$

$$ln|cx(1-v)|+v=0$$

Replacing $v\,by\,\dfrac{y}{x}$ in the above equation, we have

$$\ln|c(x-y)|+\frac{y}{x}=0$$

which is the required solution.

Example: Solve the initial value problem $\left(x^2+3y^2\right)dx-2xy\,dy=0$,

$y(2)=6.$

Solution: Here $$\frac{dy}{dx}=\frac{x^3+3y^2}{2xy}=\frac{3\left(\frac{y}{x}\right)^2+1}{2\left(\frac{y}{x}\right)} \quad (1)$$

which shows that the equation is homogeneous. Putting $y=vx$ into (1), we have

$$v+x\frac{dv}{dx}=\frac{3v^2+1}{2v}$$

$$x.\frac{dv}{dx}=\frac{3v^2+1}{2v}-v=\frac{v^2+1}{2v}$$

$$\frac{2v}{1+v^2}dv=\frac{dx}{x}$$

On integrating, we get

$$\ln\left(1+v^2\right)=\ln|x|+\ln|c|$$

$$1+v^2=|cx|$$

Replacing $v\,by\,\frac{y}{x},$ we obtain

$$1+\frac{y^2}{x^2}=|cx|$$

$$x^2+y^2=|cx|\,x^2$$

Now if $x\geq 0,$ we can write this as

$$x^2+y^2=cx^3$$

$$y^2=cx^3-x^2$$

$$y=\pm\sqrt{cx^3-x^2}$$

Applying the initial condition, we get

$$y(2)=6=\pm\sqrt{8c-4}$$

$$8c-4=36\, i.e., c=5$$

Hence $y=\sqrt{5x^3-x^2}$ is the required solution. We take the plus sign in the radical since $y(2)$ is positive.

8.11. DIFFERENTIAL EQUATION REDUCIBLE TO HOMOGENEOUS FORM

Some differential equations are not homogeneous. But they can be reduced to homogeneous form.

The differential equation $(a_1x+b_1y+c_1)dx+(a_2x+b_2y+c_2)dy=0$ is not homogeneous. But it can be reduced to a homogeneous form as illustrated below:

Case 1. If $\frac{a_1}{a_2}\neq\frac{b_1}{b_2}$, then make the transformations

$$x=X+h, y=Y+k.$$

The given equation becomes

$$(a_1X+b_1Y+a_1h+b_1k+c_1)dX+(a_2X+b_2Y+a_2h+b_2k+c_2)dY=0 \quad (1)$$

Let $h\, and\, k$ be the solution of the system of equations

$$\left.\begin{aligned} a_1h+b_1k+c_1&=0 \\ a_2h+b_2k+c_2&=0 \end{aligned}\right\}$$

Then for these values of $h\, and\, k$, (1) reduces to the homogenous equation

$$(a_1X+b_1Y)dX+(a_2X+b_2Y)dY=0$$

in the variables $X\, and\, Y$.

Case II. If $\frac{a_1}{a_2}=\frac{b_1}{b_2}$, then put $z=a_1x+b_1y$ and the given equation will reduce to a separable equation in the variable $x\, and\, z$.

Example: Solve $\frac{dy}{dx}=\frac{2y-x+5}{2x-y-4}$ (1)

Solution: Putting $x=X+h, y=Y+k$

So, the given equation becomes

$$\frac{dY}{dX}=\frac{2Y-X+2K-h+5}{2X-Y+2hXk-4} \quad (2)$$

The solution of the system of equations

$$\left.\begin{aligned}-h+2k+5=0\\2h-k-4=0\end{aligned}\right\}$$

is $h=1, k=-2.$

For these values of $h\,and\,k$, (2) reduce to

$$\frac{dY}{dX}=\frac{2Y-X}{2X-Y} \quad (3)$$

which is homogeneous.

Putting $Y=vX$ into (3), we have $v+X\frac{dv}{dX}=\frac{2v-1}{2-v}$ which is separable. Therefore,

$$\left(\frac{2-v}{v^2-1}\right)dv=\frac{dX}{X}.$$

Integrating, we obtain

$$\ln\left|\frac{v-1}{v+1}\right|-\frac{1}{2}\ln\left|v^2-1\right|=\ln\left|X\right|+\ln\left|c_0\right|$$

$$\ln\left|\frac{v-1}{v+1}\right|^2=\ln\left|X\right|^2+\ln\left|c_0\right|^2+\ln\left|v^2-1\right|$$

$$\ln\left|\frac{v-1}{v+1}\right|^2=\ln\left\{\left|c_0X\right|^2.\left|v^2-1\right|\right\}$$

$$\left|\frac{v-1}{v+1}\right|^2=c\left|X\right|^2.\left|v-1\right|.\left|v+1\right|, where\, c=c_0^2$$

$$|v-1|=c|X|^2.|v+1|^3$$

Replacing $v \, by \frac{Y}{X}$, this becomes

$$\left|\frac{Y-X}{X}\right|=c|X|^2.\left|\frac{Y+X}{X}\right|^3$$

$$|Y-X|=c|Y+X|^3 \quad (4)$$

But $X=x-1, Y=y+2$. Hence (4) takes the form

$$|y-x+3|=c|y+x+1|^3$$

which is the required solution.

Example: Solve $(2x+y+1)dx+(4x+2y-1)dy=0$ (1)

Solution: Since $\frac{a_1}{a_2}=\frac{2}{4}=\frac{b_1}{b_2}=\frac{1}{2}$, we put $2x+y=z$. then (1) becomes

$$(z+1)dx+(2z-1)(dz-2dx)=0$$

$$(z+1-4z+2)dx+(2z-1)dz=0$$

i.e., $3(1-z)dx+(2z-1)dz=0$ which is separable.

Dividing by $1-z$, we have

$$3dx+\frac{2z-1}{1-z}dz=0$$

$$3.dx+\left(-2+\frac{1}{1-z}\right)dz=0$$

Integrating, we obtain

$$3x-2z-\ln|1-z|=c_0$$

Replacing z by $2x+y$, it reduces to

$$3x-2(2x+y)-\ln|1-2x-y|=c_0$$

$$-x-2y-\ln|2x+y-1|=c_0$$

$$\ln|2x+y-1|+x+2y=c, where\, c=-c_0$$

is the required solution.

8.12. EXACT EQUATION

Definition. The expression $M(x,y)\,dx+N(x,y)\,dy$ (1)

is called an exact differential if there exists a continuously differentiable function $f(x,y)$ of two real variables x and y such that the expression equals the total differential df. We know from calculus that

$$df=\frac{\partial f}{\partial x}dx+\frac{\partial f}{\partial y}dy.$$

Thus, if (1) is exact then

$$M(x,y)=\frac{\partial f}{\partial x}=f_x \text{ and } N(x,y)=\frac{\partial f}{\partial y}=f_y.$$

If (1) is an exact differential then the differential equation

$$M(x,y)\,dx+N(x,y)\,dy=0$$

is called an exact equation.

Theorem. The differential equation $M(x,y)\,dx+N(x,y)\,dy=0$ (1)

is an exact equation if and only if $\frac{\partial M}{\partial y}=\frac{\partial N}{\partial x}$, where the functions $M(x,y)$ and $N(x,y)$ have continuous first order partial derivatives.

Proof. Suppose that the equation (1) is exact so that $M\,dx+Mdy$ is an exact differential. By definition, there exists a function $f(x,y)$ such that

$$M(x,y)=\frac{\partial f}{\partial x}=f_x \text{ and } N(x,y)=\frac{\partial f}{\partial y}=f_y$$

Then

$$M_y = \frac{\partial M}{\partial y} = f_{xy} = \frac{\partial^2 f}{\partial y \partial x}$$

$$N_x = \frac{\partial N}{\partial x} = f_{yx} = \frac{\partial^2 f}{\partial x \partial y}$$

Since $M \; and \; N$ possess continuous first order partial derivatives, we have $f_{xy} = f_{yx}$ and, therefore $\frac{\partial M}{\partial y} = \frac{\partial N}{\partial x}$ as desired.

8.13. SOLUTION OF AN EXACT EQUATION

If $M(x, y)dx + N(x, y)dy = 0$ (1)

is an exact equation, then there exists a function $f(x, y)$ such that

$$df = \frac{\partial f}{\partial x} dx + \frac{\partial f}{\partial y} dy$$

$$= M(x, y)dx + N(x, y)dy$$

Therefore $\frac{\partial f}{\partial x} = M \; and \; \frac{\partial f}{\partial y} = N$

Integrating $\frac{\partial f}{\partial x} = M$ with respect to x, we have

$f(x, y) = \int M \, dx + h(y).$ (2)

The constant of integration $h(y)$ is an arbitrary function of y since it must vanish under differentiation w.r.t x. Differentiating (2) $w.r.t. y$, we get

$$\frac{\partial f}{\partial y} = \frac{\partial}{\partial y}\left(\int M dx\right) + h'(y)$$

$$\frac{\partial f}{\partial y} = N = \frac{\partial}{\partial y}\left(\int M dx\right) + h'(y)$$

$$h'(y)=N-\frac{\partial}{\partial y}\left(\int Mdx\right)$$

Integrating the above equation $w.r.t\ y,$ we obtain $h\,and$ hence $f(x,y)=c$ is the required solution of (1).

Example: Solve $\left(3x^2y+2\right)dx+\left(x^3+y\right)dy=0$ (1)

Solution: Here $M=3x^2y+2$ and $N=x^3+y$

$$\frac{\partial M}{\partial y}=3x^2,\frac{\partial N}{\partial x}=3x^2$$

Thus $\frac{\partial M}{\partial y}=\frac{\partial N}{\partial x}$ and so the equation is exact.

To find the solution of (1), we note that the left-hand side of the equation is an exact differential. Therefore, there exists a function $f(x,y)$ such that

$$\frac{\partial f}{\partial x}=3x^2y+2 \quad (2)$$

$$\frac{\partial f}{\partial y}=x^3+y \quad (3)$$

Integrating (2) $w.r.t\ x$, we have

$$f(x,y)=x^3y+2x+h(y)$$

where $h(y)$ is the constant of integration. Differentiating the above equation $w.r.t\ y$ and using (3), we obtain

$$\frac{\partial f}{\partial y}=x^3+h'(y)=x^3+y$$

$$h'(y)=y.$$

Integrating, we have

$$h(y)=\frac{y^2}{2}$$

Thus $f(x,y)=x^3y+2x+\dfrac{y^2}{2}$

Hence the general solution of (1) is

$$x^3y+2x+\frac{y^2}{2}=c$$

8.14. ALTERNATIVE METHOD

Integrating (2) and (3) $w.r.t\, x\, and\, y$, respectively, we have

$$f(x,y)=x^3y+2y+h(y)$$

$$f(x,y)=x^3y+\frac{y^2}{2}+g(x).$$

Thus

$h(y)=\dfrac{y^2}{2}$ and $g(x)=2x$.

So, the general solution is

$$x^3y+2x+\frac{y^2}{2}=c.$$

Example: Solve the initial value problem

$$(2y\sin x\cos x+y^2\sin x)dx+(\sin^2 x-2y\cos x)dy=0, y(0)=3$$

Solution: Here $M=2y\sin x\cos x+y^2\sin x$ and

$$N=\sin^2 x-2y\cos x$$

$$\frac{\partial M}{\partial y}=2\sin x\cos x+2y\sin x$$

$$\frac{\partial N}{\partial x}=2\sin x\cos x+2y\sin x$$

Thus $\dfrac{\partial M}{\partial y}=\dfrac{\partial N}{\partial x}$ showing that the given equation is exact.

Hence there exists a function $f(x,y)$ such that

$$\frac{\partial f}{\partial x} = 2y\sin x\cos x + y^2\sin x \quad (1)$$

$$\frac{\partial f}{\partial y} = \sin^2 x - 2y\cos x \quad (2)$$

Integrating (1) $w.r.t x$, we have

$$f(x,y) = y\sin^2 x - y^2\cos x + h(y).$$

Differentiating this equation $w.r.t\ y$ and using (2), we get

$$\sin^2 x - 2y\cos x + h'(y) = \sin^2 x - 2y\cos x$$

i.e., $h'(y) = 0\ and\ so\ h(y) = c_1$.

Hence the general solution of the given equation is

$$f(x,y) = c_2$$

i.e., $y\sin^2 x - y^2\cos x + c_1 = c_2$

$$y.\sin^2 x - y^2.\cos x = c_2 - c_1 = c$$

Applying the initial condition that when $x = 0, y = 3,$ we have

$$-9 = c$$

Hence $y^2\cos x - y\sin^2 x = 9$ is the required solution.

8.15. INTEGRATING FACTORS

Definition. If the differential equation $M(x,y)dx + N(x,y)dy = 0$ (1)

is not exact but when it is multiplied by a function $\mu(x,y)$ and the resulting equation $\mu(x,y)M(x,y)dx + \mu(x,y)N(x,y)dy = 0$ is exact, then $\mu(x,y)$ is called an integrating factor (I.F) of the differential equation (1). The number of integrating factors of an equation may be infinite. We list below (without proofs) some rules to find the integrating factors of equations of special types.

Rule 1. If $M(x,y)dx+N(x,y)dy=0$ (1)

is not exact and $\frac{M_{y-N_x}}{N}=P,$ where P is a function of x only then (1) has an integrating factor $\mu(x)$ which also depends on $x.\mu(x)$ is solution of the differential equation

$$\frac{d\mu}{dx}=P\mu$$

i.e., $\mu(x)=\exp\int P\,dx$

Note that $M_y=\frac{\partial M}{\partial y}, N_x=\frac{\partial N}{\partial x}.$

Rule II. If $\frac{N_x-M_y}{M}=Q,$ where Q is a function of y only, then the differential equation $Mdx+Ndy=0$ has an integrating factor

$$\mu(y)=\exp\int Q\,dy.$$

Rule III. If $M\,dx+Ndy=0$ (1)

is homogeneous and $xM+yN\neq 0,$ then $\frac{1}{xM+yN}$ is an I.F of (1).

Rule IV. If $M\,dx+Ndy=0$ is of the form $y\,f(x,y)dx+x\,g(x,y)dy=0$

and $xM-yN\neq 0,$ then $\frac{1}{xM-yN}$ is an I.F of (1).

The following differential formulas are useful in the calculation of certain exact equations:

i. $d\left(\frac{y}{x}\right)=\frac{x\,dy-y\,dx}{x^2}$

ii. $d\left(\frac{x}{y}\right)=\frac{y\,dx-x\,dy}{y^2}$

iii. $d(xy) = x\,dy + y\,dx$

iv. $d(x^2 + y^2) = 2(x\,dx + y\,dy)$

v. $d\left(\ln\frac{x}{y}\right) = \frac{y\,dx - x\,dy}{xy}$

vi. $d\left(arc\tan\frac{x}{y}\right) = \frac{y\,dx - x\,dy}{x^2 + y^2}$

Example: Solve $y\,dx + (x^2y - x)\,dy = 0$

Solution. The equation is not exact. Rearranging the equation, we have

$$y\,dx - x\,dy + x^2y\,dy = 0$$

$$\frac{y\,dx - x\,dy}{x^2} + y\,dy = 0$$

Now it is an exact equation and may be written as $-d\left(\frac{y}{x}\right) + y\,dy = 0$

Integrating, we have

$$-\frac{y}{x} + \frac{y^2}{2} = c$$

$xy^2 - 2y = cx$ is the general solution.

Example: Solve $(x^2 - 2x + 2y^2)\,dx + 2xy\,dy = 0$ (1)

Solution. Here $\frac{M_y - N_x}{N} = \frac{4y - 2y}{2xy} = \frac{1}{x}$

Therefore, I.F $\mu(x, y)$ is the solution of $\frac{d\mu}{dx} = \frac{\mu}{x}$ or $\mu = x$ is an I.F. Multiplying the equation by x, we have

$$(x^3 - 2x^2 + 2xy^2)\,dx + 2x^2y = 0.$$

This equation is exact. So, we can easily find that its solution is

$\frac{x^4}{4}-\frac{2x^3}{3}+x^2y^2=c_0$ or $3x^4-8x^3+12x^2y^2=c$ is the solution of (1).

Example: Solve $dx+\left(\frac{x}{y}-\sin y\right)dy=0$.

Solution. Here, by rule II,

$$\frac{Nx-M_y}{M}=\frac{\frac{1}{y}-0}{1}=\frac{1}{y}$$

is a function of y only. Therefore, $\mu(y)=\exp\int\frac{dy}{y}=e^{\ln y}=y$ is an I.F. Multiplying the equation by y, we have

$$y\,dx+(x-y\sin y)dy=0$$

$$y\,dx+x\,dy-y\sin y\,dy=0$$

$$d(xy)-y\sin y\,dy=0.$$

Integrating, we get $xy+y\cos y-\sin y=c$ which is the required solution.

Example: Solve $(x^2y-2xy^2)dx-(x^3-3x^2y)dy=0$ (1)

Solution: The equation is homogeneous but not exact. We have

$$xM+yN=x^3y-2x^2y^2-x^3y+3x^2y^2=x^2y^2\neq 0$$

So, using Rule III, $\frac{1}{xM+yN}=\frac{1}{x^2y^2}$ is an I.F. Therefore, multiplying (1) by $\frac{1}{x^2y^2}$, we obtain

$$\left(\frac{1}{y}-\frac{2}{x}\right)dx-\left(\frac{x}{y^2}-\frac{3}{y}\right)dy=0$$

This equation is exact. Integrating, we get

$$\frac{x}{y}-2\ln|x|+3\ln|y|=c$$

is the required solution.

Note that, here the term $\frac{x}{y}$, involving both $x\, and\, y$, has been taken only once.

Example: Solve $y\left(x.y+2x^2y^2\right)+x\left(xy-x^2y^2\right)dy=0$ (1)

Solution: The equation is of the form $y\,f\left(xy\right)dx+x\,g\left(xy\right)dy=0$

Now, $xM-yN=x^2y^2+2x^3y^3-x^2y^2+x^3y^3=3x^3y^3\neq 0.$

Therefore, $\frac{1}{x^3y^3}$ is an I.F.

Multiplying (1) by $\frac{1}{x^3y^3}$, we have

$$\left(\frac{1}{x^2y}+\frac{2}{x}\right)dx+\left(\frac{1}{xy^2}-\frac{1}{y}\right)dy=0.$$

This is an exact equation. Integrating, we get

$$-\frac{1}{xy}+2\ln|x|-\ln|y|=c$$

as the required solution.

8.16. LINEAR EQUATIONS

A first order ordinary differential equation (ODE) is linear in the dependent variable y and the independent variable x if it is or can be written in the form

$$\frac{dy}{dx}+P(x)y=Q(x), \quad (1)$$

where $P\, and\, Q$ are functions of x.

8.17. SOLUTION OF A LINEAR EQUATION

The linear equation $\frac{dy}{dx}+P(x)=Q(x)$ may be written as

$$\left[P(x)y-Q(x)\right]dx+dy=0 \quad (2)$$

which is of the form $M\,dx+N\,dy=0,$ where

$$M=P(x)y-Q(x)\,and\;N=1.$$

$$\frac{\partial M}{\partial y}=P(x)\,and\,\frac{\partial N}{\partial x}=0$$

Thus (2) is not exact unless $P(x)=0$ in which case (1) is separable. However, an integrating factor (depending only on x) of (2) may be easily found. Let $\mu(x)$ be an I.F. of (2). Then multiplying (2) by $\mu(x),$ we get

$$\left[\mu(x)P(x)y-\mu(x)Q(x)\right]dx+\mu(x)dy=0 \quad (3)$$

Now (3) is an exact equation if and only if

$$\frac{\partial}{\partial y}\left[\mu(x)P(x)y-\mu(x)Q(x)\right]=\frac{\partial}{\partial x}\left[\mu(x)\right]$$

This condition reduces to $\mu(x)P(x)=\dfrac{d}{dx}\left[\mu(x)\right]$

i.e., $\mu P(x)=\dfrac{d\mu}{dx}$

$$\frac{d\mu}{\mu}=P(x)dx,$$

Integrating, we obtain

$$\ln|\mu|=\int P(x)dx$$

$$\mu=\exp[\int P(x)dx\,]>0.$$

Thus $e^{\int P\,dx}$ *or* $[\exp\int P(x)dx\,]$ is an I.F. of the linear equation (1). Multiplying (1) by $e^{\int P\,dx}$, we have $e^{\int P\,dx}\dfrac{dy}{dx}+e^{\int P\,dx}P(x)y=Q(x)e^{\int P\,dx}$ or

$$\frac{d}{dx}\left[ye^{\int P\,dx}dx\right]=Q(x)e^{\int P\,dx}$$

Integrating this we obtain the solution of (1) in the form

$$ye^{\int P\,dx} = \int [Q(x)e^{\int P\,dx}]dx + c$$

Example: Solve $(x-1)^3 \frac{dy}{dx} + 4(x-1)^2 y = x+1$ (1)

Solution: We write the equation in the standard form

$$\frac{dy}{dx} + \frac{4}{x-1}y = \frac{x+1}{(x-1)^3}$$

Here $P(x) = \frac{4}{x-1}$. Therefore, an I.F. of (1) is

$$\exp[\int \frac{4\,dx}{}] = \exp\left[\ln(x-1)^4\right] = (x-1)^4$$

Multiplying (1) by this I.F., we get

$$(x-1)^4 \frac{dy}{dx} + 4(x-1)^3 y = x^2 - 1$$

$$\frac{d}{dx}\left[y(x-1)^4\right] = x^2 - 1$$

Integrating, we obtain

$$y(x-1)^4 = \frac{x^3}{3} - x + c$$

which is the required solution.

Example: Solve $(x + 2y^3)\frac{dy}{dx} = y$

Solution: We have $\frac{dy}{dx} = \frac{y}{x+2y^3}$

This equation is clearly not linear in y. But in the first order differential equation, the roles of x and y are interchangeable in the sense that either variable may be regarded as dependent variable. Let us regard x as dependent variable and y as independent variable. The equation may be written as

$$\frac{dx}{dy} - \frac{1}{y}x = 2y^3 \qquad (1)$$

which is linear in x with

$$I.F. = \exp\left[\int\left(-\frac{1}{y}\right)dy\right] = \exp\left[\ln\frac{1}{y}\right] = \frac{1}{y}$$

Multiplying (1) by $\frac{1}{y}$, we get

$$\frac{1}{y}\frac{dx}{dy} - \frac{1}{y^2}x = 2y$$

$$\frac{d}{dx}\left(\frac{x}{y}\right) = 2y$$

Integrating, we have $\frac{x}{y} = y^2 + c$ or $x = y\left(y^2 + c\right)$ is the required solution.

Example: Solve the initial value problem

$$\frac{dr}{d\theta} + r\tan\theta = \cos^2\theta, r\left(\frac{\pi}{4}\right) = 1$$

Solution. The equation is linear in r with

$$\pm\ . = \exp\left[\int \tan\theta \quad \theta\right] = \exp\left[\ln\sec\theta\right] = \sec\theta,$$

Multiplying the given equation by $\sec\theta$, we have

$$\sec\theta\frac{dr}{d\theta} + r\sec\theta\tan\theta = \cos\theta$$

$$\frac{d}{d\theta}\left[r\sec\theta\right] = \cos\theta$$

Integrating, we obtain

$$r\sec\theta = \sin\theta + c$$

Applying the initial condition, we have

$$1.\sec\left(\frac{\pi}{4}\right)=\sin\left(\frac{\pi}{4}\right)+c$$

$$c=\sqrt{2}-\frac{1}{\sqrt{2}}=\frac{1}{\sqrt{2}},$$

Therefore,

$$r\sec\theta=\sin\theta+\frac{1}{\sqrt{2}}$$

$$r=\sin\theta\cos\theta+\frac{\cos\theta}{\sqrt{2}}$$

$$2r=\sin 2\theta+\sqrt{2}\cos\theta$$

is the required solution.

8.18. THE BERNOULLI EQUATION

Definition. An equation of the form $\frac{dy}{dx}+P(x)y=Q(x)y^n$ (1)

is called the Bernoulli differential equation. This equation is linear if $n=0 or 1$. If n is not zero or 1, then (1) is reducible to a linear equation.

Dividing by y^n, (1) becomes

$$y^{-n}\frac{dy}{dx}+P(x)y^{1-n}=Q(x) \quad (2)$$

In (2), put $v=y^{1-n}$ then it reduces to

$$\frac{dv}{dx}+(1-n)P(x)v-(1-n)Q(x)$$

which is linear in v.

Note. Consider the equation

$$f'(y)\frac{dy}{dx}+P(x)f(y)=Q(x)$$

Letting $v=f(y)$, this equation becomes $\frac{dv}{dx}+P(x)v=Q(x)$ which is linear in v.

Example: Solve $\frac{dy}{dx}+\frac{xy}{1-y^2}=xy^{\frac{1}{2}}$ (1)

Solution: Dividing by $y^{\frac{1}{2}}$, (1) becomes

$$y^{-\frac{1}{2}}\frac{dy}{dx}+\frac{x}{1-x^2}y^{\frac{1}{2}}=x \quad (2)$$

Put $y^{\frac{1}{2}}=v$

$$\frac{1}{2}y^{-\frac{1}{2}}\frac{dy}{dx}=\frac{dv}{dx}$$

then (2) reduces to

$$\frac{dv}{dx}+\frac{x}{2\left(1-v^2\right)}.v=\frac{x}{2} \quad (3)$$

this is linear in v.

$$I.F.=\exp\left[\int\frac{x}{2\left(1-x^2\right)}dx\right]=\exp\left[-\frac{1}{4}\ln\left(1-x^2\right)\right]=\left(1-x^2\right)^{-\frac{1}{4}}$$

Multiplying (3) by $\left(1-x^2\right)^{-\frac{1}{4}}$, we get

$$\left(1-x^2\right)^{-\frac{1}{4}}\frac{dv}{dx}+\frac{x}{2\left(1-x^2\right)^{\frac{5}{4}}}.v=\frac{x}{2\left(1-x^2\right)^{\frac{1}{4}}}$$

$$\frac{d}{dx}\left[\left(1-x^2\right)^{-\frac{1}{4}}v\right]=-\frac{1}{4}\left[-2x\left(1-x^2\right)^{-\frac{1}{4}}\right]$$

Integrating, we have

$$v\left(1-x^2\right)^{-\frac{1}{4}}=-\frac{1}{4}\frac{\left(1-x^2\right)^{\frac{3}{4}}}{\frac{3}{4}}+c$$

$$v=c\left(1-x^2\right)^{\frac{1}{4}}-\frac{1-x^2}{3}$$

$$\pm^{\frac{1}{2}} \aleph \left(1 \quad ^2\right)^{-\frac{1}{4}} \quad \frac{1-x^2}{3}$$

is the required solution of (1).

8.19. EQUATION SOLVABLE FOR P

Example: Solve $x^2p^2+xp-y^2-y=0.$ (1)

Solution: We factorize the left hand side of (1) to obtain

$$\left(x^2p^2-y^2\right)+\left(xp-y\right)=\left(xp-y\right)\left(xp+y+1\right)=0$$

Therefore, either

$xp-y=0$ (2)

$xp+y+1=0$ (3)

(2) gives $x\frac{dy}{dx}=y$ or $\frac{dy}{dx}=\frac{dx}{x}$

$y=cx$ (4)

From (3), we have $xp=-\left(y+1\right)$ or $x\frac{dy}{dx}=-\left(y+1\right)$ or $\frac{dy}{y+1}=-\frac{dx}{x}$ which yields

$x\left(y+1\right)=c.$ (5)

Combining (4) and (5), the required solution of (1) is

$$\left(y-cx\right)\left(xy+x-c\right)=0.$$

Example: Solve $xp^3-\left(x^2+x+y\right)p^2+\left(x^2+xy+y\right)p-xy=0$ (1)

Solution: By inspection, we find that $p-1$ is a factor of left hand side of (1). Thus, the given equation is

$$\left(p-1\right)\left[xp^2-\left(x^2+y\right)p+xy\right]=0$$

$$(p-1)(xp-y)(p-x)=0.$$

Therefore, either

$p-1=0$ (2)

$xp-y=0$ (3)

$p-x=0$ (4)

From (2), we have

$\frac{dy}{dx}=1 \, or \, y=x+c$ (5)

From (3), we get

$x\frac{dy}{dx}=y \, or \, y=cx$ (6)

From (4), we obtain

$$\frac{dy}{dx}=x$$

$y=\frac{x^2}{2}+c \, or \, x^2+2(y-c)=0.$ (7)

The general solution of (1) is obtained by combining (5), (6) and (7). Thus

$$(y-x-c)(y-cx)(x^2-2y+2c)=0$$

is the general solution of (1).

8.20. EQUATIONS SOLVABLE FOR Y

Example: Solve $y=p^2x+p$ (1)

Solution. Differentiating (1) $w.r.t.x,$ we have

$$\frac{dy}{dx}=p=p^2+2xp\frac{dp}{dx}+\frac{dp}{dx}$$

$$\frac{dp}{dx}(2px+1)+p^2-p=0$$

$$\frac{dp}{dx}=\frac{-p(p-1)}{2px+1}$$

$$\frac{dx}{dp}=\frac{-2px-1}{p(p-1)}$$

i.e., $\frac{dx}{dp}+\frac{2x}{p-1}=-\frac{1}{p(p-1)}$ (2)

which is linear in x.

$$I.F.=\exp\left[\int\frac{2}{p-1}dp\right]=\exp\left[\ln(p-1)^2\right]=(p-1)^2.$$

Multiplying (2) by $(p-1)^2$ and integrating, we get

$x(p-1)^2=c-p+\ln p$ (3)

$$x=\frac{c-p+\ln p}{(p-1)^2}$$

Substituting this value of x into (1), we have

$$y=p^2.\frac{c-p+\ln p}{(p-1)^2}+p$$

$y(p-1)^2=p^2(c-p+\ln p)+p$ (4)

Thus, (3) and (4) constitute the required solution of (1) with p as a parameter.

Example: Solve $y+px=p^2x^4$ (1)

Solution. From (1), we get

$$y=p^2x^4-px$$

Differentiating the above equation $w.r.t.x,$ we have

$$\frac{dy}{dx}=p=4x^3p^2+2px^4\frac{dp}{dx}-p-x\frac{dp}{dx}$$

$$2p-4x^3p^2-2x^4p\frac{dp}{dx}+x\frac{dp}{dx}=0$$

$$2p\left(1-2px^3\right)+x\left(1-2px^3\right)\frac{dp}{dx}=0$$

$$\left(1-2px^2\right)\left(2p+x\frac{dp}{dx}\right)=0$$

Hence, either $1-2px^3=0$

$$2p+x\frac{dp}{dx}=0 \qquad (2)$$

The equation (2) gives

$$\frac{dp}{p}+2\frac{dx}{x}=0$$

$$\ln p+2\ln x=\ln c$$

$px^2=c$ or $p=\frac{c}{x^2}$

Substituting this value of p into (1), we obtain

$$y=c^2-\frac{c}{x}$$

$$xy-c^2x+c=0$$

is the required solution.

We have yet to deal with relation (2). If we eliminate p from (1) and (2), we get

$$y=-\frac{1}{4x^2} \qquad (3)$$

It is easy to check that (3) is also a solution of (1) and this solution does not involve any constant.

8.21. EQUATION SOLVABLE FOR

Example: Solve $xp=1+p^3$ (1)

Solution: We have $x = \frac{1}{p} + p$ (2)

Differentiating (1) $w.r.t.y,$ we have

$$\frac{dx}{dy} = \frac{1}{p} = -\frac{1}{p^2}\frac{dp}{dy} + \frac{dp}{dx}$$

$$\frac{1}{p} = \left(1 - \frac{1}{p^2}\right)\frac{dp}{dy}$$

$$1 = \left(p - \frac{1}{p}\right)\frac{dp}{dy}$$

which is a separable equation. Therefore,

$$dy = \left(p - \frac{1}{p}\right)dp$$

Integrating, we get $c + y = \frac{p^2}{2} - \ln p$

i.e., $2y = p^2 - 2\ln p - 2c$ (3)

Thus, (2) and (3) constitute the solution of (1).

8.22. CLAIRAUT'S EQUATION

Definition. The equation $y = xp + f(p),$ (1)

is known as Clairaut's equation.

Differentiating both sides of (1), $w.r.t.x,$ we get

$$p = p + xp' + f'(p).p'$$

Cancelling like terms, we have

$$xp' + p'f'(p) = 0$$

$$p'\left[x + f'(p)\right] = 0$$

Since one of the factors must be zero, two different solution arise.

i. If $p' = 0,$ then $p = c$ and substitution of this value into (1) yields the general solution

$$y = cx + f(c)$$

ii. If $x + f'(p) = 0,$ then $x = -f'(p)$ and (1) can be rewritten as $y = -pf'(p) + f(p)$

Thus $x and y$ are both expressed as functions of p and we obtain the parametric equations

$$\left.\begin{aligned} x &= -f'(p) \\ y &= f(p) - pf'(p) \end{aligned}\right\} \qquad (2)$$

Of a curve, representing a solution of (1). The solution (2) is called the singular solution. This solution is not deducible from the general solution p may be eliminated between the two equations in (2) to get a relation in $x and y$ involving no constant.

Example: Find the general solution and singular solution of $y = xp + \frac{1}{4}p^4$ (1)

Solution. The general solution of the equation is $y = cx + \frac{1}{4}c^4.$

Differentiating (1) $w.r.t. x,$ we have

$$p = p + xp' + p^3 p'$$

$$x = -p^3 \qquad (2)$$

We eliminate p from (1) and (2) and get

$$y = x(-x)^{\frac{1}{3}} + \frac{1}{4}(-x)^{\frac{4}{3}} = -\frac{3}{4}x(-x)^{\frac{1}{3}}$$

$64y^3 + 27x^4 = 0$ is the singular solution.

Example: Solve $x^2(y - px) = yp^2$ (1)

Solution: This is not Clairaut's equation. Let us write

$x^2 = u$ and $y^2 = v$

Then, $2x\,dx = du$ and $2y\,dy = dv$.

Hence $\dfrac{y}{x}\dfrac{dy}{dx} = \dfrac{dv}{du}$

$$p = \frac{dy}{dx} = \frac{x}{y}\frac{dv}{du}$$

Substituting this value of p into (1), we get

$$x^2\left(y - \frac{x^2}{y}\frac{dv}{du}\right) = \frac{x^2}{y}\left(\frac{dv}{du}\right)^2$$

$$y^2 - x^2\frac{dv}{du} = \left(\frac{dv}{du}\right)^2$$

$$v - u\frac{dv}{du} = \left(\frac{dv}{du}\right)^2$$

$$v = u\frac{dv}{du} + \left(\frac{dv}{du}\right)^2 \qquad (2)$$

This is Clairaut's equation. Hence its general solution is

$$v = cu + c^2$$

$$y^2 = cx^2 + c$$

is the general solution of (1)

Differentiating (2) $w.r.t\,u,$ we get

$$\frac{dv}{du} = u\frac{d^2v}{du^2} + \frac{dv}{du} + 2\frac{dv}{du}\frac{d^2v}{du^2}$$

$\dfrac{d^2v}{du^2}\left(u + 2\dfrac{dv}{du}\right) = 0,$ then $v = cu + c^2,$

which is the general solution.

If $u + 2\frac{dv}{du} = 0,$ then

$$u = -2\frac{dv}{du} \qquad (3)$$

and

$$v = u\frac{dv}{du} + \left(\frac{dv}{du}\right)^2 \qquad (4)$$

give the singular solution.

Eliminating $\frac{dv}{du}$ from (3) and (4), we obtain

$v = -\frac{u^2}{4}$ or $u^2 + 4v = 0$

is the singular solution.

Replacing $u, v\, by\, x^2 and\; y^2$, respectively, we have

i. $y^2 = cx^2 + c^2$ as the general solution.

ii. $x^4 + 4y^2 = 0$ as the general solution.

8.23. ENVELOPE

Let $f(x, y, c) = 0$ be a one-parameter family of curves. Suppose all members of the family of curves are drawn for various values of the parameter c arranged in order of magnitude. The two curves that correspond to two consecutive values of c will be designated as neighboring curves. The locus of the ultimate points of intersection of neighboring curves is called the envelope of the family $f(x, y, c) = 0.$

Let A, B, C represents three neighboring intersecting members of the family. Let P be the point of intersection of $A\, and\, B\, and\, Q$ be the point of intersection of $B\, and\, C$. By definition $P\, and\, Q$ are points on the envelope. Thus, the curve B and the envelope have two contiguous points common, and therefore, have ultimately a common tangent. Hence B and the envelope touch each other. In the same way, we may show that the envelope touches any other member of the family. Thus, a curve E which, at each of its points, touches someone of the curves of the family is the envelope of the family.

For the family of circles with center on the x-axis and radius 1, that is, the family of circles with equation $(x-1)^2 + y^2 = 1$.

The pair of lines $y = \pm 1$ is the envelope. It is easy to see that the lines $y = \pm 1$ touch each member of the family of the circles.

8.23.1. Equation of the Envelope

To find an equation of the envelope of the family $f(x, y, c) = 0,$ consider two neighboring curves and

$$\left.\begin{array}{r} f(x,y,c)=0 \\ f(x,y,c+h)=0 \end{array}\right\} \qquad (1)$$

Find the intersection of these two curves and let $h \to 0$. The point of intersection then must approach the point of contact of the curve $f(x, y, c) = 0$ with the envelope. At the point of intersection of the equation

$$\frac{f(x,y,c+h) - f(x,y,c)}{h} = 0 \qquad (2)$$

is true as well as (1). Letting $h \to 0,$ we have from (2)

$$f_c(x, y, c) = 0 \quad (3)$$

and from (1) $f(x, y, c) = 0$ (4)

If we eliminate c from (3) and (4), we obtain an equation of the envelope. The eliminate is called the c-discriminant of the family $f(x, y, c) = 0$. The c-discriminant may contain loci other than the envelope.

Example: The family of parabolas $(x-c)^2 - 2y = 0$ has the x-axis as envelope.

8.24. SINGULAR SOLUTION

Let $f(x, y, p) = 0$ (1)

be a nonlinear first order differential equation in which the left-hand side is a polynomial in p. The general solution of this differential equation will be a one-parameter family

$f(x, y, c) = 0$ (2)

The envelope E of (2) is a curve which, at each of its points, touches some one member of the family (2). At a point of contact P of the envelope and a member of (2), the values of x, y, p are the same. But the values of x, y, p for the curve at P satisfy (1). Thus, the envelope of family (2) is a solution of the differential equation (1).

Example: Consider $p^2 - xp + y = 0.$

This is Clairaut's equation and its general solution (replacing simply p by c) is

$y = cx + c^2$ (1)

Now we find the envelope of the family (1)

$f(x, y, c) \equiv y - cx + c^2 = 0$ (2)

$f_c \equiv \frac{\partial f}{\partial c} = -x + 2c = 0$ (3)

Substituting the value of c from (3) into (2), we have

$$y - x\left(\frac{x}{2}\right) + \left(\frac{x}{2}\right)^2 = 0$$

i.e., $$y - \frac{x^2}{2} + \frac{x^2}{4} = 0$$

$$4y = x^2.$$

(4) is an equation of the envelope of the family (1).

We check whether (4) is a solution of the given differential equation. Differentiating (4) $w.r.t. x,$ we get

$$4\frac{dy}{dx} = 2x or 2p = x. \qquad (5)$$

Substituting (4) and (5) into $p^2 - px + y,$ we have

$$\frac{x^2}{4} - \frac{x^2}{2} + \frac{x^2}{4} = 0$$

Thus (4) is a solution of the given differential equation. This solution does not involve any constant and it cannot be derived from the general solution by giving particular values to c, such a solution is called a singular solution (S.S).

Definition: A solution of a differential equation $f(x, y, p) = 0$ is called a singular solution (S.S) if

i. It is not derived from the general solution by giving any particular value to the arbitrary constants

ii. At each of its points, it is tangent to some member of the one-parameter family of curves represented by the general solution.

We have seen that the envelope, if any, of the one-parameter family of curves represented by the general solution of a differential equation is a singular solution. Thus, the c- discriminant equation may contain singular solution, if any.

The singular solution may also be obtained from the differential equation directly without finding the general solution. Since at ultimate point of intersection of neighboring curves the $p's$ for the intersecting curves become equal, and thus the locus of the points where $p's$ have equal roots will include the envelope. If we eliminate p from

$$f(x, y, p) = 0 \, and \, \frac{\partial f}{\partial p} = 0$$

The resulting equation is called the p-discriminant. The p-discriminant represents the locus for each point of which $f(x, y, p) = 0$ has equal roots.

If envelope of the general solution of $f(x, y, p) = 0$ exists, it will be contained in the p-discriminant. Thus, the p-discriminant equation may contain (i) singular solution (ii) Solutions that are not singular and (iii) such loci that are not solutions at all.

Example: Solve $xp^2 - 2yp + 4x = 0$ (1)

Solution: $2yp = xp^2 + 4x$

$$2y = xp + \frac{4x}{p}$$

Differentiating $w.r.t.x,$ we get

$$2p = x\frac{dp}{dx} + p + \frac{4}{p} - \frac{4x}{p^2}\frac{dp}{dx}$$

$$p - \frac{4}{p} = \left(x - \frac{4x}{p^2}\right)\frac{dp}{dx}$$

$$\frac{p^2 - 4}{p} = \frac{x\left(p^2 - 4\right)}{p^2}\frac{dp}{dx}$$

$$\frac{dp}{p} = \frac{dx}{x}$$

$p = cx$ (2)

Eliminating p from (1) and (2), we obtain

$$x\left(c^2x^2\right) - 2ycx + 4x = 0$$

$c^2x^2 - 2yc + 4 = 0$ (3)

From (3), we obtain the c - discriminant as

$4y^2 - 16x^2 = 0$ or $y^2 = 4x^2$

The p-discriminant is from (1)

$4y^2 = 16x^2$ or $y^2 = 4x^2$

Since the c-discriminant and p-discriminant are the same viz., $y^2 = 4x^2$ and it satisfies the given differential equation, it is the singular solution.

Example: Solve $x^2p^2 + yp(2x + y) + y^2 = 0$ (1)

By making substitution $y = u, xy = v$ and find the singular solutions.

Solution: Let $y = u, x = \frac{v}{y} = \frac{v}{u}$

Thus $\frac{dy}{du} = 1, \frac{dx}{du} = \frac{u.\frac{dv}{du} - v}{u^2}$

$$p = \frac{dy}{dx} = \frac{dy}{du}.\frac{du}{dx} = \frac{u^2}{u.\frac{dv}{du} - v}$$

Substituting into (1), we have

$$\frac{v^2}{u^2}\left[\frac{u^2}{u.\frac{dv}{du} - v}\right]^2 + \left(2v + u^2\right).\frac{u^2}{u.\frac{dv}{du} - v}$$

$$v^2 + \left(2v + u^2\right)\left(u\frac{dv}{du} - v\right) + \left(u.\frac{dv}{du} - v\right)^2 = 0$$

$$v^2 + 2uv\frac{dv}{du} - 2v^2 + u^3.\frac{dv}{du} - u^2v + u^2\left(\frac{dv}{du}\right)^2 - 2uv\frac{dv}{du} + v^2 = 0$$

$$v = u\frac{dv}{du} + \left(\frac{dv}{du}\right)^2$$

which is a Clairaut's equation. Its solution is

$$v = cu + c^2$$

i.e., $xy = cy + c^2$ (2)

is the solution of (1).

From (2), the c-discriminant is

$y^2 + 4xy = 0$ i.e., $y\left(y + 4x\right) = 0$

Therefore, $y = 0, y + 4x = 0$ (3)

Differentiating (3), we have $p = -4.$ Substituting $p = -4$ into the left-hand member of (1) and using (3), we obtain

$$16x^2-4x(-4)(-2x)+(-4x)^2=32x^2-32x^2=0$$

Thus $y+4x=0$ is also a singular solution.

Example: Solve $(x^2-1)p^2-2xyp-x^2=0$ (1)

Moreover, find the singular solution, if any.

Solution: Solving the equation for y, we have

$$2y=\frac{(x^2-1)p^2-x^2}{xp}=xp-\frac{p}{x}-\frac{x}{p}.$$

Differentiating $w.r.t.x$, we get

$$2p=xp'+p-\frac{xp'-p}{x^2}-\frac{p-xp'}{p^2},$$ where $p'=\frac{dp}{dx}$

$$p=p'\left(x-\frac{1}{x}+\frac{x}{p^2}\right)+\frac{p}{x^2}-\frac{1}{p}$$

$$p\left(1-\frac{1}{x^2}+\frac{1}{p^2}\right)=xp'\left(1-\frac{1}{x^2}+\frac{1}{p^2}\right)$$

Therefore, $\frac{dp}{p}=\frac{dx}{x} or\ p=cx.$

Substituting this value of p into (1), we get

$$(x^2-1)c^2x^2-2x^2yc-x=0$$

$(x^2-1)c^2-2yc-1=0$ (2)

is the general solution of (1).

From (2), the c-discriminant is

$$4y^2+4(x^2-1)=0$$

i.e., $x^2+y^2=1$

which is also envelope of the family (2). We also note that $p-discriminat$ is

$$4x^2y^2 + 4x^2\left(x^2 - 1\right) = 0$$

i.e., $x^2 + y^2 = 1$

Since the envelope and the p-discriminant are same, the singular solution is

$$x^2 + y^2 = 1$$

Example: Find the singular solution of

$$x^3p^2 + x^2yp + a^3 = 0 \qquad (1)$$

Solution: The $p-discriminant$ of (1) is

$$x^4y^2 - 4x^3a^3 = 0\, or\, x^3\left(y^2x - 4a^3\right) = 0 \qquad (2)$$

The parts of (2) that satisfy (1) are singular solutions of the given equation. From (2), we have

$$x = 0\, or\, y^2x - 4a^3 = 0 \quad (3)$$

We rewrite (1) as

$$x^3 + x^2y\frac{dx}{dy} + a^3\left(\frac{dx}{dy}\right)^2 = 0 \qquad (4)$$

Clearly, $x = 0$ satisfies (4).

Thus $x = 0$ is a singular solution of (1).

Differentiating $y^2x - 4a^3 = 0\, w.r.t\, x,$ we have

$$2xyp + y^2 = 0\, or\, y\left(2xp + y\right) = 0$$

i.e., $p = -\dfrac{y}{2x}.$

Putting this value of p into the left-hand member of (1) and using (3), we obtain

$$x\left(-\frac{y}{2x}\right)^2 + x^2y\left(-\frac{y}{2x}\right) + a^3 = -\frac{xy^2}{4} + a^3 = -\left(xy^2 - 4a^3\right) = 0$$

Thus $xy^2 - 4a^3 = 0$ and its derivative satisfy (1) and so it is a solution of (1)

Hence $x = 0 \, and \, xy^2 - 4a^3 = 0$ are singular solutions of (1).

8.25. THE RICATTI EQUATION

Definition. We have already studied first order linear differential equation

$$y' + P(x)y = R(x) \qquad (1)$$

If we add the term $q(x)y^2$ to the left-hand member of (1), we obtain a nonlinear differential equation

$$y' + P(x)y + Q(x)y^2 = R(x) \qquad (2)$$

(2) is called the Ricatti equation.

Note: In many cases, the solution of (2) cannot be expressed in terms of elementary functions. However, the Ricatti equation $y' + Py + Qy^2 = R$ (1)

can be reduced to a linear equation by the substitution $y = y_1 + \frac{1}{u}$, where y_1 is a particular solution of (1) and u is an unknown non-zero function of x.

Proof: Let $y = y_1 + \frac{1}{u}$ be as given. Differentiating $w.r.t.x$, we have

$$y' = \frac{dy}{dx} = \frac{dy_1}{dx} - \frac{1}{u^2}\frac{du}{dx}$$

Substituting for $y \, and \, y'$ into (1), we get

$$\frac{dy_1}{dx} - \frac{1}{u^2}\frac{du}{dx} + P\left(y_1 + \frac{1}{u}\right) + Q\left(y_1^2 + \frac{2y_1}{u} + \frac{1}{u^2}\right) = R$$

$$\frac{dy_1}{dx} + Py_1 + Qy_1^2 - R - \frac{1}{u^2}\left(\frac{du}{dx} - Pu - 2Qy_1u - Q\right) = 0 \qquad (2)$$

Since y_1 is a solution of (1), we have

$$\frac{dy_1}{dx} + Py_1 + Qy_1^2 - R = 0$$

and so (2) reduces to $\frac{du}{dx}-\left(P+2Qy_1\right)u=Q$ which is a linear equation. Thus, if a particular solution of (1) is known then its general solution can be found.

Example: Solve $\frac{dy}{dx}-y^2=-1, y(0)=3,$ given that $y_1=1$ is a particular solution of the given equation.

Solution: Here $P=0, Q=-1, R=-1.$

Writing $y=1+\frac{1}{u},$ the given equation reduces to

$$\frac{du}{dx}-\left(0+2(-1)(1)\right)=-1$$

i.e., $\frac{du}{dx}+2u=-1,$ (1)

which is a linear equation.

I.F. of (1) is $e^{\int 2\,dx}=e^{2x}.$

Multiplying (1) by the I.F., we get

$$\frac{du}{dx}e^{2x}+2u\,e^{2x}=-e^{2x}$$

$$\frac{d}{dx}\left(ue^{2x}\right)=-e^{2x}$$

Integrating, we obtain

$$u\,e^{2x}=-\int e^{2x}\,dx+c=-\frac{e^{2x}}{2}+c$$

$$u=-\frac{1}{2}+\frac{c}{e^{2x}} \quad (2)$$

From $y=1+\frac{1}{u},$ we get by the initial condition,

$$y(0)=3=1+\frac{1}{u(0)} or\, u(0)=\frac{1}{2}$$

Hence from (2), we have

$$u(0)=\frac{1}{2}=-\frac{1}{2}+c\, or\, c=1$$

Thus $u=-\frac{1}{2}+\frac{1}{e^{2x}}=\frac{2-e^{2x}}{2e^{2x}}$

Required solution is $y=1+\frac{1}{u}=1+\frac{2e^{2x}}{2-e^{2x}}=\frac{2+e^{2x}}{2-e^{2x}}$.

Example: Solve $\frac{dy}{dx}-\frac{y}{x}-x^3y^2=-x^5$, (1)

by finding a particular solution.

Solution: It is a Ricatti equation with $P=-\frac{1}{x}, Q=-x^3\, and\, R=-x^5$.

An obvious solution of (1) is $y_1=x.$ Substituting $y=x+\frac{1}{u}$ into (1), we have

$$\frac{du}{dx}-\left[-\frac{1}{x}+2\left(-x^3\right)x\right]u=-x^3$$

$$\frac{du}{dx}+\left(\frac{1}{x}+2x^4\right)u=-x^3, \qquad (2)$$

which is a linear equation.

Its $I.F.=e^{\int\left(\frac{1}{x}+2x^4\right)dx}$

$$=e^{\ln x}.e^{\frac{2}{5}x^5}=xe^{\frac{2}{5}x^5}$$

Multiplying (2) by the I.F. and integrating, we have

$$uxe^{\frac{2}{5}x^5}=-\int x^4.e^{\frac{2}{5}x^5}dx+c'=-\frac{1}{2}e^{\frac{2}{5}x^5}+c'$$

$$u = \frac{c' - \frac{1}{2} e^{\frac{2}{5}x^5}}{x e^{\frac{2}{5}x^5}}$$

Hence the general solution of (1) is

$$y = x + \frac{x e^{\frac{2}{5}x^5}}{c' - \frac{1}{2} e^{\frac{2}{5}x^5}} = \frac{x\left(c' + \frac{1}{2} e^{\frac{2}{5}x^5}\right)}{c' - \frac{1}{2} e^{\frac{2}{5}x^5}}$$

$$\frac{y}{x} = \frac{c' + \frac{1}{2} e^{\frac{2}{5}x^5}}{c' - \frac{1}{2} e^{\frac{2}{5}x^5}}$$

$$\frac{y - x}{y + x} = \frac{e^{\frac{2}{5}x^5}}{2c'} = c e^{\frac{2}{5}x^5}$$

is the required solution.

SUMMARY AND DISCUSSION

In this chapter, we discussed some elementary aspects of differential equation. A very important topic in real life problem.

The study of differential equations is a wide sphere in pure and applied mathematics, physics, and engineering. All of these disciplines are concerned with the estate of differential equations of various types. Pure sum focuses on the life and singularity of solutions, while applied tally emphasizes the rigorous occasion of the methods for approximating solutions. Differential equations pleasure an important role in modeling virtually every physical, technical, or biological process, from celestial motion, to bridge design, to interactions between neurons. Differential equations such as those used to solve real-life problems may not necessarily be directly solvable, i.e., do not have closed pattern solutions. Instead, solutions can be approximated using numerical methods.

Applications

Many fundamental jurisprudences of physics and chemistry can be formulated as differential equations. In biology and economics, differential equations are used to mold the behavior of complex systems. The mathematical opinion of differential equations first developed together with the sciences where the equations had originated and where the results found application. However, diverse problems, sometimes originating in quite distinct scientific fields, may give rise to identical differential equations.

Whenever this happens, mathematical plan seats the equations can be viewed as a unifying principle seat diverse phenomena. As an example, consider dissemination of light and sound in the atmosphere, and of waves on the surface of a pond. All of them may be described by the same second-order partial differential equation, the breaker equation, which allows ourselves to think of brightening and sound as forms of waves, scads like familiar cascade in the water. Conduction of heat, the theory of which was developed by Joseph Fourier, is governed by another second-order partial differential equation, the heat equation. It turns out that many scattering processes, while seemingly different, are described by the same equation; the Black-Scholes equation in finance is, for instance, related to the heat equation.

Euler-Lagrange equation in classical mechanics, Hamilton's equations in classical mechanics, Radioactive decay in nuclear physics Newton's statute of cooling in thermodynamics, The breaker equation, The heat equation in thermodynamics, Laplace's equation, which defines harmonic functions, Poisson's equation, The geodesic equation, The Navier-Stokes equations in fluid dynamics, The Diffusion equation in stochastic processes, The Convection-diffusion equation in fluid dynamics, The Cauchy-Riemann equations in complex analysis, The Poisson-Boltzmann equation in molecular dynamics, The shallow water equations and Universal differential equation.

8.2. Differential Equation

An equation involving one dependent variable and its derivatives with respect to one or more independent variables is called differential equation. For example

vi. $$\frac{dy}{dx} + y\cos x = \sin x$$

8.2.1. Ordinary Differential Equation (O. D. E)

A differential equation, in which ordinary derivatives of the dependent variable with respect to a single independent variable occur, is called an ordinary differential equation (O. D. E).

8.2.2. Partial Differential Equation

A differential equation involving partial derivatives of the dependent variable with respect to more than one independent variable is called a partial differential equation.

$$x\frac{\partial z}{\partial x}+y\frac{\partial z}{\partial y}=nx$$

8.3. Order of Differential Equation

The order of a differential is the order of the highest derivative that occurs in the equation.

8.4. Degree of a Differential Equation

The degree of a differential equation is the greatest exponent of the highest order derivative that appears in the equation. (The dependent variable and its derivatives should be expressed in a form free from radicals and fractions).

i. $\frac{dy}{dx}+y\cos x=\sin x$ (order 1, degree 1)

ii. $\frac{d^2y}{dx^2}+xy\left(\frac{dy}{dx}\right)^2=0$ (order 2, degree 1)

iii. $\left[1+\left(\frac{dy}{dx}\right)^2\right]^{\frac{3}{2}}=\frac{d^2y}{dx^2}$ (order 2, degree 2)

8.5. Linear Differential Equation

An ordinary differential equation $F\left(x,y,\frac{dy}{dx},\frac{d^2y}{dx^2},\ldots,\frac{d^ny}{dx^n}\right)=0$ is said to be linear if F is a linear function of the variables $x,\frac{dy}{dx},\frac{d^2y}{dx^2},\ldots,\frac{d^ny}{dx^n}$

It should be carefully noted that in a linear ordinary differential equation

i. The dependent variable y and its derivative are all of degree one.

ii. No product of y or any of its derivative appear.

iii. No transcendental function of y and/or its derivative occur.

Nonlinear Differential Equation. A differential equation that is not linear is called a nonlinear differential equation. Differential equations occur in the mathematical formulation of many problems in science and engineering. Some such problems are

a. Determining the motion of projectile, rocket, satellite or planet.

b. Finding the charge or current in an electric circuit.

c. Study of chemical reactions.

d. Determination of curves with given geometrical properties.

8.6. Solution of a Differential Equation

A solution (or integral) of a differential equation is a relation between the variables, not containing derivatives, such that this relation and the derivatives obtained from it satisfy the given differential equation identically.

8.8.1. Initial Condition

It is often required to find the solution of a differential equation subject to certain conditions. If the conditions relate to one value of the independent variable such as $y = y_0$ at $x = x_0$ (written as $y(x_0) = y_0$) and $\frac{dy}{dx} = y'(x_0)$ at $x = x_0$, where x_0 belongs to some interval $]\alpha, \beta[$ then they are called initial conditions (or one- point boundary conditions) and x_0 is called the initial point.

8.8.2. Boundary Conditions

The problem of finding the solution of a differential equation such that all the associated constraints relate to two different values of the independent variable is called a two-point boundary value problem (or simply a boundary value problem). The associated supplementary boundary conditions are called two-point boundary conditions.

8.9. Separable Equations

Definition. A differential equation of the type

$$F(x)G(x)dx + f(x)g(y)dy = 0 \quad (1)$$

is called an equation with separable variables or simply a separable equation. Equation (1) may be written as

$$\frac{F(x)}{f(x)}dx + \frac{g(y)}{G(y)}dy = 0$$

which can be easily integrated.

8.10. Homogenous Equations

Definition. A function $f(x,y)$ is called homogeneous of degree n if $f(tx,ty) = t^n f(x,y)$, Where t is a non-zero real number. Thus, $\sqrt{xy}, \frac{x^{10}+y^{10}}{x^2+y^2}$ *and* $\sin\left(\frac{x}{y}\right)$ are homogeneous function of degree 1,8 and 0, respectively.

Exact Equation. The expression $M(x,y)dx + N(x,y)dy$ is called an exact differential if there exists a continuously differentiable function $f(x,y)$ of two real variables x and y such that the expression equals the total differential df. We know from calculus that $df = \frac{\partial f}{\partial x}dx + \frac{\partial f}{\partial y}dy.$ Thus, if it is exact then $M(x,y) = \frac{\partial f}{\partial x} = f_x andN(x,y) - \frac{\partial f}{\partial y} = f_y.$

8.15. Integrating Factors

If the differential equation $M(x,y)dx + N(x,y)dy = 0$ is not exact but when it is multiplied by a function $\mu(x,y)$ and the resulting equation $\mu(x,y)M(x,y)dx + \mu(x,y)N(x,y)dy = 0$ is exact, then $\mu(x,y)$ is called an integrating factor (I.F) of the differential equation. The number of integrating factors of an equation may be infinite.

We list below (without proofs) some rules to find the integrating factors of equations of special types.

Rule 1. If $M(x,y)\,dx+N(x,y)\,dy=0$ (1)

is not exact and $\frac{M_{y-N_x}}{N}=P,$ where P is a function of x only then (1) has an integrating factor $\mu(x)$ which also depends on $x.\mu(x)$ is solution of the differential equation

$$\frac{d\mu}{dx}=P\mu$$

i.e., $\mu(x)=\exp\int Pdx$

Note that $M_y=\frac{\partial M}{\partial y},N_x=\frac{\partial N}{\partial x}.$

Rule II. If $\frac{N_x-M_y}{M}=Q,$ where Q is a function of y only, then the differential equation $Mdx+Ndy=0$ has an integrating factor,

$$\mu(y)=\exp\int Qdy.$$

Rule III. If $Mdx+Ndy=0$ (1)

is homogeneous and $xM+yN\neq 0,$ then $\frac{1}{xM+yN}$ is an I.F of (1).

Rule IV. If $Mdx+Ndy=0$ is of the form $yf(x,y)\,dx+xg(x,y)\,dy=0$ and $xM-yN\neq 0,$ then $\frac{1}{xM-yN}$ is an I.F of (1).

The following differential formulas are useful in the calculation of certain exact equations:

i. $d\left(\frac{y}{x}\right)=\frac{xdy-ydx}{x^2}$

ii. $d\left(\frac{x}{y}\right)=\frac{ydx-xdy}{y^2}$

iii. $d(xy)=xdy+ydx$

iv. $d\left(x^2+y^2\right)=2(xdx+ydy)$

v. $d\left(\ln\frac{x}{y}\right)=\frac{ydx-xdy}{xy}$

vi. $d\left(arc\tan\frac{x}{y}\right)=\frac{ydx-xdy}{x^2+y^2}$

8.16. Linear Equations

A first order ordinary differential equation (ODE) is linear in the dependent variable y and the independent variable x if it is or can be written in the form

$$\frac{dy}{dx}+P(x)y=Q(x), \quad (1)$$

where $P and Q$ are functions of x.

8.18. The Bernoulli Equation

Definition. An equation of the form $\frac{dy}{dx}+P(x)y=Q(x)y^n$ (1)

is called the Bernoulli differential equation. This equation is linear if $n=0 or 1$. If n is not zero or 1, then (1) is reducible to a linear equation.

8.22. Clairaut's Equation

Definition. The equation

$$y=xp+f(p), \quad (1)$$

is known as Clairaut's equation.

8.23. Envelope

Let $f(x, y, c) = 0$ be a one-parameter family of curves. Suppose all members of the family of curves are drawn for various values of the parameter c arranged in order of magnitude. The two curves that correspond to two consecutive values of c will be designated as neighboring curves. The locus of the ultimate points of intersection of neighboring curves is called the envelope of the family $f(x, y, c) = 0$.

8.24. Singular Solution

A solution of a differential equation $f(x, y, p) = 0$ is called a singular solution (S.S) if

i. It is not derived from the general solution by giving any particular value to the arbitrary constants.

ii. At each of its points, it is tangent to some member of the one-parameter family of curves represented by the general solution.

8.25. The Ricatti Equation

Definition. We have already studied first order linear differential equation

$$y' + P(x)y = R(x) \quad (1)$$

If we add the term $q(x)y^2$ to the left-hand member of (1), we obtain a nonlinear differential equation

$$y' + P(x)y + Q(x)y^2 = R(x) \quad (2)$$

Equation (2) is called the Ricatti equation.

CHAPTER 9

DIFFERENTIAL EQUATIONS OF HIGHER ORDER

CONTENTS

9.1. INTRODUCTION

In previous chapter, we studied methods of solving special types of linear and nonlinear differential equations of the first order. In this chapter, we shall consider systematic methods for the solution of certain classes of differential equations of order more than one.

9.2. LINEAR DIFFERENTIAL EQUATIONS

Definition. A linear differential equation of order n in the dependent variable y and the independent variable x is of the form

$$a_0(x)\frac{d^n y}{dx^n}+a_1(x)\frac{d^{n-1} y}{dx^{n-1}}+\ldots+a_{n-1}\frac{dy}{dx}+a_n(x)y=F(x), \qquad (1)$$

where $a_0(x), a_1(x), \ldots, a_{n-1}(x), a_n(x) and\, F(x)$ are functions of the independent variable x only and $a_0(x)$ is not identically zero. Using primes, (1) is also written as

$$a_0 y^n + a_1 y^{n-1} + \ldots + a_{n-1} y' + a_n y = F(x) \qquad (2)$$

where $a_0, a_1, \ldots, a_{n-1}, a_n$ are real constants.

Equation (1) is with variable coefficients while (2) is with constant coefficients. In order to solve (2), we shall first consider the equation

$$a_{\ddot{u}} y^n + a\ y^{n-1} + \ldots \aleph a_{n-}\ y'\ \ a_n y\ \ 0 \qquad (3)$$

The coefficient of y^n may be made 1 by dividing throughout by a_0. The differential equation (3) is called homogeneous linear differential equation of order n. The use of the word homogeneous here is quite different from the one already mentioned in previous chapter. If $F(x)$ is not identically zero then (2) is called non-homogeneous and (3) is called the associated homogeneous equation of (2).

Definition. If $y_1(x), y_2(x), \ldots, y_m(x)$ are m functions of an independent variable x and $c_1, c_2, \ldots, c_m$ are constants, then the expression $c_1 y_1(x) + c_2 y_2(x) + \ldots + c_m y_m(x)$ is called a linear combination of $y_1(x), y_2(x), \ldots, y_m(x)$. We usually write $y_1, y_2, \ldots, y_m$ instead of

$y_1(x), y_2(x), \ldots, y_m(x)$ when it is clear from the context that $y_1, y_2, \ldots, y_m$ are functions of x.

As in vector spaces, the m non-zero functions $y_1, y_2, \ldots, y_m$ are called linearly dependent if and only if, there exist constants $c_1, c_2, \ldots, c_m$ at least one of which is non-zero, such that

$$c_1 y_1 + c_2 y_2 + \ldots + c_m y_m = 0.$$

The functions $y_1, y_2, \ldots, y_m$ are called linear independent if and only if, they are not linearly dependent, i.e., if and only if $c_1 y_1 + c_2 y_2 + \ldots + c_m y_m = 0$ implies $c_1 = c_2 = \ldots = c \ \ = 0$.

Now before investigating a solution of (2), we state the following facts:

i. Every homogeneous linear $nth-order$ differential equation $a_0 y^n + a_1 y^{n-1} + \ldots + a_{n-1} y' + a_n y = 0$ has n linearly independent solutions $y_1, y_2, \ldots, y_n$. If $y_1, y_2, \ldots, y\_m$ are n linearly independent solutions of (3), then any linear combination $y_c = c_1 y_1 + c_2 y_2 + \ldots + c_n y_n$ of $y_1, y_2, \ldots, y_n$ is the general solution of (3), $c_1, c_2, \ldots, c_m$ being arbitrary constants. Let y_p be any particular solution of (2). i.e., y_p does not contain any constant, then $y_c + y_p$ is the general solution of (2).(ii) and (iii) can be easily checked by actual substitutions into (3) and (2).Thus, to find the general solution of (2), we have to find a linearly independent set of n solutions of (3). So as to determine y_c and a particular solution y_p of (2) and then obtain

$$y = y_c + y_p \qquad (4)$$

as the general solution of (2). In the general solution (4), y_c is called the Complementary Function (C.F) and is called the Particular Integral (P.I) of (2).

These statements are also true for the equation (1) with variable coefficients.

9.3. HOMOGENEOUS LINEAR EQUATION

Consider the equation $a_0\dfrac{d^n y}{dx^n}+a_1\dfrac{d^{n-1}y}{dx^{n-1}}+\ldots+a_{n-1}\dfrac{dy}{dx}+a_n\,y,=0$
(1)

where $a_0,a_1,\ldots,a_{n-1}$ are real constants. To find a solution of (1) we shall try a judicious guess. The differential equation (1) requires a function y with the property that if y and its successive derivatives are each multiplied by constants $a_j, j=n,n-1,\ldots,1,0,$ the resulting products are then added, the sum should equal zero. This can only happen if a function is such that its various derivatives are constant multiples of itself. The exponential function $y=e^{mx}$, m being a constant, has such properties. Here we have

$$\frac{dy}{dx}=me^{mx},\frac{d^2y}{dx^2}=m^2e^{mx},\ldots,\frac{d^ny}{dx^n}=m^ne^{mx}.$$

Substituting into (1) we have

$$a_0m^ne^{mx}+a_1m^{m-1}e^{mx}+\ldots+a_{n-1}me^{mx}+a_ne^{mx}=0$$

$$e^{mx}\,(a_0m^n+a_1m^{n-1}e^{mx}+\ldots+a_{n-1}me^{mx}+a_ne^{mx}=0$$

Since $e^{mx}\neq 0,$ we have

$a_0m^n+a_1m^{m-1}+\ldots+a_{n-1}m+a_n=0$ (2)

Thus, $y=e^{mx}$ is a solution of (1) if and only if m is a solution of (2). Equation (2) is called the characteristics (or auxiliary) equation of the given differential equation (1). Observe that (2) can be obtained from (1) by merely replacing the kth derivative in (1) by $m^k\left(k=n,n-1,\ldots,3,2,1\right).$ Three cases arise according as the roots of (2) are:

i. real and distinct;

ii. real and repeated; and

iii. complex.

9.3.1. Case I. Distinct Real Roots

Let $m_1,m_2,\ldots,m_n$ be n distinct real roots of (2). Then $e^{m_1x},e^{m_2x},\ldots,e^{m_nx}$ are n distinct solutions of (1). These n solutions are linearly independent. Hence the general solution of (2) is

$$y = c_1 e^{m_1 x} + c_2 e^{m_2 x} + \ldots + c_n e^{m_n x},$$

where $c_1, c_2, \ldots, c_n$ are arbitrary constants.

9.3.2. Case II. Repeated Real Roots

In equation (1), writing $D \equiv \dfrac{d}{dx}$, we have $\left(a_0 D^n + a_1 D^{n-1} + \ldots + a_{n-1} D + a_n\right) y = 0$ or $\left[f(D)\right] y = 0,$

where $f(D) = a_0 D^n + a_1 D^{n-1} + \ldots + a_{n-1} D + a_n.$

Note that if we write D for m in the characteristic equation (2) then it is the same as $f(D) = 0.$ If $m_1, m_2, \ldots, m_n$ are the roots of $f(D) = 0,$ then (1) may be written as

$$(D - m_1)(D - m_2)\ldots(D - m_n) y = 0.$$

Let the root m_1 be repeated twice say $m_2 = m_1.$

The part of the general solution of (1) corresponding to the twice repeated root m_1 of (2) is solution of

$$(D - m_1)^2 y = 0$$

i.e., of $(D - m_1)(D - m_1) y = 0$ (3)

Let $(D - m_1) y = V.$ Then (3) becomes

$$(D - m_1) V = 0$$

$$\frac{dV}{dx} - m_1 V = 0 \qquad (4)$$

Separating the variables, we have

$$\frac{dV}{V} = m_1 dx.$$

Therefore, $\ln V = m_1 x + k,$ where k is a constant $.V = c_2 e^{m_1 x}$, where c_2 is a constant.

Replacing V in (4), we obtain

$$(D - m_1) y = c_2 e^{m_1 x}$$

$$\frac{dy}{dx} - m_1 y = c_2 e^{m_1 x} \qquad (5)$$

which is a linear equation of order one. Its I.F. $= \exp\left(\int(-m_1)dx\right) = e^{-m_1 x}$.

Multiplying (5) by $e^{-m_1 x}$, we get

$$\frac{d}{dx}\left(ye^{-m_1 x}\right) = c_2$$

$$ye^{-m_1 x} = c_2 x + c_1$$

$y = (c_1 + c_2 x)e^{m_1 x}$ is the part of the general solution corresponding to the twice repeated m_1. The general solution of (1) is

$$y = (c_1 + c_2 x)e^{m_1 x} + c_3 e^{m_3 x} + \ldots + c_n e^{m_n x}$$

In the same manner, if the characteristic equation (2) has the triple repeated root m_1, the corresponding part of the general solution of (1) is the solution of

$$(D - m_1)^3 y = 0$$

Proceeding as before, we can easily find

$$y = \left(c_1 + c_2 x + c_3 x^2\right)e^{m_1 x}$$

As the part of the general solution corresponding to this triple repeated root m_1. If the characteristics equation (2) has the real root m_1 occurring k times then the part of the general solution of (1) corresponding to the $k - fold$ repeated root m_1 is

$$y = \left(c_1 + c_2 x + c_3 x^2 + \ldots + c_k x^{k-1}\right)e^{m_1 x}$$

9.3.3. Case III. Complex Roots

Suppose the characteristic equation has the complex number $a+ib$ as a non-repeated root. Since coefficients of (1) are real, the conjugate complex number $a-ib$ is also a non-repeated root. The corresponding part of the general solution is $y = k_1 e^{(a+ib)x} + k_2 e^{(a-ib)x}$, where $k_1 \; and \; k_2$ are arbitrary constants.

$$y = e^{ax}\left[k_1 e^{ibx} + k_2 e^{-ibx}\right]$$

$$= e^{ax}\left[k_1(\cos bx + i\sin bx) + k_2(\cos bx - i\sin bx)\right]$$

$$= e^{ax}\left[(k_1+k_2)\cos bx + i(k_1-k_2)\sin bx\right]$$

$$= e^{ax}\left[c_1 \sin bx + c_2 \cos bx\right]$$

where $c_1 = i(k_1-k_2), c_2 = k_1 + k_2$ are two arbitrary constants.

If $a+ib \, and \, a-ib$ are conjugate complex roots, each repeated k times, the corresponding part of the general solution of (1) may be written as

$$y = \left[(c_1 + c_2 x + c_3 x^2 + \ldots + c_k x^{k-1})\sin bx + (c_{k+1} + c_{k+2} x + \ldots + c_{2k} x^{k-1})\cos bx\right]$$

In the examples given below we use these concepts to solve homogeneous linear differential equations.

Example: Solve $(D^2 + 4D + 3)y = 0$

Solution: The characteristic equation is $D^2 + 4D + 3 = 0$ with roots $D = -1, -3$.

Hence the general solution of the given equation is $y = c_1 e^{-x} + c_2 e^{-3x}$.

Example: Solve $(D^3 - 5D^2 + 7D - 3)y = 0$.

Solution. The characteristic equation is $D^3 - 5D^2 + 7D - 3 = 0$ (1)

By inspection, $D = 1$ is a solution of this equation. The other two roots can be found from the quadratic factor of $(D-1)(D^2 - 4D + 3) = 0$

Hence all roots of (1) are $D = 1, 1, 3$.

The general solution is $y=(c_1+c_2x)e^x+c_3e^{3x}$.

Example. Solve $(D^3-D^2+D-1)y=0$.

Solution. The characteristic equation is $D^3-D^2+D-1=0$.

By inspection, $D=1$ is a root of this equation.

Now, $D^3-D^2+D-1=(D-1)(D^2+1)=0$.

Hence the other two roots are $D=\pm i=a+ib\, with\, a=0, b=1$.

The general solution is

$$y=c_1e^x+e^{0x}(c_2\sin x+c_3\cos x)$$

$$y=c_1e^x+c_2\sin x+c_3\cos x.$$

Example: Solve $(D^2+D-12)y=0, where\, y(2)=2, y'(2)=0$.

Solution. The characteristic equation is $D^2+D-12=0$ with roots $D=3,-4$.

Hence the general solution of the given equation is $y=c_1e^{3x}+c_2e^{-4x}$ (1)

We now determine the co-efficient (constants) c_1,c_2 in (1) from the initial conditions as follows. Since the initial conditions are given at $x=2$, we rewrite the general solution in the form

$$y=k_1e^{3(x-2)}+k_2e^{-4(x-2)} \qquad (2)$$

where $k_1=c_1e^6, k_2=c_2e^{-8}$.

Applying initial condition, (i.e., replacing c_1,c_2 by the new constants k_1e^6, k_2e^{-8}, respectively in (1), we get

$$y(2)=2=k_1+k_2 \qquad (3)$$

Differentiating (2) $w.r.t.x$, we have

$$y' = \frac{dy}{dx} = 3k_1 e^{3(x-2)} - 4k_2 e^{-4(x-2)}$$

Therefore, $y'(2) = 0 = 3k_1 - 4k_2$ (4)

Solving (3) and (4), we obtain $k_1 = \frac{8}{7}, k_2 = \frac{6}{7}.$

Hence the solution (2) of the differential equation satisfying the given conditions is

$$y = \frac{8}{7} e^{3(x-2)} + \frac{6}{7} e^{-4(x-2)}.$$

Example. Solve $(D^2 + 4D + 5) = 0, y(0) = 1, y'(0) = 0.$

Solution. The characteristic equation is $D^2 + 4D + 5 = 0$ which has roots $D = -2 \pm i.$

The general solution is $y = e^{-2x}(c_1 \sin x + c_2 \cos x).$ (1)

Applying the given conditions, we have from (1)

$$y(0) = 1 = c_2$$

Differentiating (1) $w.r.t.x,$ we get

$$y' = -2e^{-2x}(c_1 \sin x + c_2 \cos x) + e^{-2x}(c_1 \cos x - c_2 \sin x)$$

Therefore, $y'(0) = 0 = -2c_2 + c_1,$ giving $c_1 = 2.$

Substituting the values of c_1 and c_2 into (1), the required solution is

$$y = e^{-2x}(2\sin x + \cos x).$$

Example. Solve $(D^3 - 3D^2 + 4)y = 0, y(0) = 1, y'(0) = -8, y''(0) = -4.$

Solution. The characteristic equation is $D^3 - 3D^2 + 4 = 0.$

$$(D+1)(D^2 - 4D + 4) = 0.$$

Therefore, $D = -1, 2, 2$

The general solution is $y = c_1 e^{-x} + e^{2x}(c_2 + c_3 x)$ (1)

Now, from (1) $y' = -c_1 e^{-x} + 2e^{2x}(c_2 + c_3 x) + c_3 e^{2x}$

$$y'' = c_1 e^{-x} + 4e^{2x}(c_2 + c_3 x) + 4c_3 e^{2x}$$

Applying initial conditions, we get from the above three equations

$y(0) = 1 = c_1 + c_2$ (2)

$y'(0) = -8 = -c_1 + 2c_2 + c_3$ (3)

$y''(0) = -4 = c_1 + 4c_2 + 4c_3$ (4)

Multiplying (3) by –4 and adding to (4), we have

$28 = 5c_1 - 4c_2$ (5)

Multiplying (2) by 4 and adding to (5), we obtain

$$9c_1 = 32 \, or \, c_1 = \frac{32}{9}.$$

Therefore, $c_2 = 1 - \frac{32}{9} = -\frac{23}{9} \, and \, c_3 = \frac{6}{9}.$

The required solution is

$$y = \frac{32}{9} e^{-x} + e^{2x}\left(-\frac{23}{9} + \frac{6}{9}x\right) = \frac{1}{9}\left(32e^{-x} - 23e^{2x} + 6xe^{2x}\right).$$

9.4. DIFFERENTIAL OPERATORS

Let $T: V \to V$, where V is a vector space over a field F, be a linear transformation (or an operator) from $V \, to V$.

Linear transformation S, T, U defined on V have the following properties.

i. $S + T = T + S$ where $(S+T)v = S(v) + T(v), v \in V$,

ii. $(S+T)+U = S+(T+U)$

iii. $S(T+U) = ST + SU$

iv. $(ST)U = S(TU)$

v. $ST \neq TS$ in general.

Here $ST : V \to V$ is a linear transformation defined by:

$(ST)(v) = S(Tv)$ for all $v \in V$.

Let X be the vector space of all real (or complex) valued real (or complex) functions possessing a kth order derivative for every $k = 1, 2, 3, \ldots$ For each $g \in X, y = g(x) \in R(or \in C).D^k$ which associates with each $y = g(x)$, its kth derivative, is a linear operator on X. We write $D^k y \, as \dfrac{d^k y}{d^k x}$. Since the sum and product of the linear operator and scalar multiple of a linear operator are also linear operators, the linear combination

$$f(D) = a_0 D^n + a_1 D^{n-1} + \ldots + a_{n-1} D + a_n I \qquad (1)$$

Of $D^n, D^{n-1}, \ldots, D, I$ is also a linear operator. Here, I is the identity linear operator defined by $I(y) = y \; for \; all \; y = g(x)$ and $a_0, a_1, \ldots, a_n$ are scalars. The image of a $y = g(x)$ under $f(D)$ is written as

$$f(D)y = a_0 D^n y + a_1 D^{n-1} y + \ldots + a_{n-1} Dy + a_n y$$

$$= a_0 \frac{d^n y}{dx^n} + a_1 \frac{d^{n-1} y}{dx^{n-1}} + \ldots + a_{n-1} \frac{dy}{dx} + a_n y.$$

The equation $f(D)y = 0.$ is then a linear differential equation of order n.

The operator $D, D^2, \ldots, D^n \; and \; f(D),$ given by (1), are called differential operators. We have written $D \; for \dfrac{d}{dx}, D^2 \; for \dfrac{d^2}{dx^2}$ and so on $D^n \; for \dfrac{d^n}{dx^n}$.

Thus, $Dy = \dfrac{dy}{dx}, D^2 y = \dfrac{d^2 y}{dx^2} \; and \; D^n y = \dfrac{d^n y}{dx^n}$

The expression $A = f(D) = a_0 D^n + a_1 D^{n-1} + \ldots + a_{n-1} D + a_n I$ is called a differential operator of order n. This may be thought of as an operator which, when applied to any function y, result in

$$Ay = f(D)y = a_0\frac{d^n y}{dx^n} + a_1\frac{d^n - 1y}{dx^{n-1}} + \ldots + \frac{a_{n-1}dy}{dx} + a_n Iy.$$

9.5. NONHOMOGENEOUS LINEAR EQUATIONS

In this section we discuss the solution of nonhomogeneous linear equation $a_0 y^n + a_1 y^{n-1} + \ldots + a_{n-1}y' + a_n y = F(x)$.

9.5.1. Solution of the Equation

$$\boldsymbol{f(D)y = F(x)} \qquad (1)$$

where $f(D) = a_0 D^n + a_1 D^{n-1} + \ldots + a_{n-1}D + a_n$.

We have already discussed that the general solution of (1) consist of two parts, namely,

i. Complementary Function (C.F);
ii. Particular Solution (P.I).

The C.F is the solution of the homogeneous equation $f(D)y = 0$ and we have described different cases of its solution. Now to find the $P.I$ of (1), we can write

$$y = \frac{1}{f(D)}F(x)$$

and try to evaluate $y = \frac{1}{f(D)}F(x)$.

A real number m is said to be a zero of $f(D)$ if $f(m) = 0$. Suppose $f(D)$ has n distinct zeros $m_1, m_2, \ldots, m_n$. Then

$$\frac{1}{f(D)}F(x) = \frac{1}{(D - m_1)(D - m_2)\ldots(D - m_n)}F(x).$$

Suppose $\frac{1}{D - m_n}F(x) = y^*$, then $(D - m_n)y^* = F(x)$ or $\frac{dy}{dx} - m_n y^* = F(x)$ which is a linear equation of the first order. Its

$$I.F=\exp\left(\int(-m_n)\right)dx=\exp(-m_n x).$$

Hence $\frac{d}{dx}\left(ye^{-m_n x}\right)=F(x)e^{-m_n x}$

$$y^*=e^{m_n x}\int F(x)e^{-m_n x}dx,$$

Omitting the constant of integration, since we are looking for a particular solution, Therefore

$$\frac{1}{f(D)}F(x)=\frac{1}{(D-m_1)\ldots(D-m_{n-1})}e^{m_n x}\int e^{-m_n x}F(x)dx.$$

Next, we evaluate $\frac{1}{D-m_{n-1}}e^{m_n x}\int e^{-m_n x}F(x)dx$ as before and continue the process. This is the required P.I of the equation (1).

Alternatively, we may also write

$$\frac{1}{f(D)}F(x)=\frac{1}{(D-m_1)(D-m_2)\ldots(D-m_n)}F(x)$$

$$=\left[\frac{A_1}{D-m_1}+\frac{A_2}{D-m_2}+\ldots+\frac{A_n}{D-m_n}\right]F(x)$$

after resolving $\frac{1}{f(D)}$ into partial fractions

$$=A_1e^{m_1 x}\int e^{-m_1 x}F(x)dx+A_2e^{m_2 x}\int e^{-m_2 x}F(x)dx+\ldots+A_ne^{m_n x}\int e^{-m_n x}F(x)dx$$

which is the required P.I of the equation (1).

Example: Solve $(D^3-D)y=e^x$.

Solution. The characteristic equation is $D^3-D=0$ with roots $0,1,-1$

$$C.F=y_c=c_1+c_2e^x+c_3e^{-x}.$$

P.I = $y_p = \dfrac{e^x}{D(D+1)(D-1)}$

Now $\dfrac{1}{D-1}e^x = e^x \int e^{-x}.e^x dx = xe^x,$ by the method already described. Thus

$$\frac{1}{D(D+1)(D-1)}e^x = \frac{1}{D(D+1)}xe^x$$

Again,

$$\frac{1}{D+1}\left(xe^x\right) = e^{-x}\int e^x.xe^x dx = \frac{xe^x}{2} - \frac{e^x}{4}$$

Hence

$$\frac{1}{D(D+1)}\left(xe^x\right) = \frac{1}{D}\left(\frac{xe^x}{2} - \frac{e^x}{4}\right) = \int \frac{xe^x}{2}dx - \int \frac{e^x}{4}dx$$

$$= \frac{xe^x}{2} - \frac{3}{4}e^x$$

Therefore, the general solution of the equation is

$$y = y_c + y_p = c_1 + c_2e^x + c_3e^{-x} + \frac{xe^x}{2} - \frac{3}{4}e^x.$$

$$= c_1 + \left(c_2 - \frac{3}{4}\right)e^x + c_3e^{-x} + \frac{xe^x}{2}$$

$$= c_1 + c_2'e^x + c_3e^{-x} + \frac{xe^x}{2}$$

9.5.2. Alternative Method

Consider $f(D)^{-1}$ as the inverse of the operator $f(D)$, so that

$$y_p = \frac{1}{D(D+1)(D-1)}e^x$$

$$=\left(-\frac{1}{D}+\frac{1}{2}.\frac{1}{1+D}+\frac{1}{2}.\frac{1}{D-1}\right)e^{x}.$$

$$=-\int e^{x}dx+\frac{1}{2}e^{-x}\int e^{x}.e^{x}\,dx+\frac{1}{2}e^{x}\int e^{-x}.e^{x}\,dx$$

$$=-e^{x}+\frac{1}{4}e^{x}+\frac{1}{2}xe^{x}$$

$=\frac{xe^{x}}{2}$, since e^{x} already occurs in the C.F and so the terms involving e^{x} are omitted.

9.6. WORKING RULES FOR FINDING P.I

The method of evaluating $\frac{1}{f(D)}F(x)$ involves successive integrations and is quite unwieldy. We list below short methods for finding the P.I. when $F(x)$ is a function of a particular type. To evaluate $\frac{1}{f(D)}e^{ax}$, where $f(D)=a_0D^n+a_1D^{n-1}+\ldots+a_{n-1}D+a_n$. We know that $D^k e^{ax}=a^k e^{ax}$, where k is a positive integer. Hence

$$f(D)e^{ax}=\left(a_0D^n+a_1D^{n-1}+\ldots+a_{n-1}D+a_n\right)e^{ax}$$

$$=a_0.D^n e^{ax}+a_1.D^{n-1}e^{ax}+\ldots+a_{n-1}.De^{ax}+a_n.e^{ax}$$

$$=a_0.a^n e^{ax}+a_1.a^{n-1}e^{ax}+\ldots+a_{n-1}.a\,e^{ax}+a_n.e^{ax}.$$

$-f(a)e^{ax}.$ (1)

Applying $\frac{1}{f(D)}$ to both sides of (1), we obtain

$$\frac{f(D)}{f(D)}e^{ax}=\frac{1}{f(D)}f(a)e^{ax}$$

$$e^{ax} = f(a).\frac{1}{f(D)}e^{ax}$$

i.e., $\frac{1}{f(D)}e^{ax} = \frac{e^{ax}}{f(a)}, if\ f(a) \neq 0$

if $f(a)=0,$ then a is a root of $f(D)=0.$

Let a be a k-fold root of $f(D)=0.$ so that $(D-a)^k$ is a factor of $f(D).$ Then

$$f(D)=(D-a)^k \varnothing(D), where \varnothing(a) \neq 0.$$

We first check the effect of $(D-a)^k \, on e^{ax} p(x),$ for a polynomial $p(x)$ in x. We have $(D-a)(e^{ax} p(x))=D(e^{ax} p(x))-a e^{ax} p(x)=e^{ax} Dp(x)$.

Continuing in this way, we get

$(D-a)^k (e^{Ax} p(x))=e^{ax} D^k p(x).$ (3)

Setting $p(x)=x^k$ in (3), we are led to

$$(D-a)^k (x^k e^{ax})=e^{ax} D^k x^k = k!e^{ax}$$

$$\varnothing(D)(D-a)^k n(x^k e^{ax})=\varnothing(D)k!e^{ax}$$

$=k!\varnothing(D)e^{ax} = k!\varnothing(a)e^{ax},$ by (1)

Operating on both sides by $\frac{1}{\varnothing(D)(D-a)^k}, we\, obtain$

$$x^k e^{ax} = \frac{1}{\varnothing(D)(D-a)^k} k!\varnothing(a)e^{ax}$$

$\frac{1}{\varnothing(D)(D-a)^k} e^{ax} = \frac{x^k e^{ax}}{k!\varnothing(a)}$ (4)

9.7. PRINCIPLE OF SUPERPOSITION

$$f(D)y = \left(a_0 D^n + a_1 D^{n-1} + \ldots + a_{n-1} D + a_n\right) y = F(x) \qquad (1)$$

Be a linear differential equation of order n. if $F(x) = F_1(x) + F_2(x)$, then particular integral of (1) is the sum of particular integrals of

$$F(D)y = F_1(x) \qquad (2)$$

and $F(D)y \quad F(x)$ (3)

Proof: Let y_1 *and* y_2 be particular integrals of (2) and (3), respectively. Then

$$f(D)y_1 = F_1(x)$$

and $f(D)y_2 = F_2(x)$.

Suppose $y = y_1 + y_2$. Setting this value of y into (1), we obtain

$$f(D)y = f(D)(y_1 + y_2)$$

$$= f(D)y_1 + f(D)y_2$$

$$= F_1(x) + F_2(x) = F(x)$$

Showing that y is a solution of (1).

Example: Solve $\left(D^3 + 1\right)y = 1 + e^{-x} + e^{2x}$.

Solution. The characteristic equation is

$D^3 + 1 = 0$ with roots $-1, \dfrac{1 \pm i\sqrt{3}}{2}$

Therefore, $C.F$ is

$$y_c = c_1 e^{-x} + e^{\frac{x}{2}} \left(c_2 \sin \frac{\sqrt{3}}{2} x + c_3 \cos \frac{\sqrt{3}}{2} x \right).$$

The P/I is given by

$$y_p = \frac{1}{D^3 + 1}\left(1 + e^{-x} + e^{2x}\right)$$

$$= \frac{1}{D^3+1}e^{0x} + \frac{1}{D^3+1}e^{2x} + \frac{1}{(D+1)(D^2-D+1)}e^{-x}$$

$$= 1 + \frac{e^{2x}}{9} + \frac{1}{3(D+1)}e^{-x}, \quad \text{using } \frac{1}{f(D)}e^{ax} = \frac{1}{f(a)}e^{ax},$$

for $\frac{1}{D^2-D+1}e^{-x}, with\, D = -1\, when\, f(-1) \neq 0$

$$= 1 + \frac{e^{2x}}{9} + \frac{1}{3}xe^{-x},$$

Hence the general solution of the equation is

$$y = y_c + y_p$$

$$= c_1 e^{-x} + e^{\frac{x}{2}}\left(c_2 \sin\frac{\sqrt{3}}{2}x + c_3 \cos\frac{\sqrt{3}}{2}x \right) + 1 + \frac{e^{2x}}{9} + \frac{1}{3}x.e^{-x}.$$

To Find the P.I when $F(x) = \sin ax \ or \cos ax.$

Here we have to evaluate $\frac{1}{f(D)}\sin ax\, and\, \frac{1}{f(D)}\cos ax.$

From Euler's theorem, we have $e^{iax} = \cos ax + i\sin ax$. Thus $\frac{1}{f(D)}\sin ax\, and\, \frac{1}{f(D)}\cos ax$ are, respectively imaginary and real parts of $\frac{1}{f(D)}e^{iax}.$

If $f(D)$ contains only even powers of $D, say\, f(D) = f(D^2),$ it is easy to see that

$$\frac{1}{f(D^2)}\begin{Bmatrix} \sin ax \\ \cos ax \end{Bmatrix} = \frac{1}{f(-a^2)}\begin{Bmatrix} \sin ax \\ \cos ax \end{Bmatrix}, \ provided - a^2\, is\, not\, a\, zero\, of\, f(D^2).$$

Example: Solve $(D^2 - 5D + 6)y = \sin 3x$

Solution: The roots of the characteristic equation

$D^2-5D+6=0$ are $2\, and\, 3$.

$$C.F=y_c=c_1e^{2x}+c_2e^{3x}$$

$$P.I=y_p=\frac{1}{D^2-5D+6}\sin 3x$$

which is imaginary part of $\frac{1}{(D-2)(D-3)}e^{3ix}$ .Now

$$\frac{1}{(D-2)(D-3)}e^{3ix}=\frac{e^{3ix}}{(3i-2)(3i-3)},$$

$$=\frac{-1+5i}{78}e^{3ix}$$

$$=\left(-\frac{1}{78}+\frac{5}{78}i\right)(\cos 3x+i\,\sin 3x)$$

Its imaginary part is $-\frac{1}{78}\sin 3x+\frac{5}{78}\cos 3x=y_p$. Hence the required general solution is

$$y=c_1e^{2x}+c_2e^{3x}-\frac{1}{78}\sin 3x+\frac{5}{78}\cos 3x$$

Theorem. If a is not zero of $f(D)$, then $\frac{1}{f(D-a)}e^{ax}F(x)=e^{ax}\frac{1}{f(D)}F(x).$ This replacing of $D\, by\, D+a$ is known as exponential shift.

Proof. We have already shown in previous section that $(D-a)^k.(e^{ax}p(x))=e^{ax}D^k p(x)$

Using linearity of differential operators, we conclude that, when $f(D)$ is a polynomial (with constant coefficients), then

$$f(D-a)(e^{ax}p(x))\quad e^{ax}f(D)p(x) \qquad (1)$$

Suppose $f(D)p(x)=F(x)$. Then

$$p(x)=\frac{1}{f(D)}F(x) \qquad (2)$$

From (1) and (2), we have

$$f(D-a)e^{ax}\frac{1}{f(D)}F(x)=e^{ax}F(x)$$

Operating on both sides by $\frac{1}{f(D-a)}$, we get

$$e^{ax}\frac{1}{f(D)}F(x)=\frac{1}{f(D-a)}e^{ax}F(x)$$

$\frac{1}{f(D-a)}e^{ax}F(x)=e^{ax}\frac{1}{f(D)}F(x),$ provided that a is not a zero of $f(D)$.

Example: Solve $\left(D^3+D^2-4D-4\right)y=e^{2x}\cos 3x.$

Solution. The characteristics equation is $D^3+D^2-4D-4=0.$

By inspection $D=2$ is a root of this equation. The other roots are $D=-2,-1$

$$C.F.=y_c=c_1e^{-x}+c_2e^{-2x}+c_3e^{2x}$$

$$P.I=y_p=\frac{1}{(D-2)(D+2)(D+1)}e^{2x}\cos 3x$$

$$=e^{2x}\frac{1}{D(D+4)(D+3)}\cos 3x, \textit{by the exponential shift}$$

$$=e^{2x}\frac{1}{\left(D^3+7D^2+12D\right)}\cos 3x$$

$$=e^{2x}\frac{1}{-9D-63+12D}\cos 3x,\ \textit{putting}\ D^2=-3^2$$

$$=e^{2x}\frac{1}{3(D-21)}\cos 3x$$

$$= e^{2x} \frac{(D+21)}{3(D^2 - 441)} \cos 3x$$

$$= e^{2x} . \frac{-1}{3 \times 450} [-3 \sin 3x + 21 \cos 3x]$$

$$= \frac{e^{2x}}{450} (\sin 3x - 7 \cos 3x)$$

Alternative Method. Here, for $\frac{1}{D(D+4)(D+3)} \cos 3x,$ we proceed as follow:

$$\frac{\cos 3x}{(D^3 + 7D^2 + 12D)} = Re \frac{1}{(D^3 + 7D^2 + 12D)} e^{3ix},$$

$$= Re \frac{1}{(-27i - 63 + 36i)}$$

$$= Re \frac{e^{3ix}}{9(i-7)}$$

$$= Re \frac{(i+7) e^{3ix}}{-450}$$

$$= Re \frac{(i+7)(\cos 3x + i \sin 3x)}{-450}$$

$$= -\frac{7}{450} \cos 3x + \frac{1}{450} \sin 3x.$$

Therefore, $y_p = \frac{e^{2x}}{450} (\sin 3x - 7 \cos 3x)$ as before.

The general solution is $y = y_c + y_p$

$$= c_1 e^{-x} + c_2 e^{-2x} + c_3 e^{2x} - \frac{7}{450} \cos 3x + \frac{1}{450} \sin 3x.$$

$$= c_1 e^{-x} + c_2 e^{-2x} + c_3 e^{2x} + \frac{e^{2x}}{450}\left(\sin 3x - 7\cos 3x\right).$$

9.8. SERIES EXPANSION OF POLYNOMIALS OR EXPRESSIONS INVOLVING OPERATORS

This method is useful when $F(x)$ is a polynomial in x.

Let $f(D)y = F(x)$ be such that $F(x)$ is a polynomial in x. To evaluate the particular integral

$$y_p = \frac{1}{f(D)} F(x)$$

It is often useful to expand $\frac{1}{f(D)}$ in a series in D by the Binomial Theorem for negative exponent so that

$$y_p = \left(1 + a_1 D + a_2 D^2 + a_3 D^3 + \ldots\right) F(x)$$

The derivatives on the right will vanish after certain stage since

$$D^n x^r = 0 \; if \; n > r.$$

The following binomial expansions will be useful in this connection:

i. $\frac{1}{1-x} = 1 + x + x^2 + x^3 + \ldots.$

ii. $\frac{1}{1+x} = 1 - x + x^2 - x^3 + \ldots.$

Example: Solve $\left(D^3 - 2D + 1\right) y = 2x^3 - 3x^2 + 4x + 5$

Solution. The characteristic equation is $D^3 - 2D + 1 = 0$

$$\left(D - 1\right)\left(D^2 + D - 1\right) = 0$$

$$D = 1, \frac{-1+\sqrt{5}}{2}, \frac{-1-\sqrt{5}}{2}$$

$$y_c = c_1 e^x + c_2 e^{\frac{-1+\sqrt{5}}{2}x} + c_3 e^{\frac{-1-\sqrt{5}}{2}x}$$

Next,

$$y_p = \frac{1}{1-2D+D^3}\left(2x^3 - 3x^2 + 4x + 5\right)$$

Now

$$\frac{1}{1-2D+D^3} = \left[1-\left(2D-D^3\right)\right]^{-1}$$

$$= 1 + \left(2D - D^3\right) + \left(2D - D^3\right)^2 + \left(2D - D^3\right)^3 + \ldots$$

$$= 1 + 2D - D^3 + 4D^2 + 8D^3$$

Neglecting D^4 and higher powers of D in view of the degree 3 of the polynomial $p(x)$ in x therefore

$$y_p = \left(1 + 2D + 4D^2 + 7D^3\right)\left(2x^3 - 3x^2 + 4x + 5\right)$$

$$= 2x^3 - 3x^2 + 4x + 5 + 2\left(6x^2 - 6x + 4\right) + 4\left(12x - 6\right) + 7\left(12\right)$$

$$= 2x^3 + 9x^2 + 40x + 73.$$

The general solution is $y = y_c + y_p$

$$= c_1 e^x + c_2 e^{\frac{-1+\sqrt{5}}{2}x} + c_3 e^{\frac{-1-\sqrt{5}}{2}x} + 2x^3 + 9x^2 + 40x + 73.$$

9.9. THE METHOD OF UNDETERMINED COEFFICIENTS (U. C. METHOD)

We have studied some special cases in which particular integral can be evaluated by the inverse operator. Now we consider the method of undetermined coefficients which can prove simpler in finding the particular integral of the equation $f(D)y = F(x)$ when $F(x)$ is

a. An exponential function: e^{ax}

b. A polynomial: $\left(b_0 x^n + b_1 x^{n-1} + \ldots + b_n\right)$

c. Sinusoidal function: $\sin ax \; or \; \cos bx$

d. The more general case in which $F(x)$ is sum of a product of terms of the above types, such as

$$F(x)=e^{ax}\left(b_0x^n+b_1x^{n-1}+\ldots+b_n\right)\begin{Bmatrix}\sin ax\\ \cos bx\end{Bmatrix}$$

The P.I., i.e., y_p will be constructed according to the following table:

$F(x)$ *is of the form*	*Take* y_p *as*
$1.a$	$A.x^k$
$2.ax^n\,(n\,is\,a+ve\,integer)$	$x^k\left(A_0x^n+A_1x^{n-1}+\ldots+A_n\right)$
$3.ax^ne^{rx}\,(n\,is\,a+ve\,integer)$	$x^k\left(A_0x^n+A_1x^{n-1}+\ldots+A_{n-1}x+A_n\right)e^{rx}$
$4.Cx^n\cos ax$	$x^k\left(A\cos ax+B\sin ax\right)$
$5.Cx^n\sin ax$	
$6.C.x^n.e^{rx}\cos ax$	$x^k\,[\left(A_0x^n+A_1x^{n-1}+\ldots+A_{n-1}x+A_n\right)e^{rx}\cos ax\,+$
$7.Cx^ne^{rx}\sin ax$	$\left(B_0x^n+B_1x^{n-1}+\ldots+B_{n-1}x+B_n\right)e^{rx}\sin ax$

In x^k, k is the smallest nonnegative integer, which will ensure that no terms in y_p is already in the C.F. If $F(x)$ is sum of several terms, write y_p for each term individually and then add up all of them. The y_p and its derivatives will be substituted into the equation $f(D)y=F(x)$ and coefficients of like terms on the left hand and right-hand sides will be equated to determine the $U.C.A_0, A_1, \ldots, A_n, B_0, B_1, \ldots, B_n.$ The method is illustrated by examples as follows.

Example: Solve $y''-3y'+2y=2x^3-9x^2+6x$ (1)

Solution: C.F. is easily found as $y_c = c_1 e^x + c_2 e^{2x}$

For a particular solution, we assume

$$y_p = x^k \left(Ax^3 + Bx^2 + Cx + D \right).$$

Since no term of the C.F. is present in y_p, we take $k = 0$ and so

$$y_p = Ax^3 + Bx^2 + Cx + D$$

$$y_p' = 3Ax^2 + 2Bx + C$$

$$y_p'' = 6Ax + 2B$$

Substituting for y_p, y_p' *and* y_p'' into (1), we have

$$\left(6Ax + 2B\right) - 3\left(3Ax^2 + 2Bx + C\right) + 2\left(Ax^3 + Bx^2 + Cx + D\right) = 2x^3 - 9x^2 + 6x$$

$$2Ax^3 + x^2\left(2B - 9A\right) + x\left(6A - 6B + 2C\right) + 2B - 3C + 2D = 2x^3 - 9x^2 + 6x.$$

Equating coefficients of like terms, we obtain

Coeff. of $x^3 : 2A = 2 \, or \, A = 1.$

Coeff. of $x^2 : 2B - 9A = -9 \, or \, 2B = 9A - 9 \, giving \, B = 0$

Coeff. of $x : 6A - 6B + 2c = 6 \, or \, C = 0.$

Coeff. of $x^0 : 2B - 3C + 2D = 0 \, or \, D = 0.$

So $y_p = x^3$

The required general solution is $y = c_1 e^x \mid c_2 e^{2x} + x^3.$

Example: Solve $y'' \text{Ñü} y' \quad y \quad x^2 e^x$ (1)

Solution. The C.F. is $y_c = c_1 e^x + c_2 e^{2x}.$

To construct a particular solution, we use (3) of the table

$$y_p = x^k \left(Ax^2 + Bx + C \right) e^x.$$

Since e^x is already in $C.F.$, we take $k=1$ so that no terms of C.F. is in y_p. Hence the modified P.I. is

$$y_p = \left(Ax^3 + Bx^2 + Cx\right)e^x$$

$$y_p' = \left(Ax^3 + Bx^2 + Cx\right)e^x + \left(3Ax^2 + 2Bx + C\right)e^x$$

$$y_p'' = \left(Ax^3 + Bx^2 + Cx\right)e^x + 2\left(3Ax^2 + 2Bx + C\right)e^x + \left(6Ax + 2B\right)e^x.$$

Substituting for y_p, y_p' *and* y_p'' into (1), we have

$$\left(Ax^3 + Bx^2 + Cx\right)e^x + 2\left(3Ax^2 + 2Bx + C\right)e^x + \left(6Ax + 2B\right)e^x + \left(6Ax + 2B\right)e^x$$

$$-3\left(Ax^3 + Bx^2 + Cx\right)e^x - 3\left(3Ax^2 + 2Bx + C\right)e^x + 2\left(Ax^3 + Bx^2 + Cx\right)e^x = x^3e^x$$

$$\left[-3Ax^2 + \left(6A - 2B\right)x + C\right]e^x = x^3e^x$$

Equating coefficients of like terms, we obtain

$$Co-eff.of\ x^3 : -3A = 1 \, or \, A = -\frac{1}{3}$$

$$Co-eff.of\ x : 6A - 2B = 0 \, or \, B = -1$$

$$Co-eff.of\ x^0 : C = 0$$

So $y \; = - - x^3e^x - x^2e^x.$

The required solution is

$$y = c_1e^x + c_2e^{2x} - \frac{1}{3}x^3e^x - x^2e^x.$$

Example: Solve $y'' + 4y = xe^x + x\sin 2x$ (1)

Solution. The C.F., as readily found above, is $y_c = c_1 \sin 2x + c_2 \cos 2x.$

For the P.I of (1), we find the P.I. of $y'' + 4y = xe^x$ (2)

and $y'' + 4y = x\sin 2x$ (3)

Separately. Their sum will be the P.I of (1) (by the Principle of superposition). For a particular solution of (2), we have

$$y_p = x^k\left(Ax+B\right)e^x.$$

Since no term of C.F. is in the y_p, we must take $k=0$. Therefore,

$$y_p = \left(Ax+B\right)e^x$$

$$y_p' = \left(Ax+B\right)e^x + Ae^x$$

$$y_p'' = \left(Ax+B\right)e^x + 2Ae^x$$

Substituting for y_p, y_p' and y_p'' into (2), we have

$$\left(Ax+B\right)e^x + 2Ae^x + 4\left(Ax+B\right)e^x = xe^x$$

$$5Axe^x + \left(2A+5B\right)e^x = xe^x.$$

Equating coefficients of like terms, we obtain

$$Co-efficient\ of\ xe^x : 5A = 1\ or\ A = \frac{1}{5}$$

$$Coefficient\ of\ e^x : 2A+5B = 0\ or\ B = -\frac{2}{25}$$

So, P.I of (2) is $y_p = \frac{1}{5}xe^x - \frac{2}{25}e^x$.

$$y_p = x^k\left[\left(Cx+D\right)\sin 2x + \left(Ex+F\right)\cos 2x\right].$$

Since $\sin 2x$ and $\cos 2x$ are already in the C.F., we must take $k=1$, so that the P.I. is

$$y_p = \left(Cx^2+Dx\right)\sin 2x + \left(Ex^2+Fx\right)\cos 2x$$

$$y_p' = \left(Cx^2+Dx\right)\left(2\cos 2x\right) + \left(2Cx+D\right)\sin 2x$$

$$+\left(Ex^2+Fx\right)\left(-2\sin 2x\right) + \left(2E\left(x\right)+F\right)\cos 2x$$

$$y_p'' = (Cx^2 + Dx)(-4\sin 2x) + (2Cx + D)(2\cos 2x) + (2Cx + D)(2\cos 2x)$$

$$+(2Cx + D)(2\cos 2x) + 2C\sin 2x + (Ex^2 + Fx)(-4\cos 2x)$$

$$+(2Ex + F)(-2\sin 2x) + (2Ex + F)(-2\sin 2x) + 2E\cos 2x.$$

Substituting into (3), we have

$$(Cx^2 + Dx)(-4\sin 2x) + 4(2Cx + D)\cos 2x + 2C\sin 2x + (Ex + Fx)(-4\cos 2x)$$

$$-4(2Ex + F)\sin 2x + 2E\cos 2x + 4(Cx^2 + Dx)\sin 2x + 4(Ex^2 + Fx)\cos 2x = x\sin 2x.$$

Equating the coefficients of like terms, we get

Coeff. of $x\sin 2x : -8E = 1 \, or \, E = -\frac{1}{8}$

Coeff. of $x\cos 2x : 8C = 0 \, or \, C = 0$

Coeff. of $\sin 2x : 2C - 4F = 0 \, or \, F = 0$

Coeff. of $\cos 2x : 2E + 4D = 0 \, or \, D = \frac{1}{16}.$

Hence P.I of (3) is

$$y_p = \frac{1}{16}x\sin 2x - \frac{1}{8}x^2\cos 2x \qquad (5)$$

Sum of (4) and (5) is the required P.I of (I), i.e., $y_p = \frac{1}{5}xe^x - \frac{2}{25}e^x + \frac{1}{16}x\sin 2x - \frac{1}{8}x^2\cos 2x$ is the particular solution of (1).

Hence the general solution of (1) is

$$y = c_1\sin 2x + c_2\cos 2x + \frac{1}{5}xe^x - \frac{2}{25}e^x + \frac{1}{16}x\sin 2x - \frac{1}{8}x^2\cos 2x.$$

9.10. THE CAUCHY-EULER EQUATION

An equation of the form

$$a_0 x^n \frac{d^n y}{dx^n} + a_1 x^{n-1} \frac{d^{n-1} y}{dx^{n-1}} + \ldots + a_{n-1} x \frac{dy}{dx} + a_n y = F(x)$$

is called Cauchy-Euler (or equidimensional) equation. $a_0, a_1, \ldots, a_{n-1}, a_n$ are real constants. The equation can be reduced to a linear differential equation with constant coefficients by the transformation $x = e^t \; or \, t = \ln x$. Then

$$\frac{dy}{dx} = \frac{dy}{dt} \cdot \frac{dt}{dx} = \frac{1}{x} \frac{dy}{dt}$$

$$\frac{d^2 y}{dx^2} = \frac{d}{dx}\left(\frac{1}{x}\frac{dy}{dt}\right) = \frac{1}{x}\frac{d}{dx}\left(\frac{dy}{dt}\right) + \frac{dy}{dt} \cdot \frac{d}{dx}\left(\frac{1}{x}\right)$$

$$= \frac{1}{x}\frac{d}{dt}\left(\frac{dy}{dt}\right) \cdot \frac{dt}{dx} + \frac{dy}{dt}\left(-\frac{1}{x^2}\right) = \frac{1}{x^2}\left(\frac{d^2 y}{dt^2} - \frac{dy}{dt}\right)$$

$$\frac{d^3 y}{dx^3} = \frac{d}{dx}\left[\frac{1}{x^2}\frac{d^2 y}{dt^2} - \frac{1}{x^2}\frac{dy}{dt}\right]$$

$$= \frac{1}{x^2}\frac{d}{dx}\left(\frac{d^2 y}{dt^2}\right) + \frac{d^2 y}{dt^2} \cdot \frac{d}{dx}\left(\frac{1}{x^2}\right) - \frac{1}{x^2}\frac{d}{dx}\left(\frac{dy}{dt}\right) + \frac{dy}{dt} \cdot \frac{d}{dx\left(-\frac{1}{x^2}\right)}$$

$$= \frac{1}{x^2}\frac{d}{dt}\left(\frac{d^2 y}{dt^2}\right) \cdot \frac{dt}{dx} - \frac{2}{x^3}\frac{d^2 y}{dt^2} - \frac{1}{x^2}\frac{d}{dt}\left(\frac{dy}{dt}\right) \cdot \frac{dt}{dx} + \frac{2}{x^3}\frac{dy}{dt}$$

$$= \frac{1}{x^3}\frac{d^3 y}{dt^3} - \frac{3}{x^3}\frac{d^2 y}{dt^2} + \frac{2}{x^3}\frac{dy}{dt}$$

$$= \frac{1}{x^3}\left(\frac{d^3 y}{dt^3} - 3\frac{d^2 y}{dt^2} + 2\frac{dy}{dt}\right)$$

If we write $D = \frac{d}{dx} \, and \, \Delta = \frac{d}{dt}, then$

$$xD = \Delta$$
$$x^2D^2 = \Delta^2 - \Delta = \Delta(\Delta-1)$$
$$x^3D^3 = \Delta^3 - 3\Delta^2 + 2\Delta = \Delta(\Delta-1)(\Delta-2)$$
$$\vdots$$
$$x^nD^n = \Delta(\Delta-1)(\Delta-2)\ldots(\Delta-n+1).$$

Substituting these values of $xD, x^2D^2, \ldots$ into (1), we obtain an equation of nth order with constant coefficients having t as the independent variable. This equation can be solved by the previous methods.

Example: Solve $x^2\dfrac{d^2y}{dx^2} - 2x\dfrac{dy}{dx} + 2y = x^3.$ (1)

Solution: Let $x = e^t$. Then we have $t = \ln x$

$$xD = \Delta.$$
$$x^2D^2 = \Delta(\Delta-1).$$

Substituting into (1), we get

$$\left[\Delta(\Delta-1) - 2\Delta + 2\right]y = e^{3t}$$

$$\left[\Delta^2 - 3\Delta + 2\right]y = e^{3t} \qquad (2)$$

The characteristic equation has roots $2 \, and \, 1$. Therefore, C.F. of (2) is

$$y_c = c_1e^t + c_2e^{2t}$$

P.I. of equation (2) is $y_p = \dfrac{1}{(\Delta-1)(\Delta-2)}e^{3t} = \dfrac{1}{2}e^{3t}$. The general solution of (2) is

$$y = c_1e^t + c_2e^{2t} + \frac{1}{2}e^{3t}.$$

Replacing t by $\ln x\left(or\, e^t by\, x\right)$, we have $y = c_1x + c_2x^2 + \dfrac{1}{2}x^3$ as the general solution of (1).

Example: Solve $x^4\dfrac{d^4y}{dx^4} + 6x^3\dfrac{d^3y}{dx^3} + 9x^2\dfrac{d^2y}{dx^2} + 3x\dfrac{dy}{dx} + y = (1+\ln x)^2.$
(1)

Solution: We let $x = e^t$ so that $t = \ln x$

$$xD = \Delta$$

$$x^2D^2 = \Delta(\Delta - 1) = \Delta^2 - \Delta.$$

$$x^3D^3 = \Delta(\Delta - 1)(\Delta - 2) = \Delta^3 - 3\Delta^2 + 2\Delta$$

$$x^4D^4 = \Delta(\Delta - 1)(\Delta - 2)(\Delta - 3) = \Delta^4 - 6\Delta^3 + 11\Delta^2 - 6\Delta.$$

Substituting into (1), we get

$$\left[\Delta^4 - 6\Delta^3 + 11\Delta^2 - 6\Delta + 6\left(\Delta^3 - 3\Delta^2 + 2\Delta\right) + 9\left(\Delta^2 - \Delta\right) + 3\Delta + 1\right]y = (1+t)^2$$

$$\left(\Delta^4 + 2\Delta^2 + 1\right)y = 1 + 2t + t^2 \qquad (2)$$

The roots of the characteristic equation of (2) are $\pm i, \pm i$. Therefore, C.F. of (2) is

$$y_c = \left(c_1 + c_2 t\right)\sin t + \left(c_3 + c_4 t\right)\cos t.$$

To find a particular integral of (2), we use the method of $U.C.$ Let the particular integral of (2) be

$$y_p = At^2 + Bt + C$$

$$y_p' = 2At + B$$

$$y_p'' = 2A.$$

Substituting into (2), we obtain

$$4A + At^2 + Bt + C = 1 + 2t + t^2$$

$$At^2 + Bt + \left(4A + C\right) = t^2 + 2t + 1.$$

Equating coefficients of like terms, we have

$$A = 1, B = 2, 4A + C = 1 \, or \, C = -3.$$

Therefore, a particular integral of (2) is

$$y_p = t^2 + 2t - 3,$$

Hence the general solution of (2) is

$$y = \left(c_1 + c_2 t\right)\sin t + \left(c^3 + c_4 t\right)\cos t + t^2 + 2t - 3 \qquad (3)$$

Replacing t by $\ln x$ in (3), we get

$$y = (c_1 + c_2 \ln x)\sin(\ln x) + (c_3 + c_4 \ln x)\cos(\ln x) + (\ln x)^2 + 2\ln x - 3$$

as the general solution of (1).

9.11. REDUCTION OF ORDER

If one solution of the second order linear equation $\frac{d^2y}{dx^2} + P\frac{dy}{dx} + Qy = 0$ (1)

(where P, Q are not necessarily constants and may be functions of x) is known, then we can use it to find the general solution of $\frac{d^2y}{dx^2} + P\frac{dy}{dx} + Qy = F(x)$
(2)

The procedure, due to D' Alembert, is known as the method reduction of order.

Suppose it is known that $y = y_1$ is a solution of (1). we assume that

$y = vy_1$ (3)

is a solution of (2), where v is a function of to be determined.

From (3), we get

$$\frac{dy}{dx} = v\frac{dy_1}{dx} + y_1\frac{dv}{dx},$$

$$\frac{d^2y}{dx^2} = v\frac{d^2y_1}{dx^2} + 2\frac{dv}{dx}\frac{dy_1}{dx} + y_1\frac{d^2v}{dx^2}$$

Substituting into (2), we have

$$v\frac{d^2y_1}{dx^2} + 2\frac{dv}{dx}\frac{dy_1}{dx} + y_1\frac{d^2v}{dx^2} + Pv\frac{dy_1}{dx} + Py_1\frac{dv}{dx} + Qvy_1 = F(x)$$

$$y_1\frac{d^2v}{dx^2} + \left(2\frac{dy_1}{dx} + Py_1\right)\frac{dv}{dx} + \left(\frac{d^2y_1}{dx^2} + P\frac{dy_1}{dx} + Qy_1\right)v = F(x) \quad (4)$$

Since $y = y_1$ is a solution of (1)

$$\frac{d^2y_1}{dx^2} + P\frac{dy_1}{dx} + Qy_1 = 0$$

and, therefore, equation (4) reduces to

$$y_1 \frac{d^2v}{dx^2} + \left(2\frac{dy_1}{dx} + Py_1 \right)\frac{dv}{dx} = F(x) \qquad (5)$$

Setting $\frac{dv}{dx} = u$ in (5), we obtain

$$y_1 \frac{du}{dx} + \left(2\frac{dy_1}{dx} + Py_1 \right)u = F(x)$$

which is a linear equation of the first order in u and can be solved for u.
From $\frac{dv}{dx} = u,$ we determine v and hence the solution $y = vy_1.$ The method is illustrated by means of example.

Example: Solve $\frac{d^2y}{dx^2} + y = \csc x.$ (1)

Solution. The C.F, of (1) is $y_c = c_1 \sin x + c_2 \cos x.$

We may take any special value of $y_c\ as\ y_1.$ Let us take y_1 to be the value of $y_c\ when\ c_1 = 1\ and\ c_2 = 0$. Then assume that $y = v\sin x$ (2)

is a solution of (1).
From (2), we get

$$\frac{dy}{dx} = \sin x \frac{dv}{dx} + v\cos x$$

$$\frac{d^2y}{dx^2} = \sin x \frac{d^2v}{dx^2} + 2\cos x \frac{dv}{dx} - v\sin x.$$

Substituting into (1), wc obtain

$$\sin x \frac{d^2v}{dx^2} + 2\cos x \frac{dv}{dx} - v\sin x + v\sin x = \csc x$$

$$\frac{d^2v}{dx^2} + 2\cot x \frac{dv}{dx} = \csc^2 x.$$

Setting $u = \frac{dv}{dx},$ the above equation becomes

$$\frac{du}{dx}+2\cot x\,u=\csc^2 x \qquad (3)$$

which is linear equation of first order. An integrating factor of (3) is

$$exp\left[\int 2\cot x dx\right]=\exp\left[\ln\sin^2 x\right]=\sin^2 x$$

Multiplying (3) by $\sin^2 x$ and integrating, we have

$$u\sin^2 x=x,$$

Neglecting constant of integration since we seek only a particular solution.

$$u\,x\csc^2 x$$

$$\frac{dv}{dx}=x\csc^2 x$$

$$v=\int x\csc^2 x\,dx=-x\cot x+\ln|\sin x| \text{ (on integrating by parts)}$$

Hence a particular solution of (1) is

$$y_p=v\sin x=-x\cos x+\sin x\ (\ln|\sin x|).$$

The general solution of (1) is

$$y=c_1\sin x+c_2\cos x-x\cos x+\sin x(\ln|\sin x|).$$

Example: Solve $(x+2).\frac{d^2y}{dx^2}-(2x+5).\frac{dy}{dx}+2y=(x+1)e^x \qquad (1)$

Solution: We note that $y=e^{2x}$ makes the left-hand side of (1) zero. Therefore, we put

$$y=ve^{2x}$$

$$\frac{dy}{dx}=\left(\frac{dv}{dx}+2v\right)e^{2x}$$

$$\frac{d^2y}{dx^2}=\left(\frac{d^2v}{dx^2}+4\frac{dv}{dx}+4v\right)e^{2x}$$

Substituting into (1), we obtain

$$(x+2)\left(\frac{d^2v}{dx^2}+4\frac{dv}{dx}+4v\right)e^{2x}-(2x+5)\left(\frac{dv}{dx}+2v\right)e^{2x}+2ve^{2x}=(x+1)e^{x}.$$

$$(x+2)\frac{d^2v}{dx^2}e^{2x}+\left[4(x+2)-(2x+5)\right]\frac{dv}{dx}e^{2x}+\left[4(x+2)-2(2x+5)+2\right]ve^{2x}=(x+1)e^{x}$$

$$(x+2)\frac{d^2v}{dx^2}+(2x+3)\frac{dv}{dx}=(x+1)e^{-x}.$$

Writing $u=\frac{dv}{dx}$, the above equation becomes

$$(x+2)\frac{du}{dx}+(2x+3)u=(x+1)e^{-x}$$

$$\frac{du}{dx}+\frac{3x+3}{x+2}u=\frac{x+1}{x+2}e^{-x} \qquad (2)$$

which is linear equation of first order. An integrating factor of (2) is

$$exp\left[\int\frac{2x+3}{x+2}dx\right]=\exp\left[\int\left(2-\frac{1}{x+2}\right)dx\right]$$

$$=\exp\left[2x-\ln(x+2)\right]=\frac{e^{2x}}{x+2}$$

Multiplying (2) by $\frac{e^{2x}}{x+2}$, we have

$$\frac{d}{dx}\left(u\frac{e^{2x}}{x+2}\right)=\frac{x+1}{(x+2)^2}.e^{x}$$

$$u\frac{e^{2x}}{x+2}=\int\frac{x+1}{(x+2)^2}.e^{x}\,dx+c_1$$

$$=\int\frac{e^{x}}{x+2}dx-\int\frac{e^{x}}{(x+2)^2}dx+c_1$$

$$=\frac{e^{x}}{x+2}+c_1$$

$$u = e^{-x} + c_1(x+2)e^{-2x}$$

Therefore, $\frac{dv}{dx} = e^{-x} + c_1(x+2)e^{-2x}.$

Integrating, we get

$$v = -e^{-x} - \frac{1}{4}c_1(2x+5)e^{-2x} + c_2$$

Hence $y = ve^{2x} = -e^{x} - \frac{1}{4}c_1(2x+5) + c_2e^{2x}$ is the general solution of (1).

Note. It is easy to see that $y'' + Py' + Qy = 0$ is satisfied by $y = e^{\pm x}$. If $1 \pm P + Q = 0$ and by $y = x$ if $P + Qx = 0.$ These can be used to find a solution $y'' + Py' + Q = 0$ by inspection.

9.12. THE WRONSKIAN

Definition. If y_1, y_2 are two differentiable functions of x on $I = [a,b]$ then their Wronskian, denoted by $W = W[y_1, y_2]$, is obtained by

$$W = W[y_1, y_2] = y_1y_2' - y_1'y_2 = \begin{vmatrix} y_1 & y_2 \\ y_1' & y_2' \end{vmatrix}$$

Similarly, the Wronskian of three differentiable functions y_1, y_2, y_3 on $I = [a,b]$ is defined by

$$W = W[y_1, y_2, y_3] = \begin{vmatrix} y_1 & y_2 & y_3 \\ y_1' & y_2' & y_3' \\ y_1'' & y_2'' & y_3'' \end{vmatrix}$$

The definition can be extended in a similar manner for the Wronskian of n differentiable functions on $I = [a,b]$.

Theorem. Let y_1, y_2 be two solutions of the homogeneous linear equation

$\frac{d^2y}{dx^2}+P\frac{dy}{dx}+Qy=0$, (1) where P, Q are functions of x and are continuous on $I=[a,b]$. Then their Wronskian $W=W[y_1,y_2]$ is either identically zero or is never zero on I.

Proof. We have $W=y_1y_2'-y_1'y_2$

Differentiating $w.r.t.x,$ we get

$$W'=y_1y_2''+y_1'y_2'-y_1'y_2'-y_1''y_2$$

$$=y_1y_2''-y_1''y_2.$$

Since y_1,y_2 are solutions of (1), we have

$$y_1''+Py_1'+Qy_1=0 \qquad (2)$$

and

$$y_2''+Py_2'+Qy_2=0 \qquad (3)$$

Multiply (2) by y_2 and (3) by y_1 to have an equivalent system

$$y_1''y_2+P\,y_1'y_2+Qy_1y_2=0 \qquad (4)$$

$$y_1y_2''+P\,y_2'y_1+Qy_1y_2=0 \qquad (5)$$

Subtracting (4) from (5), we obtain

$$\left(y_1y_2''-y_2y_1''\right)+P\left(y_1y_2'-y_1'y_2\right)=0.$$

$$\frac{dW}{dx}+PW=0.$$

The general solution of the above equation is readily found as $W=c\exp(-\int Pdx)$. Since the exponential function is never zero, W is identically zero if $c=0$ or is never zero when $c\neq 0.$

Theorem. If y_1,y_2 are two solutions of $\frac{d^2y}{dx^2}+P\frac{dy}{dx}+Qy=0,$ then their Wronskian $W=W[y_1,y_2]=0$ if and only if y_1,y_2 are linearly dependent.

Proof. If one of the two solutions is identically zero, then the theorem is obviously true. Assume that y_1, y_2 are both non-zero and let y_1, y_2 be linearly dependent. Then $y_1 = cy_2,$ where c is a constant.

i.e., $\dfrac{y_1}{y_2} = c$

$$\frac{d}{dx}\left(\frac{y_1}{y_2}\right) = \frac{y_2 y_1' - y_1 y_2'}{y_2^2} = \frac{dc}{dx} = 0$$

i.e., $y_1 y_2' - y_1' y_2 = 0$

$$W[y_1, y_2] = 0.$$

Conversely, if $W[y_1, y_2] = 0,$ then $\dfrac{d}{dx}\left(\dfrac{y_1}{y_2}\right) = \dfrac{W}{y_2^2} = 0, \dfrac{y_1}{y_2} = c,$ where c is a constant. Thus $y_1 = cy_2$ and so y_1, y_2 are linearly dependent.

Corollary. $W[y_1, y_2] \neq 0$ if and only if y_1, y_2 are linearly independent.

Above theorems can only be applied to check the linear dependence or linear independence of differentiable functions which are solutions of a linear homogeneous $O.D.E.$ Arbitrary differentiable functions exist with their Wronskian zero but they are linearly independent.

Consider the functions $y_1 = x^3, y_2 = |x^3|$ on $]-\infty, \infty[$. Here $W(y_1, y_2) = 0$ but y_1 *and* y_2 are linearly independent.

9.13. VARIATION OF PARAMETERS

We found the solution of the equation $\dfrac{d^2y}{dx^2} + P\dfrac{dy}{dx} + Qy = F(x)$ (1)

where P, Q are functions of $x,$ in the previous section by reduction of order of (1). The solution of (1) can be determined by a procedure known as the method of variation of parameters. This method can be applied even to equations of higher order.

Suppose that linearly independent solutions of $\frac{d^2y}{dx^2}+P\frac{dy}{dx}+Q=0.$ (2)

are given by $y=y_1(x)$ and $y=y_2(x).$ Then the complementary function of (1) is

$$y_c=c_1y_1+c_2y_2$$

where $c_1\ and\ c_2$ are arbitrary constants. We replace the arbitrary constants $c_1\ and\ c_2$ by unknown functions $u_1(x)\ and\ u_2(x)$ and require that

$$y_p=u_1y_1+u_2y_2 \qquad (3)$$

be a particular solution of (1). In order to determine the two functions $u_1\ and\ u_2$ we need two conditions. One condition is that (3) must satisfy (1). A second condition can be imposed arbitrarily.

Differentiating (3) $w.r.t.x,$ we have

$$y_p'=\left(u_1'y_1+u_2'y_2\right)+u_1y_1'+u_2y_2'.$$

If we differentiate the above equation, y_p'' will contain $u_1''\ and\ u_2''.$ To avoid second derivatives of $u_1\ and\ u_2.$ we put

$$u_1'y_1+u_2'y_2=0 \qquad (4)$$

With this condition, we have

$$y_p'=u_1y_1'+u_2y_2'.$$

So that

$$y_p''=u_1'y_1'+u_2'y_2'+u_1y_1''+u_2y_2''.$$

Substituting for y_p,y_p',y_p'' into equation (1), we get

$$\left(u_1'y_1'+u_2'y_2'+u_1y_1''+u_2y_2''\right)+P\left(u_1y_1'+u_2y_2'\right)+Q\left(u_1y_1+u_2y_2\right)=F(x)$$ or

$$u_1\left(y_1''+Py_1'+Qy_1\right)+u_2\left(y_2''+Py_2'+Qy_2\right)+u_1'y_2'=F(x).$$

Expressions within the parenthesis are zero since $y_1\ and\ y_2$ are solutions of

(2). Hence

$u_1'y_1' + u_2'y_2' = F(x)$ (5)

Taking (4) and (5) together, we have two equations in the two unknowns u_1' *and* u_2';

$$u_1'y_1 + u_2'y = 0$$

and $u_1'y_1' + u_2'y_2' = F(x)$,

Solving these, we have

$$\left.\begin{aligned} u_1' &= \frac{-y_2F(x)}{y_1y_2' - y_1'y_2} \\ u_2' &= \frac{y_1F(x)}{y_1y_2' - y_1'y_2} \end{aligned}\right\} \qquad (6)$$

In (6), the Wronskian of y_1, y_2 namely $W(y_1, y_2) = y_1y_2' - y_1'y_2 \neq 0,$ since y_1, y_2 are linearly independent solutions of (2). Integrating (6), we find u_1 *and* u_2 as

$$u_1 = \int \frac{-y_2F(x)}{y_1y_2' - y_1'y_2}dx = \int \frac{-y_2F(x)}{W}dx \qquad (7)$$

$$u_2 = \int \frac{y_1F(x)}{y_1y_2' - y_1'y_2}dx = \int \frac{y_1F(x)}{W}dx \qquad (8)$$

where W is the Wronskian of y_1, y_2.

Thus y_p is completely determined. In numerical problems, instead of performing the complete process, formulas (7) will be directly applied to evaluate u_1 *and* u_2.

Example. Find the general solution of $\frac{d^2y}{dx^2} - 2\frac{dy}{dx} + y = xe^x \ln x$

(1)

Solution. The C.F. of (1) is

$$y_c = c_1 e^x + c_2 x e^x$$

Let $y_p = u_1 e^x + c_2 x e^x$

Here $y_1 = e^x, y_2 = x e^x, F(x) = x e^x \ln x$

$$W = W(y_1, y_2) = y_1 y_2' - y_1' y_2 = e^x\left(e^x + xe^x\right) - xe^x . e^x = e^{2x}.$$

By the formulas of (7) and (8), we have

$$u_1 = \int \frac{-y_2 F(x)}{W} dx = \int \frac{-xe^x . xe^x \ln x}{e^{2x}} dx$$ and

$$u_2 = \int \frac{y_1 F(x)}{W} dx = \int \frac{e^x . xe^x \ln x}{e^{2x}} dx$$

i.e., $$u_1 = -\int x^2 \ln x dx = -(\ln x).\frac{x^3}{3} + \int \frac{x^3}{3} dx = \frac{x^3}{9} - \frac{x^3}{3} \ln x$$

$$u_2 = \int x \ln x \, dx = \frac{x^2}{2} \ln x - \int \frac{x}{2} dx = -\frac{x^2}{4} + \frac{x^2}{2} \ln x.$$

$$y_p = \left(\frac{x^3}{9} - \frac{x^3}{3} \ln x\right) e^x + \left(-\frac{x^2}{4} + \frac{x^2}{2} \ln x\right) x e^x$$

$$= e^x \left(\frac{x^3}{9} - \frac{x^3}{3} \ln x - \frac{x^3}{4} + \frac{x^3}{2} \ln x\right)$$

$$= e^x \left(-\frac{5x^3}{36} + \frac{x^3}{6} \ln x\right).$$

The general solution of (1) is

$$y = y_c + y_p = c_1 e^x + c_2 x e^x = e^x \left(-\frac{5x^3}{36} + \frac{x^3}{6} \ln x\right).$$

Example. Find the general solution of $x^2 \frac{d^2 y}{dx^2} - x(x+2)\frac{dy}{dx} + (x+2) y = x^3$ (1)

Given that $y = xe^x$ is a solution of the associated homogeneous equation

$$x \frac{d\ y}{dx} - x(x+2)\frac{dy}{dx} + (x+2)y = 0 \qquad (2)$$

Solution. The given equation in the standard form is $\frac{d^2y}{dx^2} - \frac{x+2}{x}\frac{dy}{dx} + \frac{x+2}{x^2}y = x.$

Since $P + Qx = -\frac{x+2}{x} + x\frac{x+2}{x^2} = 0, y = x$ is also a solution of (2). The two solutions, $y \quad x$ and $y = xe^x$ are linearly independent. The complementary function of (1) is

$$y_c = c_1 x + c_2 x e^x.$$

We assume that

$$y_p = u_1 x + u_2 x e^x \qquad (3)$$

is a particular solution of (1).

Here $y_1 = x, y_2 = xe^x, F(x) = x$

$$W = W(y_1, y_2) = y_1 y_2' - y_1' y_2 = x(xe^x + e^x) - xe^x = x^2 e^x$$

By the formulas of (7) and (8), we have

$$u_1 = \int \frac{-y_2 F(x)}{W} dx \quad \text{and} \quad u_2 = \int \frac{y_1 F(x)}{W} dx$$

i.e., $u_1 = -\int \frac{x^2 e^x}{x^2 e^x} dx = -x$ and $u_2 = \int \frac{x^2}{x^2 e^x} dx = -e^{-x}.$

Substituting for u_1, u_2 into (3), a particular solution of (1) is

$$y = y_c + y_p = c_1 x + c_2 x e^x - x^2 - x.$$

Example. Explain how the method of variation of parameters can be applied to find a particular solution of a non-homogeneous linear third order differential equation whose complementary function is known. Apply the method to find a particular solution of

$$\frac{d^3y}{dx^3} + \frac{dy}{dx} = \cos x.$$

Solution. Suppose the C.F. of a linear third order differential equation

$$\frac{d^3y}{dx^3}+P\frac{d^2y}{dx^2}+Q\frac{dy}{dx}+Ry=F(x) \qquad (1)$$

is known to be $y_c=c_1y_1+c_2y_2+c_3y_3; y_1,y_2,y_3$ being linearly independent solutions of the associated homogeneous equation of (1).We assume that $y_p=u_1y_1+u_2y_2=u_3y_3,$ where u_1,u_2,u_3 are functions of x.

$$y_p'=u_1y_1'+u_3y_3'+u_1'y_1+u_2'y_2+u_3'y_3.$$

We set $u_1'y_1+u_2'y_2+u_3'y_3=0.$ Then

$$y_p'=u_1y_1'+u_2y_2'+u_3y_3'$$

$$y_p''=u_1y_1''+u_2y_2''+u_3y_3''$$

$$y_p'''=u_1y_1'''+u_2y_2'''+u_3y_3'''+u_1'y_1''+u_2'y_2''+u_3'y_3''.$$

Substituting for y_p,y_p',y_p'',y_p''' into equation (1), we have

$$\left(u_1y_1'''+u_2y_2'''+u_3y_3'''+u_1'y_1''+u_2'y_2''+u_3'y_3''\right)+P\left(u_1y_1''+u_2y_2''+u_3y_3''\right)$$

$$+Q\left(u_1y_1'+u_2y_2'+u_3y_3'\right)+R\left(u_1y_1+u_2y_2+u_3y_3\right)=F(x)$$

$$u_1\left(y_1'''+Py_1''+Qy_1'+Ry_1\right)+u_2\left(y_2'''+Py_2''+Qy_2'+Ry_2\right)$$

$$+u_2\left(y_2'''+Py_2''+Qy_2'+Ry_2\right)$$

$$+u_3\left(y_3'''+Py_3''+Qy_3'+Ry_3\right)+u_1'y_1''+u_2'y_2''+u_3'y_3''=F(x) \qquad (4)$$

From the system of equations (2), (3) and (4), we find $u_1',u_2'\,and\,u_3'$ and on integration we get $u_1,u_2 and\, u_3.$ Now we apply this method to find P.I. of $\frac{d^3y}{dx^3}+\frac{dy}{dx}=\csc x.$

The C.F. of the equation is obtained by solving the equation $D\left(D^2+1\right)=0.$ Thus

$y_c = c_1 + c_2 \cos x + c_3 \sin x$ with $y_1 = 1, y_2 = \cos x, y_3 = \sin x.$

For the particular integral, let $y_p = u_1 y_1 + u_2 y_2 + u_3 y_3,$

$$= u_1 + u_2 \cos x + u_3 \sin x.$$

Substituting for y_1, y_2, y_3 and their derivatives into equations (2), (3) and (4) above, we have the following system of linear equations in u_1', u_2', u_3'.

$$u_1' + u_2' \cos x + u_3' \sin x = 0$$

$$u_1'.0 - u_2' \sin x + u_3' \cos x = 0$$

$$u_1'.0 - u_2' \cos x - u_3' \sin x = \csc x.$$

We solve this system by the Gaussian Elimination method. Augmented matrix of the system is

$$\left[\begin{array}{ccc|c} 1 & \cos x & \sin x & 0 \\ 0 & -\sin x & \cos x & 0 \\ 0 & -\cos x & -\sin x & \csc x \end{array}\right]$$

$$\left[\begin{array}{ccc|c} 1 & 0 & 0 & \csc x \\ 0 & -\sin x & \cos x & 0 \\ 0 & -\cos x & -\sin x & \csc x \end{array}\right] by\, R_1 + R_3$$

$$\left[\begin{array}{ccc|c} 1 & 0 & 0 & \csc x \\ 0 & -\sin x \cos x & \cos^2 x & 0 \\ 0 & -\cos x \sin x & -\sin^2 x & 1 \end{array}\right] \begin{array}{l} by\,(\cos x) R_2 \\ and\,(\sin x) R_3 \end{array}$$

$$\left[\begin{array}{ccc|c} 1 & 0 & 0 & \csc x \\ 0 & -\sin x \cos x & \cos^2 x & 0 \\ 0 & 0 & -1 & 1 \end{array}\right]$$

Therefore,

$-u_3' = 1\, or\, u_3' = -1,$ from R_3

$(-\sin x \cos x) u_2' + (\cos^2 x) u_3' = 0$ from R_2

or $u_2' = -\cot x.$

$u_1' = \csc x,$ from R_1

Hence $u_1 = \int \csc x \, dx = \ln|\csc x - \cot x|$

$$u_2 = \int \frac{-\cos x}{\sin x} dx = -\ln|\sin x|.$$

and $u_3 = \int -dx = -x.$

$$y_p = u_1 + u_2 \cos x + u_3 \sin x.$$

$$= \ln|\csc x - \cot x| - (\cos x) ln|\sin x| - x\sin x$$

is the required particular solution.

9.14. APPLICATION OF DIFFERENTIAL EQUATIONS

Differential equations can be used in the solution of many problems in physical, biological and social sciences. In this section, we shall study such problems which can be formulated in terms of first and second order differential equations. The solution of the differential equation will lead to the solution of the given problem.

The following examples illustrate the techniques.

Example. A stone weighting 8 lbs falls from rest toward the earth from a great height. As it falls it is acted upon by air resistance that is numerically $\frac{1}{2}v(\textit{in pounds})$ where v is the velocity (in feet per second). Find the velocity and distance fallen at time t.

Solution. We use Newton's second law $F = ma$ to formulate mathematical model of this problem. The forces acting on the body are

i. F_1, the weight of the stone acting downward and hence is positive.

ii. F_2, the air resistance, numerically equal to $\frac{1}{2}v$, which acts upward and is therefore negative.

Newton's second law becomes $m\dfrac{dv}{dt} = F_1 + F_3$

i.e., $\dfrac{1}{4}\dfrac{dv}{dt} = 9 - \dfrac{1}{2}v, taking\ g = 32.$

$$\frac{dv}{dt} = 32 - 2v$$

$$\frac{dv}{dt} + 2v = 32$$

which is a linear equation. We now solve this differential equation. Here $I.F = e^{\int 2\,dt} = e^{2t}$.

And so, $\dfrac{d}{dt}\left(ve^{2t}\right) = 32e^{2t}$

Integrating, we get

$$ve^{2t} = 16e^{2t} + c\ or\ v = 16 + ce^{-2t}$$

Now $v(0) = 0.$

Hence $0 = 16 + c\ or\ c = -16$ Therefore,

$v = 16\left\{1 - e^{-2t}\right\}$ (1)

is the velocity after time t. (1) may be written as

$\dfrac{dx}{dt} = 16\left(1 - e^{-2t}\right)$ (2)

From (2), we have $dx = 16\left(1 - e^{-2t}\right)dt$ or $x = 16t + 8\,e^{-2t} + k$. Applying the initial condition $x(0) = 0,$ we find $k = -8.$ Hence $x = 16t + 8e^{-2t} - 8$ is the distance fallen after time t.

Example. A body of constant mass is projected upward from the earth's surface with an initial velocity v_0. Assuming there is no air resistance, but taking into consideration the variation of the earth's gravitational field with altitude, find the smallest initial velocity for which the body will not return to the earth (This is the so-called escape velocity).

Solution. The general expression for the weight $w(x)$ of a body of mass m is obtained from Newton's inverse- square law of gravitational attraction. If R is the radius of the earth and x is the altitude above sea level, then $w(x)=\frac{k}{(R+x)^2}, k$ being constant. At $x=0, w=mg,$ hence $k=mgR^2$ and $w(x)=\frac{mg\,R^2}{(R+x)^2}.$ The only force acting on the body is its weight which acts downward. Thus, the equation of motion is $m\frac{dv}{dt}=\frac{-mg\,R^2}{(R+x)^2}$ or $v\frac{dv}{dt}=-\frac{mg\,R^2}{(R+x)^2}.$

Separating the variables and integrating, we have

$\frac{1}{2}v^2=\frac{gR^2}{R+x}+c.$ But

$v=v_0\ at\,t=0$ i.e., at $x=0.$ Therefore,

$c=\frac{1}{2}v_0^2-gR$ and so

$$v^2=v_0^2-2gR+\frac{2gR^2}{R+x} \quad (1)$$

The escape velocity is found by requiring that v given by (1) remains positive for all (positive) value of $x.$ Thus we must have

$$v_0^2 \geq n\,2g\,R.$$

Hence the escape velocity is

$v_0=\sqrt{2gR}=6.9 miles/sec.taking\ R=4000\,miles$.

SUMMARY AND DISCUSSION

In this chapter we discussed higher order differential equation in detail. Also, different methods to solve higher order differential equation.

9.2. Linear Differential Equations

Definition. A linear differential equation of order n in the dependent variable y and the independent variable x is of the form

$$a_0(x)\frac{d^n y}{dx^n}+a_1(x)\frac{d^{n-1}y}{dx^{n-1}}+\ldots+a_{n-1}\frac{dy}{dx}+a_n(x)y=F(x), \qquad (1)$$

where $a_0(x),a_1(x),\ldots,a_{n-1}(x),a_n(x) and F(x)$ are functions of the independent variable x only and $a_0(x)$ is not identically zero. Using primes, (1) is also written as

$$a_0y^n+a_1y^{n-1}+\ldots+a_{n-1}y'+a_ny=F(x) \qquad (2)$$

where $a_0,a_1,\ldots,a_{n-1},a_n$ are real constants.

Definition. If $y_1(x),y_2(x),\ldots,y_m(x)$ are m functions of an independent variable x and $c_1,c_2,\ldots,c_m$ are constants, then the expression $c_1y_1(x)+c_2y_2(x)+\ldots+c_my_m(x)$ is called a linear combination of $y_1(x),y_2(x),\ldots,y_m(x)$.

i. Every homogeneous linear $nth-order$ differential equation $a_0y^n+a_1y^{n-1}+\ldots+a_{n-1}y'+a_ny=0$ has n linearly independent solutions $y_1,y_2,\ldots,y_n$

ii. If $y_1,y_2,\ldots,y\_m$ are n lineally independent solutions of (3), then any linear combination $y_c=c_1y_1+c_2y_2+\ldots+c_ny_n$ of $y_1,y_2,\ldots,y_n$ is the general solution of (3), $c_1,c_2,\ldots,c_m$ being arbitrary constants.

iii. Let y_p be any particular solution of (2). i.e., y_p does not contain any constant, then y_c+y_p is the general solution of (2).

9.3. Homogeneous Linear Equation

Consider the equation $a_0\frac{d^n y}{dx^n}+a_1\frac{d^{n-1}y}{dx^{n-1}}+\ldots+a_{n-1}\frac{dy}{dx}+a_n\,y,=0$ (1)

where $a_0, a_1, \ldots, a_{n-1}$ are real constants.

9.3.1. Case I. Distinct Real Roots

Let $m_1, m_2, \ldots, m_n$ be n distinct real roots of (2). Then $e^{m_1 x}, e^{m_2 x}, \ldots, e^{m_n x}$ are n distinct solutions of (1). These n solutions are linearly independent. Hence the general solution of (2) is $y = c_1 e^{m_1 x} + c_2 e^{m_2 x} + \ldots + c_n e^{m_n x}$, where $c_1, c_2, \ldots, c_n$ are arbitrary constants.

9.3.2. Case II. Repeated Real Roots

$$y = (c_1 + c_2 x) e^{m_1 x} + c_3 e^{m_3 x} + \ldots + c_n e^{m_n x}$$

In the same manner, if the characteristic equation (2) has the triple repeated root m_1, the corresponding part of the general solution of (1) is the solution of

$$(D - m_1)^3 y = 0$$

Proceeding as before, we can easily find

$$y = (c_1 + c_2 x + c_3 x^2) e^{m_1 x}$$

9.3.3. Case III. Complex Roots

Suppose the characteristic equation has the complex number $a + ib$ as a non-repeated root. Since coefficients of (1) are real, the conjugate complex number $a - ib$ is also a non-repeated root. The corresponding part of the general solution is $y = k_1 e^{(a+ib)x} + k_2 e^{(a-ib)x}$, where $k_1 \; and \; k_2$ are arbitrary constants.

$$y = e^{ax}\left[k_1 e^{ibx} + k_2 e^{-ibx}\right]$$

$$= e^{ax}\left[k_1(\cos bx + i \sin bx) + k_2(\cos bx - i \sin bx)\right]$$

$$= e^{ax}\left[(k_1 + k_2)\cos bx + i(k_1 - k_2)\sin bx\right]$$

$$= e^{ax}\left[c_1 \sin bx + c_2 \cos bx\right]$$

where $c_1 = i\left(k_1 - k_2\right), c_2 = k_1 + k_2$ are two arbitrary constants.

If $a + ib \, and \, a - ib$ are conjugate complex roots, each repeated k times, the the corresponding part of the general solution of (1) may be written as

$$y = \left[\left(c_1 + c_2 x + c_3 x^2 + \ldots + c_k x^{k-1}\right)\sin bx + \left(c_{k+1} + c_{k+2} x + \ldots + c_{2k} x^{k-1}\right)\cos bx\right]$$

Solution of the Equation

i. Complementary Function (C.F)

ii. Particular Solution (P.I)

The C.F is the solution of the homogeneous equation $f(D)y = 0$ and we have described different cases of its solution.

Now to find the $P.I$ of (1), we can write

$$y = \frac{1}{f(D)} F(x)$$

and try to evaluate $y = \dfrac{1}{f(D)} F(x).$

To Find the P.I when $F(x) = \sin ax \; or \cos ax.$

Here we have to evaluate $\dfrac{1}{f(D)} \sin ax \, and \, \dfrac{1}{f(D)} \cos ax.$ From Euler's theorem, we have

$$e^{iax} = \cos ax + i \sin ax$$

Thus $\dfrac{1}{f(D)} \sin ax \, and \, \dfrac{1}{f(D)} \cos ax$ are, respectively imaginary and real parts of $\dfrac{1}{f(D)} e^{iax}.$ If $f(D)$ contains only even powers of $D, say \, f(D) = f\left(D^2\right),$ it is easy to see that

$$\frac{1}{f(D^2)}\begin{cases}\sin ax\\ \cos ax\end{cases}=\frac{1}{f(-a^2)}\begin{cases}\sin ax\\ \cos ax\end{cases},\ provided\ -a^2\ is\ not\ a\ zero\ of\ f(D^2).$$

9.8. Series Expansion of Polynomials or Expressions Involving Operators

This method is useful when $F(x)$ is a polynomial in x.

Let $f(D)y=F(x)$ be such that $F(x)$ is a polynomial in x. To evaluate the particular integral

$$y_p=\frac{1}{f(D)}F(x)$$

It is often useful to expand $\frac{1}{f(D)}$ in a series in D by the Binomial Theorem for negative exponent so that

$$y_p=\left(1+a_1D+a_2D^2+a_3D^3+\ldots\right)F(x)$$

The derivatives on the right will vanish after certain stage since

$$D^n x^r=0\ if\ n>r.$$

The following binomial expansions will be useful in this connection:

i. $\frac{1}{1-x}=1+x+x^2+x^3+\ldots.$

ii. $\frac{1}{1+x}=1-x+x^2-x^3+\ldots.$

9.9. The Method of Undetermined Coefficients (U. C. Method)

We have studied some special cases in which particular integral can be evaluated by the inverse operator. Now we consider the method of undetermined coefficients which can prove simpler in finding the particular integral of the equation $f(D)y=F(x)$ when $F(x)$ is

a. An exponential function: e^{ax}

b. A polynomial: $\left(b_0x^n+b_1x^{n-1}+\ldots+b_n\right)$

c. Sinusoidal function: $\sin ax \; or \; \cos bx$

The more general case in which $F(x)$ is sum of a product of terms of the above types, such as

$$F(x)=e^{ax}\left(b_0x^n+b_1x^{n-1}+\ldots+b_n\right)\begin{Bmatrix}\sin ax\\ \cos bx\end{Bmatrix}$$

9.10. The Cauchy-Euler Equation

An equation of the form

$$a_0x^n\frac{d^ny}{dx^n}+a_1x^{n-1}\frac{d^{n-1}y}{dx^{n-1}}+\ldots+a_{n-1}x\frac{dy}{dx}+a_ny=F(x)$$

is called Cauchy-Euler (or equidimensional) equation. $a_0,a_1,\ldots,a_{n-1},a_n$ are real constants. The equation can be reduced to a linear differential equation with constant coefficients by the transformation $x=e^t \; or \; t=\ln x$.

9.11. Reduction of Order

If one solution of the second order linear equation $\frac{d^2y}{dx^2}+P\frac{dy}{dx}+Qy=0$ (1)

(where P,Q are not necessarily constants and may be functions of x) is known, then we can use it to find the general solution of $\frac{d^2y}{dx^2}+P\frac{dy}{dx}+Qy=F(x)$ (2)

The procedure, due to D' Alembert, is known as the method reduction of order.

9.12. The Wronskian

Definition. If y_1,y_2 are two differentiable functions of x on $I=[a,b]$ then their Wronskian, denoted by $W=W[y_1,y_2]$, is obtained by

$$W=W[y_1,y_2]=y_1y_2'-y_1'y_2=\begin{vmatrix}y_1 & y_2\\ y_1' & y_2'\end{vmatrix}$$

Similarly, the Wronskian of three differentiable functions y_1, y_2, y_3 on $I = [a,b]$ is defined by

$$W = W[y_1, y_2, y_3] = \begin{vmatrix} y_1 & y_2 & y_3 \\ y_1' & y_2' & y_3' \\ y_1'' & y_2'' & y_3'' \end{vmatrix}$$

The definition can be extended in a similar manner for the Wronskian of n differentiable functions on $I = [a,b]$.

Theorem. If $y_1\ y_2$ are two solutions of $\frac{d^2y}{dx^2} + P\frac{dy}{dx} + Qy = 0,$ Then their Wronskian $W = W[y_1, y_2] = 0$ if and only if y_1, y_2 are linearly dependent.

Corollary. $W[y_1, y_2] \neq 0$ if and only if y_1, y_2 are linearly independent.

9.13. Variation of Parameters

We found the solution of the equation $\frac{d\ y}{dx} + P\frac{dy}{dx} + Qy = F(x)$ (1)

where P, Q are functions of $x,$ in the previous section by reduction of order of (1). The solution of (1) can be determined by a procedure known as the method of variation of parameters. This method can be applied even to equations of higher order.

$$u_1 = \int \frac{-y_2 F(x)}{y_1 y_2' - y_1' y_2} dx = \int \frac{-y_2 F(x)}{W} dx$$

$$u_2 = \int \frac{y_1 F(x)}{y_1 y_2' - y_1' y_2} dx = \int \frac{y_1 F(x)}{W} dx$$

where W is the Wronskian of y_1, y_2.

CHAPTER 10

BASIC CONCEPTS OF PARTIAL DIFFERENTIAL EQUATION

CONTENTS

10.1. INTRODUCTION

Many practical problems in science and engineering, when formulated mathematically gives to partial differential equation (often referred to as PDE) in order to understand the physical behavior of the mathematical model, it is necessary to have some knowledge about the mathematical character, properties, and the solution of the governing PDE. An equation which involves several independent variables (usually denoted by $x, y, z, t, \ldots$) a dependent functions u of these variables and the partial derivatives of the dependent function u with respect to the independent variables such as

$$F\left(x, y, z, t \ldots, u_x, u_y, u_z, u_1, \ldots \ldots u_{xx}, u_{yy}, \ldots u_{xy}, \ldots\right) = 0 \tag{1.1}$$

is called a partial differential equation. A few well-known examples are:

(i) $u_1 = k\left(u_{xx} + u_{yy} + u_{zz}\right)$ [Linear three-dimensional heat equation]

(ii) $u_{xx} + u_{yy} + u_{zz} = 0$ [Laplace equation in three dimensions]

(iii) $u_{tt} = c^2\left(u_{xx} + u_{yy} + u_{zz}\right)$ [Linear three-dimensional wave equation]

(iv) $u_t + uu_x = \mu u_{xx}$ [Nonlinear one-dimensional Burger equation].

In all these example, u is the dependent function and the subscript denote partial differentiation with respect to these variables.

Definition. The order of the partial differential equation is the order of the highest derivative occurring in the equation. Thus, the above examples are partial differential equation of second order, whereas

$$u_1 = u_{xx} + \sin x$$

is an example for third order partial differential equation.

10.2. CLASSIFICATION OF SECOND ORDER PDE

The most general linear second order PDE, with one dependent function u on a domain Ù of point

$X = \left(x_1, x_2, \ldots \ldots x_n\right), n > 1.$ is

$$\sum_{i,j=1}^{n} a_{ij} u_{xi} x_j + \sum_{i=1}^{n} b_1 u_{x1} + F(u) = G \tag{2}$$

The classification of a PDE depends only on the highest order derivatives present.

The classification of PDE is motivated by the classification of the quadratic equation of the form

$$Ax^2 + Bxy + Cy^2 + Dx + Ey + F = 0 \quad (3)$$

which is ellipse, parabolic or hyperbolic.

The classification of PDE is motivated by the classification of the quadratic equation of the form

$$Ax^2 + Bxy + Cy^2 + Dx + Ey + F = 0 \quad (4)$$

which is elliptic, parabolic, or hyperbolic according as the discriminant $B^2 - 4AC$ is negative, zero or positive. Thus, we have the following second order linear PDE in two variables $x\, and\, y$

$$Au_{xx} + Bu_{xy} + Cu_{yy} + Du_x + Eu_y + Fu = G$$

where the coefficient $A, B, C, \ldots$ may be functions of $x\, and\, y,$ however for the sake of simplicity we assume them to be constants. Equation (1.4) is elliptic, parabolic or hyperbolic at a point (x_0, y_0) according as the discriminant

$$B^2(x_0, y_0) - 4A(x_0, y_0)C(x_0, y_0) = 0$$

is negative, zero or positive. If this is true at all points in a domain Ù, then is said to be elliptic, parabolic in that domain. if the number of independent variables is two or three, transformation can always be found to reduce the given PDE to a canonical form (also called normal form). In general, when the number of independent variables is greater than 3, it is not always possible to find such a transformation except in certain special cases. The idea of reducing the given PDE to a canonical form is that the transformed equation assumes a simple form so that the subsequent analysis of solving the equation is made easy.

10.3. CANONICAL FORM

Consider the most general transformation of the independent variables $x\, and\, y$ of Eq. (4)

$$\xi = \xi(X, Y), \eta = \eta(x, y) \quad (5)$$

Such that the functions ξ and η are continuously differentiable and the Jacobean

$$j = \frac{\partial(\xi,\eta)}{\partial(x,y)}\begin{vmatrix}\xi_x\xi_y\\ \eta_x\eta_y\end{vmatrix} = \left(\xi_x\eta_y - \xi_y\eta_x\right) \neq 0 \tag{6}$$

In the domain $\grave{U}\, where$ Eq. (4) holds. Using the chain rule of partial differentiation. The partial derivatives becomes

$$u_x = u_\xi\xi_x + u_\eta\eta_x$$

$$u_y = u_\xi\xi_y + u_\eta\eta_y$$

$$u_{xx} = u_{\xi\xi}\xi_y^2 + 2u_{\xi\eta}\xi_x\eta_x + u_{\eta\eta}\eta_x^2 + u_\xi\xi_{xx} + u_\eta\eta_{xx}$$

$$u_{xy} = u_{\xi\xi}\xi_x\xi_y + 2u_{\xi\eta}\xi_y\eta_y + u_{\eta\eta}\eta_y^2 + u_\xi\xi_{yy} + y_\eta\eta_{yy} \tag{7}$$

Substituting these expression into the original differential equation (4) we get

$$\bar{A}u_{\xi\xi} + \bar{B}u_{\xi\eta} + \bar{C}u_{\eta\eta} + \bar{D}u_\xi + \bar{E}u_\eta + \bar{F}u = \bar{G} \tag{8}$$

$$\bar{A} = A\xi_x^2 + B\xi_x\xi_y + C\xi_y^2$$

$$\bar{B} = 2A\xi_x\eta_x + B\left(\xi_x\eta_v + \xi_y\eta_x\right) + 2C\xi_y\eta_y$$

$$\bar{C} = A\eta_x^2 + B\eta_{xy} + C\eta_y^2$$

$$\bar{D} = A\xi_{xx} + B\xi_{xy} + C\xi_{yy} + D\xi_x + E\xi_y$$

$$\bar{E} = A\eta_{xx} + B\eta_{xy} + C\eta_{yy}\; D\eta_x + E\eta_y$$

$$\bar{F} = F, \bar{G} = G \tag{9}$$

It may be noted that the transformed equation has the same as that of the original equation (4) under the general transformation.(5).

Since the classification of Eq. depend on the coefficient $A, B, and\, C,$ we can also rewrite the equation in the form

$$Au_{xx} + Bu_{xy} + Cu_{yy} = H\left(x, y, u, u_x, u_y\right) \tag{10}$$

It can be shown easily that under the transformation (5), Eq. (10) takes one of the following three canonical form:

(i) $u_{\xi\xi} - u_{\eta\eta} - \phi(\xi, \eta, u, u_\xi, u_\eta)$ (11a)

$or\, u_{\xi\eta} = \phi,(\xi, \eta, u, u_\xi, u_\eta)$ in the hyperbolic case

(ii) $u_{\xi\xi} + u_{\eta\eta} = \phi(\xi, \eta, u, u_\xi, u_\eta)$ in the hyperbolic case

(ii) $u_{\xi\xi} = \phi(\xi, \eta, u, u_\xi, u_\eta)$ (11b)

(iii) $u_{\eta\eta} = \phi(\xi, \eta, u, u_\xi, u_\eta)$ (11c)

$or\, u_{\eta\eta} = \phi(\xi, \eta, u, u_\xi, u_\eta)$ in the parabolic case.

We shall discuss in detail each of these cases separately. Using Eq. (9), it can also be verified that

$$\bar{B}^2 - 4\bar{A}\bar{C} = (\xi_x\eta_y - \xi_y\eta_x)^2 (B^2 - 4AC)$$

and therefore we conclude that the transformation of the independent variables does not modify the type of PDE.

10.3.1. Canonical Form for Hyperbolic Equation

Since the discriminant $\bar{B}^2 - 4\bar{A}\bar{C} > 0$ for hyperbolic case, we set $\bar{A} = 0$ and $\bar{C} = 0$ in Eq.(9) which will gives us the coordinates $\xi\, and\, \eta$ that reduce the given PDE to a canonical form in which the coefficient of $u_{\xi\xi}, u_{\eta\eta}\, are\, zero$. Thus, we have

$$\bar{A} = A\xi_x^2 + B\xi_x\xi_y + C\xi_y^2 = 0$$

$$\bar{C} = A\eta_x^2 + B\eta_x\eta_y + C\eta_y^2 = 0$$

which, on rewriting become

$$A\left(\frac{\xi x}{\xi y}\right)^2 + B\left(\frac{\xi x}{\xi y}\right) + C = 0$$

$$A\left(\frac{\eta x}{\eta y}\right)^2 + B\left(\frac{\eta x}{\eta y}\right) + C = 0$$

Solving these equation for $(\xi_x \xi_y) and (\eta_x \eta_y)$, we get

$$\frac{\xi x}{\xi_y} = \frac{-B + \sqrt{B^2 - 4AC}}{2A}$$

$$\frac{\eta x}{\eta_y} = \frac{-B + \sqrt{B^2 - 4AC}}{2A} \qquad (12)$$

The condition $B^2 > 4AC$ implies that the slopes of the curves $\xi(x, y) = C_1 \eta(x, y) = C_2$ are Thus, if $B^2 > 4AC$, then at any point (x, y) there exist two real directions given by two roots (12) along which the PDE (4) reduces to the canonical form. These are called Characteristic solution for each. Otherwise we will end up with same two coordinates. Thus, if $B^2 > 4AC$, then at any point (x, y) there exist two real directions given by two roots (12) along which the PDE (4) reduces to the canonical form. These are called characteristic solution for each. Otherwise we will end up with same two coordinates.

Along the curve $\xi(x, y) = c_1 . we have$

$$d\xi = \xi_x \, dx + \xi_y \, dy = 0$$

Hence,

$$\frac{dy}{dx} = \left(\frac{\xi_x}{\xi_y} \right) \qquad (13)$$

Similarly, along the curve $\eta(x, y) = c_2$, we have

$$\frac{dy}{dx} = -\left(\frac{\eta x}{\eta y} \right)$$

Integrating Eqs. (13) and (14), we obtain the equations of family of characteristics of the PDE (4). Now to obtain the canonical form the given PDE, we substitute the expressions of $\xi \, and \, \eta$ into Eq. (8) which reduce to Eq. (11a)

To make the ideas clearer let us consider the following example:

$$3u_{xx} + 10u_{xy} + 3u_{yy} = 0$$

Comparing with the standard PDE (4) we have $A = 3.B = 10, C = 3, B^2 - 4AC = 64 > 0.$ Hence the given equation is hyperbolic PDE. The corresponding are:

$$\frac{dy}{dx} = -\left(\frac{\xi_x}{\xi_y}\right) = \frac{-B + \sqrt{B^2 - 4AC}}{2A} = \frac{1}{3}$$

$$\frac{dy}{dx} = \left(\frac{\eta x}{\eta y}\right) = -\left(\frac{-B + \sqrt{B^2 - 4AC}}{2A}\right) = 3$$

To find $\xi \, and \, \eta, we$ first solve for y by integrating the above equations. Thus, we get

$$y = 3x + c_1, y = \frac{1}{3}x + c_2$$

which give the constant as

$$c_1 = y - 3x, c_2 = y - x/3$$

Therefore,

$$\xi = y - 3x = c_1, \eta = y - \frac{1}{3}x = c_2$$

These are the characteristics lines for the given hyperbolic equation. In this example, the characteristics are found to be straight lines in the $(x, y)-$ plane along which the initial data. Impulses will propagate.

To find the canonical equation, we substitute the expression for $\xi \, and \, \eta$ into Eq. (9) to get

$$\bar{A} = A\xi_x^2 + B\xi_x\xi_y + C\xi_y^2 = 3(-3)^2 + 10(-3)(1) + 3 = 0$$

$$\bar{B} = 2A\xi_x\eta_x + B(\xi_x\eta_x) + 2C\xi_y\eta_y$$

$$= 2(3)(-3)\left(-\frac{1}{3}\right) + 10\left[(-3)(1) + 1\left(-\frac{1}{3}\right)\right] + 2(3)(1)(1)$$

$$= 6 = 10\left(-\frac{10}{3}\right) + 6 = 12 - \frac{100}{3} = -\frac{64}{3}$$

$$\overline{Ac} = 0, \overline{D} = 0, \overline{E} = 0, \overline{F} = 0$$

Hence, the required canonical form is

$$\frac{64}{3}u_{\xi\eta} = 0 \, or \, u_{\xi\eta} = 0$$

On integration, we obtain

$$u(\xi,\eta) = f(\xi) + g(\eta)$$

where $f \, and \, g$ are arbitrary. Going back to the original variables, the general solution is

$$u(x,y) = f(y-3x) + g(y-x/3)$$

10.3.2. Canonical Form Parabolic Equation

For the parabolic equation, the discriminant $\overline{B}^2 - 4\overline{A}\overline{C} = 0,$ which can be true if $\overline{B} = 0$ and $\overline{A} \, or \, \overline{C}$ is equal to zero. Suppose we set first $\overline{A} = 0$ in Eq. (9). Then we obtain

$$\overline{A} = A\xi_x^2 + B\xi_x\xi_y + C\xi_y^2 = 0$$

or
$$A\left(\frac{\xi_x}{\xi_y}\right)^2 + B\left(\frac{\xi_x}{\xi_y}\right) + C = 0$$

which gives

$$\frac{\xi_x}{\xi_y} = \frac{-B + \sqrt{B^2 - 4AC}}{2A}$$

Using the condition for parabolic case, we get

$$\frac{\xi_x}{\xi_y} = -\frac{B}{2A} \qquad (15)$$

Hence, to find the function $\xi = \xi(x,y)$ which satisfies Eq. (15), we get

$$\frac{dy}{dx} = -\frac{\xi x}{\xi y} = \frac{B}{2A}$$

and get the implicit solution

$$\xi(x, y) = C_1$$

In fact, one can verify that $\bar{A} = 0\, implies\, \bar{B} = 0$ as follows:

$$\bar{B} = 2A\xi_x\eta_x + B(\xi_x\eta_y + \xi_y\eta_x) + 2C\xi_y\eta_x$$

$$= 2\left(\sqrt{A}\xi_x + \sqrt{C}\xi_y\right)\left(\sqrt{A\eta_x} + \sqrt{C}\eta_y\right)$$

However,

$$\frac{\xi_x}{\xi_y} = -\frac{B}{2a} = \frac{2\sqrt{AC}}{2A} = -\sqrt{\frac{C}{A}}$$

Hence,

$$\bar{B} = 2\left(\sqrt{A}\xi_x + \sqrt{A}\xi_y\right)\left(\sqrt{A\eta_x} + \sqrt{C}\eta_y\right) = 0$$

We therefore choose ξ in such a way that both $\bar{A}\, and\, \bar{B}$ are zero. Then η can be chosen in any way like as long as it is not parallel to the ξ coordinates. In other words, we choose η such that the Jacobian of the transformation is not zero. Thus, we can write the canonical the equation for parabolic case by simply substituting $\xi\, and\, \eta$ into equation which reduce to either of the forms. We therefore choose ξ in such a way that both $\bar{A}\, and\, \bar{B}$ are zero. Then η can be chosen in any way like as long as it is not parallel to the ξ coordinates. In other words, we choose η such that the Jacobian of the transformation is not zero. Thus, we can write the canonical the equation for parabolic case by simply substituting $\xi\, and\, \eta$ into equation which reduce to either of the forms (11 c). To illustrate the procedure, we consider the following examples:

$$x^2u_{xx} - 2xyu_{xy} + y^2u_{yy} = e^x$$

The discriminant $B^2 - 4AC = 4x^2y^2 - 4x^2\, y^2 = 0$, and hence the given PDE is parabolic everywhere. The characteristics equation is

$$\frac{dy}{dx} = -\frac{\xi_x}{\xi_y} = -\frac{B}{2a} = \frac{2xy}{2x^2} = -\frac{y}{x}$$

On integration we have

And hence $\xi = xy$ will satisfy the characteristics equation and we can choose $\eta = y.To$ find the canonical equation, we substitute the expressions for $\xi\, and\, \eta$ into Eq.(9) to get

$$\overline{A} = Ay^2 + Bxy + cx62 = x^2y^2 - 2x^2y^2 + y^2x^2 = 0$$

$$\overline{B} = 0, \overline{C} = y^2, \overline{D} = -2xy$$

$$\overline{E} = 0, \overline{F} = 0\,\overline{G} = e^x$$

Hence, the transformed equation is

$$y^2u_{\eta\eta} - 2xyu_{\xi} = e^x$$

$$y^2u_{\xi\xi} - 2xyu_{\xi} = e^{\xi/\eta}$$

The canonical form is, therefore,

$$u_{\eta\eta} = \frac{2\xi}{\eta^2}u\xi + \frac{1}{\eta^2}e^{\xi/\eta}$$

\

10.3.3. Canonical Form for Elliptic Equation

Since the discriminant $B^2 - 4AC < 0,$ for elliptic case, the characteristics equation

$$\frac{dy}{dx} = \frac{B - \sqrt{B^2 - 4AC}}{2A}$$

$$\frac{dy}{dx} = \frac{B + \sqrt{B^2 - 4AC}}{2A}$$

Gives us complex conjugate coordinates, say $\xi\, and\, \eta.$ now, we make another transformation $(\xi,\eta)to(\alpha,\beta)$ so that

$$\alpha = \frac{\xi + \eta}{2}, \beta = \frac{\xi - \eta}{2i}$$

which gives us the required canonical equation in the form (11b). To illustrate the procedure. We consider the following example:

The discriminant $B^2 - 4AC = -4x^2 < 0$. Hence, the given PDE is elliptic. The characteristics equations

$$\frac{dy}{dx} = \frac{B - \sqrt{B^2 - 4AC}}{2A} = \sqrt{-\frac{4x^2}{2}} = -ix$$

$$\frac{dy}{dx} = \frac{B + \sqrt{B^2 - 4AC}}{2A} = ix$$

Integration of these equation yields

$$iy + \frac{x^2}{2} = c_1, -iy + \frac{x^2}{2} = c_2$$

Hence, we may assume that

$$\xi = \frac{1}{2}x^2 + iy, \eta = \frac{1}{2}x^2 - iy$$

Now, introducing the second transformation

$$\alpha = \frac{\xi + \eta}{2}, \beta = \frac{\xi - n}{2i}$$

We obtain

$$\alpha = \frac{x^2}{2}, \beta = y$$

The canonical form can now be obtained by computing

$$\bar{A} = Aa_x^2 + \beta\alpha_x\alpha_y + c\alpha_y^2 = x^2$$

$$\bar{B} = 2A\alpha_x\beta_x + b(\alpha_x\beta_y + \alpha_y\beta_x) + 2c(\alpha_y\beta_y) = 0$$

$$\bar{C} = A\beta_x^2 + B\beta_x\beta_y + c\beta_y^2 = x^2$$

$$\bar{D} = Aa_{xx} + Ba_{xx} + ca_{yy} + Da_x + Ea_y = 1$$

$$\bar{E} = A\beta_{xx} + B\beta_{xy} + c\beta_{yy} + D\beta_x + E\beta_y = 0$$

$$\bar{F} = 0, \bar{G} = 0$$

Thus the required canonical equation is

$$x^2u_{aa} + x^2u_{\beta\beta} + u_a = 0$$

$$u_{aa} + u_{\beta\beta} = -\frac{u_a}{sa}$$

Example

Classify and reduce the relation

$$y^2u_{xx} - 2xyu_{xy} + x^2u_{yy} = \frac{y^2}{x}u_x + \frac{x^2}{y}u_y$$

to a canonical form and solve it.

Solution

The discriminant of the given PDE is

$$B^2 - 4AC = 4x^2y^2 - 4c^2y^2 = 0$$

Hence the given equation is of a parabolic type. The characteristics equation is

$$\frac{dy}{dx} = -\frac{\xi_x}{\xi_y} = \frac{B}{2A} = -\frac{-2xy}{2y^2}$$

Integration gives $x^2 + y^2 = c_1$. Therefore, $\xi = x^2 + y^2$ satisfies the characteristics equation. The $\eta -$ coordinates can be chosen arbitrary so that it is not parallel to ξ. i.e., the Jacobian of the transformation is not zero. Thus, we choose

$$\xi = x^2 + y^2, \eta = y^2$$

To find the canonical equation, we compute

$$\overline{A}\, A\xi_x^2 + B\xi_x\xi_y + G\xi_y^2 = 4x^2y^2 - 8x^2y^2 + 4x^2y^2 = 0$$

$$\overline{B} = 0, \overline{C} = 4x^2y^2, \overline{D} = \overline{E} = \overline{F} = \overline{G} = 0$$

Hence, the required canonical equation is

$$4x^2y^2u_{\eta\eta} = 0\, or\, u_{\eta\eta} = 0$$

To solve this equation, we integrate it twice with respect to η to get

$$u_\eta = f(\xi).u = f(\xi)\eta + g(\xi)$$

where $f(\xi) and\, g(\xi)$ are arbitrary function of ξ. Now, going back to the original independent variables, the required solution is

$$u = y^2 f(x^2 + y^2) + g(x^2 + y^2)$$

Example

Reduce the following equation to a canonical form :

$$(1 + x^2)u_{xx} + (1 + y^2)u_{yy} + xu_x + yu_y = 0$$

Solution

The discriminant of the given PDE is

$$B^2 - 4AC = -= -(1 + x^2)(1 + y^2) < 0$$

Hence the given PDE is an elliptic type. The characteristics equation are

$$\frac{dy}{dx} = \frac{B + \sqrt{B^2 - 4AC}}{2A} = -\sqrt{\frac{-4(1 + x^2)(1 + y^2)}{2(1 + x^2)}}\sqrt{\frac{1 + y^2}{1 + x^2}}$$

$$\frac{dy}{dx} = \frac{B + \sqrt{B^2 - 4AC}}{2A} = i\sqrt{\frac{1 + y^2}{1 + x^2}}$$

On integration, we get

$$\xi = In\left(x + \sqrt{x^2 + 1}\right) - i\, In\ (y + \sqrt{y^2 + 1} = c_1$$

$$\eta = In\left(x + \sqrt{x^2 + 1}\right) + i\, In\ (y + \sqrt{y^2 + 1} = c_2$$

Introducing, the second transformation

$$\alpha = \frac{\xi + \eta}{2}, \beta = \frac{\eta - \xi}{2i}$$

we obtain

$$\alpha = In\left(x + \sqrt{x^2 + 1}\right)$$

$$\beta = In\left(y+\sqrt{y^2+1} \right)$$

Then the canonical for can be obtained by computing

$$\overline{A} = A\alpha_x^2 + B\alpha_x\alpha_y + Ca_y^2 = 1, \overline{B} = 0, \overline{C} = 1, \overline{D} = \overline{E} = \overline{F} = \overline{G} = 0$$

Thus the canonical equation for the given PDE is

$$u_{aa} + u_{\beta\beta} = 0$$

Example

Reducing the following equation to a canonical form and hence solve it.

$$u_{xx} - 2\sin xu_{xy} - \cos \ \ xu_{yy} - \cos xu_y = 0$$

Solution

Computing with the general second order PDE (4), we have

$$A = 1\, B = -2\sin x, C = -\cos^2 x,$$

$$D = 0, E = -\cos x\, F = 0, G = 0$$

The discriminant $B^2 - 4AC = 4(\sin^2 x + cos^2 x = 4 > 0. Hence$ the given PDE is hyperbolic relevant characteristics equation are:

$$\frac{dy}{dx} = \frac{B - \sqrt{B^2 - 4AC}}{2A} = -\sin x - 1$$

$$\frac{dy}{dx} = \frac{B + \sqrt{B^2 - 4AC}}{2A} = 1 - \sin x$$

On integration, we get

$$y = \cos x - x + c_1, y = \cos x + x + c_2$$

Thus, we choose the characteristics lines as

$$\xi = x + y - \cos x = c_1, \eta = -x + y - \cos x = c_2$$

In order to find the canonical equation, we compute

$$\overline{A} = A\xi_x^2 + B\xi_x\xi_y + C\xi_y^2 = 0$$

$$\overline{B} = 2A\xi_x\eta_x + B\left(\xi_x\eta_y + \xi_y\eta_x\right) + 2C\xi_y\eta_y$$

$$= 2\ (\sin x + 1\ )(\sin x - 1) - 4\ \sin^2 x - 2\ \cos^2 x = -4$$

$$\bar{C} = 0\ \bar{D} = 0\ \bar{E} = 0, \bar{F} = 0, \bar{G} = 0$$

Thus, required canonical equation is

$$u_{\xi\eta} = 0$$

Integrating with respect to ξ, we obtain

$$u_\eta = f(\eta)$$

where f is arbitrary. Integrating once again with respect to η. we have

$$u = \int f(\eta)d\eta = g(\xi)$$

$$u = \psi(\eta) + g(\xi)$$

where $g(\xi)$ is another arbitrary function. Returning to the old variables $x, y,$ the solution of the given PDE is

$$u(x, y) = \psi(y - x - \cos x) + g(y + x - \cos x)$$

Example

find the characteristics of the equation

$$u_{xx} + 2u_{xy} + \sin^2(x)u_{yy} + u_y = 0$$

when it is of hyperbolic type.

Solution

The discriminant $B^2 - 4AC = 4 - 4\sin^2 x = 4\cos^2 x. Hence$ for all $x \neq (2\pi - 1)\pi / 2$ the given PDE is of hyperbolic type. The characteristics equation are

$$\frac{dy}{dx} = \frac{B - \sqrt{B^2 - 4AC}}{2A} = 1 \mp \cos x$$

On integration, we get

$$y = x - sinx + c_1, y = x + \sin x + c_2$$

Thus, the characteristics equation are

$$\xi = y - x + \sin x, \eta - x - \sin x$$

Example

Reduce the following equation to a canonical form and hence solve it:

$$yu_{xx} + (x + y)u_{xy} + xu_{yy} = 0$$

Solution

The discriminant

$$B^2 - 4AC = (x = y)^2 - 4xy = (x - y)^2 > 0$$

Hence the given PDE is hyperbolic everywhere except along the line $y = x$: whereas on the $v = x$ it is parabolic. When $y \neq x$. the characteristics equations are

$$\frac{dy}{dx} = \frac{B \mp \sqrt{B^2 - 4AC}}{2A} = \frac{(x + y) \mp (x - y)}{2y}$$

Therefore,

$$\frac{dy}{dx} = 1, \frac{dy}{dx} = \frac{x}{y}$$

On integration, we obtain

$$y = x + c_1, y^2 = x^2 + c_2$$

Hence, the characteristics equation are

$$\xi = y - x, \eta = y^2 - x^2$$

These are straight lines and rectangular hyperbola. The canonical form can be obtained by computing

$$\overline{A} = A\xi_x^2 + B\xi_x\xi_y + C\xi_y^2 = y - x - y + x = 0\,\overline{B} = -2(x - y)^2,$$

$$\overline{C} = 0, \overline{D} = 0, \overline{E} = 2(x - y), \overline{F} = \overline{G} = 0$$

Thus, the canonical equation for the given PDE is

$$-2(x - y)^2 u_{\xi\eta} + 2(x - y)u_\eta = 0$$

$$-2\xi^2 u_{\xi\eta} + 2(-\xi)u_\eta = 0$$

$$\xi u_{\xi\eta} + u_\eta = \frac{\partial}{\partial \xi}\left(\xi \frac{\partial u}{\partial \eta} \right) = 0$$

Integration yields

$$\xi \frac{\partial u}{\partial \eta} = f(\eta)$$

Again integrating with respect to η, we obtain

$$u = \frac{1}{\xi} \int f(\eta) d\eta + g(\xi)$$

Hence,

$$u = \frac{1}{y-x} \int f\left(y^2 - x^2\right) d\left(y^2 - x^2\right) + g(y-x)$$

is the general solution.

Example

Classify and transform the following equation to a canonical form::

$$\sin^2(x) u_{xx} + \sin(2x) u_{xy} + \cos^2(x) u_{yy} = x$$

Solution. The discriminant of the given PDE is

$$B^2 - 4A = \sin^2 2x - 4\sin^2 x \cos^2 x = 0$$

Hence, the given equation is of parabolic type. The characteristics equation is

$$\frac{dy}{dx} = \frac{B}{A} = \cot$$

Integration gives

$$y = in \sin x + c_1$$

Hence, the characteristics equation are :

$$\xi = y - In \sin x, \eta = y$$

η is chosen in such a way that the Jacobian of the transformation is non-zero. Now the canonical form can be obtained by computing

$$\overline{A} = 0, \overline{B} = 0, \overline{C} = \cos^2 x . \overline{D} = 1,$$

$$\overline{E}=0,\overline{F}=0,\overline{G}=x$$

Hence, the canonical equation is

$$\cos^2(x)u_{\eta\eta}+u_{\xi}=x$$

$$\left[1-e^{2(\eta-\xi)}\right]u_{\eta\eta}=\sin^{-1}\left(e^{\eta-\xi}\right)-u_{\xi}$$

Example

Show that the equation

$$u_{xx}+\frac{2N}{x}u_x=\frac{1}{a^2}u_{tt}$$

where $N\,and\,a\,are$ constant, is hyperbolic and obtain it canonical form.

Solution

Comparing with ss the general PDE and replacing $y\,by\,t.$ we have $A=1,B=0,C=-/a^2$. $D=2N\backslash x, and\ E=F=G-0.$ The discriminant $B^2-4AC=4\backslash a^2>-0.$ Hence, the given PDE is hyperbolic. The characteristics equation are

$$\frac{dt}{dx}\aleph\frac{B\mp\sqrt{B^2-4AC}}{2A}\quad\frac{\sqrt{\frac{4}{a^2}}}{2}\quad\mp\frac{1}{a}$$

Therefore,

$$\frac{dt}{dx}=-\frac{1}{a},\frac{dt}{dx}=\frac{1}{a}$$

On integration, we get

$$t=-\frac{x}{a}+c_1,t=\frac{x}{a}+c_2$$

Hence, the characteristics equations are

$$\xi=x+at,\eta=x-at$$

The canonical form can be obtained by computing

$$\overline{A}=A\xi_x^2+B\xi_x\xi_t+C\xi_t^2=0,$$

$$\bar{B} = 2A\xi_x\eta_x + B(\xi_x\eta_t + \xi_t\eta_t) + 2C\xi_t\eta_t = 4,$$

$$\bar{C} = 0, \bar{D} = D\xi_x + E\xi_t = \frac{2N}{x}, \bar{E} + D\eta_x + E\eta_t = \frac{2N}{x}$$

Thus, the canonical equation for the given PDE is

$$4u_{\xi\eta} + \frac{2N}{x}(u_\xi + u_\eta) = 0$$

Expressing x in terms of ξ *and* η. the required canonical equation is

$$u_{\xi\eta} + \frac{N}{\xi+\eta}(u_\xi + u_\eta) = 0$$

Example

Transform the following differential equation to canonical form:

$$u_{xx} + 2u_{xy} + 4u_{yy} + 2u_x + 3u_y = 0$$

Solution

The discriminant $B^2 - 4Ac = -12 < 0$: Hence, the given PDE is elliptic. The characteristics equation are

$$\frac{dy}{dx} = \frac{B - \sqrt{B^2 - 4AC}}{2A} = +1 - i\sqrt{3}$$

$$\frac{dy}{dx} = \frac{B + \sqrt{B^2 - 4AC}}{2A} = +1 + i\sqrt{3}$$

Integration of these equation yields

$$y = +\left(1 - i\sqrt{3}\right)x + c_1, y = +\left(1 + i\sqrt{3}\right)x + c_2$$

Hence, we may take the characteristics equation in the form

$$\xi = y - \left(1 - i\sqrt{3}\right)x\eta = y - \left(1 + i\sqrt{3}\right)x$$

In order, to avoid calculation with complex variables, we introduced the second transformation n

$$\alpha = \frac{\xi + \eta}{2}, \beta = \frac{\xi - \eta}{2i}$$

Therefore,

$$\alpha = y - x, \beta = \sqrt{3x}$$

The canonical form can now be obtained by computing

$$\bar{A} = A\alpha_x^2 + B\alpha_x\alpha_y + C\alpha_y^2 = 3$$

$$\bar{A} = 2A\alpha_x\beta_x + B\left(\alpha_x\beta_y + a_y\beta_x\right) + 2Ca_y\beta_y = 0$$

$$\bar{C} = A\beta_x^2 + B\beta_x\beta_y + C\beta_y^2 = 3$$

$$\bar{D} = Aa_{xx} + Ba_{xy} + Ca_{yy} + Da_x + Ea_y = 1$$

$$\bar{E} = A\beta_{xx} + B\beta_{xy} + C\beta_{yy} + D\beta_x + E\beta_y = 2\sqrt{3}$$

$$\bar{F} = 0, \bar{G} = 0$$

Thus the b required canonical form is

$$3u_{aa} + 3u_{\beta\beta} + u_a + 2\sqrt{3u_\beta} = 0$$

$$u_{aa} + u_{\beta\beta} = -\frac{1}{3}\left(u_a + 2\sqrt{3u_\beta}\right)$$

10.4. ADJOINT OPERATORS

Let

$$Lu = \phi$$

where L is a differentiate operator given by

$$L = a_0(x)\frac{d^n}{dx^n} + a_1(x)\frac{d^{n-1}}{dx^{n-1}} + \ldots + a_n(x)$$

One way of introducing the adjoint differentiate operator L^* associated with L is to form product vLu and integrate it over the interval of interest. Let

$$\int_A^B vLu\,dx = \int_A^B uL^*\,v\,dx$$

which is obtained after repeated integration by parts, Here, L^* is the operator adjoint to L, where the function $u\,and\,v$ are completely arbitrary except that $Lu\, and\, L^* v$ should exist which is obtained after repeated integration by parts, Here, L^* is the operator adjoint to $L,$ where the function $u\, and\, v$ are completely arbitrary except that $Lu\, and\, L^* v$ should exist.

Example

Let $Lu = a(x)(d^2 u / dx^2) + b(x)\left(\frac{du}{dx}\right) + c(x)u;$ construct its adjoint L^*.

Solution

Consider the equation

$$\int_A^B vLu\, dx = \int_A^B u\, v\left[a(x)\frac{d^2u}{dx^2} + b(x)\frac{du}{dx} c(x)u\right]dx$$

$$= \int_A^B (av)\frac{d^2u}{dx^2}dx + \int_A^B (bv)\frac{du}{dx}dx \int_A^B (cv)u\, dx$$

However,

$$\int_A^B (av)\frac{d^2u}{dx^2}du = \int_A^B (av)\frac{d}{dx}(u)'dx$$

$$= [u'va]_A^B - \int_A^B (av)' u'dx\, [u'av]_A^B - \left[u(av)'\right]_A^B + \int_A^B u(av)'u'dx$$

$$\int_A^B (bv)\frac{du}{dx}dx = [u(bv)]_A^B - \int_A^B u(bv)' dx$$

$$\int_A^B (cv)u\, dx = \int_A^B u(cv)dx$$

Comparing this equation with, we get

$$L^* v = (av)'' - (bv)' + (cv) = av'' + (2a' - b)v' + (a'' - b' + c)v$$

Therefore,

$$L^* = a\frac{d^2}{dx^2} + (2a' - b)\frac{d}{dx} + (a'' - b' + c)$$

Consider the partial differential equation

$$L(u) = Au_{xx} + 2Bu_{xy} + Cu_{yy} + Du_x + Eu_y + Fu = \phi \qquad (18)$$

which is valid in a region S of the $xy-plane$, where $A,B,C,\ldots,\phi$ are function of $x \, and \, y.$ In addition, linear boundary conditions of the general form

$$\alpha u = \beta u_x = f$$

are presented over the boundary curve ∂S of the region $S.$ Let

∬ vLuđ =[ ]+ ∬L $\overset{*}{\text{v}}$ dó

where the integrated part [ ] is a line integral evaluated over $\partial S,$ the boundary of $S,$ then L^* is called the adjoint operator. In general, a second order linear partial differential operator L is denoted by

$$L(u) = \sum_{1,j=1}^{n} A_{ij} \frac{\partial^2 u}{\partial x_i \, \partial x_j} + \sum_{i=1}^{n} B_i \frac{\partial u}{\partial x_1} + Cu \qquad (19)$$

Its adjoint operator is defined by

$$L^*(v) = \sum_{1,j=1}^{n} \frac{\partial^2 u}{\partial x_i \, \partial x_j}(A_{ij} v) - \sum_{i=1}^{n} \frac{\partial}{\partial x_1} + (B_i v) + Cv \qquad (20)$$

Here it is assumed that $A_{ij} \in C^2 \, and \, B_t \in C^1.$ for any pair of function $u,v \in C^2.$ it can be shown that

$$vLv(u) - uL^*(v) = \sum_{i=1}^{n} \frac{\partial}{\partial x_1}\left[\sum_{i=1}^{n} A_{ij}\left(v\frac{\partial u}{\partial x_j} - u\frac{\partial v}{\partial x_j}\right) + uv\left(B_1 - \sum_{j=1}^{n} \frac{\partial A_{ij}}{\partial x_j}\right)\right]$$

(21)

This is known as Lagrange's identity.

Example

Find the adjoint of the differential operator

$$L(u) = u_{xx} - u_t$$

Solution

Comparing with the general second order PDE, we have $A_{11} = 1, B_1 = -1$. From equation (20) the adjoint of is given by

$$L^*(v) = \frac{\partial^2}{\partial x^2}(v) - \frac{\partial}{\partial t}(-v) = v_{xx} + v_t$$

Therefore,

$$L^*(u) = u_{xx} + u_t$$

It may be noted that the diffusion operators is not a self-adjoint operator.

10.5. RIEMANN'S METHOD

In Section we have noted with interest that a linear second order PDE

$$L(u) = G(x, y)$$

is classified as hyperbolic if $B^2 > 4AC, and\, it\, has$ two families of real characteristics curves in the $xy - plane$ whose equations are

Hence (ξ, η) are the natural coordinates for the hyperbolic system. In the x y-plane the curves $\xi(x, y) - c_1\ and\ \eta(x, y) = c_2$ are the characteristics of the given PDE as shown in fig while in the $\xi\eta - plane,$ the curve $\xi = c_1\ and\ \eta = c_2$ are families of straight line parallel to the axes.

A linear second order partial differential equation in two variables, once classified as a hyperbolic equation, can always be reduced to the canonical form

$$\frac{\partial^2 u}{\partial x \partial y} = F(x, y, u, u_x, u_y)$$

In partial, consider an equation which is already reduced to its canonical form in the variables x, y

$$L(u) = \frac{\partial^2 u}{\partial x \partial y} + a\frac{\partial u}{\partial x} + b\frac{\partial u}{\partial y} + cu = F(x, y)$$

where L is a linear differential operator and a, b, c, f are functions of

$x \, and \, y$ only and are differential in some domain IR.

Let $v(x,y)\, be\, an\, arbitrary\, function$ having continuous second order partial derivative consider the adjoint operator L^* of L defined by

$$L^*(v)=\frac{\partial^2 u}{\partial x \partial y}-\frac{\partial}{\partial x}(av)-\frac{\partial}{\partial y}(bv)+cv$$

$$M_x+N_y=u_x(av)+u(av)_x-u_x=u_xv_y-uv_{xy}+u_y(bv)+u(bv)_y+v_yu_x+vu_{xy}$$

Adding and subtracting $cuv,$ we get

$$M_x+N_y=u\left[\frac{\partial^2 u}{\partial x \partial y}-\frac{\partial}{\partial x}(av)-\frac{\partial}{\partial y}(bv)+c\right]+v\left[\frac{\partial^2 u}{\partial x \partial y}+a\frac{\partial u}{\partial x}+b\frac{\partial u}{\partial y}+cu\right]$$

i.e.

This is known as Lagrange identity which all be used in the subsequent discussion. The operator L is a self –adjoint if and only if $L=L^*$. Now we shall attempt to solve Cauchy's problem which is described as follows. Let

$$L(u)=F(x,y)$$

with the condition (Cauchy data)

$(i)\, u=f(x)\, on\, \Gamma, a\, curve$ in the $xy-plane$

$$(ii)\frac{\partial u}{\partial x}=g(x)\, on\tilde{A}$$

This is $u,$ and its normal derivative are prescribed on a curve $\tilde{A}$ which is not a characteristics line. Let $\tilde{A}$ be a smooth initial curve which is also continuous as shown in. Since Eq. is in Canonical form, $x \, and \, y$ are the characteristics coordinates. We also assume that the tangent $\tilde{A}$ is nowhere parallel to the coordinates axes.

Let $P(\xi,\eta)$ be a point at which the solution to the Cauchy problems is sought. Let us draw the characteristics $PQ\, and\, PR\, throuh\, p$ to meet the curve $\tilde{A}$ at Q and R. We assume that u, u_x, u_y are presented along $\tilde{A}$

Let ∂IR be a closed contour $PQRP$ bounding IR. Since Eq. is already in canonical form, the characteristics are lines parallel to $x\,and\,y\,axes.$ Using characteristics $PQ\,and\,PR\,throuh\,p$ to meet the curve $\tilde{A}$ at Q and R. We assume that u, u_x, u_y are presented along $\tilde{A}$ Let ∂IR be a closed contour $PQRP$ bounding IR. Since Eq. is already in canonical form, the characteristics are lines parallel to $x\,and\,y\,axes.$ Using Green's theorem we have

$$\iint (M_x + N_y)dxdy = \oint_{\partial IR} (M\,dy - N\,dx \tag{29}$$

where ∂ IR is the boundary of IR. Applying this theorem to the surface integral of Eq. we obtain

$$\int_{\partial IR} \left(M\,dy - N\,dx\right) = \int_{IR}\int \left[vL(u) - uL^*(v)\right]dx\,dy \tag{30}$$

In other words, now using the fact that $dy = 0$ on $PQ\,and\,dx = 0\,on\,PR,$ we have

$$\int_{PQ} N\,dx = \int_P^Q buc\,dx + \int_P^Q vu_x\,dx$$

Integrating by parts the second term on the right-hand side and grouping. the above equation becomes

$$\int_{PQ} N\,dx = [nv]_P^Q \int_P^Q u\left(bv - v_x\right)dx$$

Substituting this result in, we obtain

And at the same time satisfy the following conditions:

(i) $v_x = bv\,when\,y = \eta, PQ$ (31)

(ii) $v_y = av\,when\,x = \xi, i.e, on\,PR$ (32)

(iii) $v = 1\,when\,x = \xi, y = \eta$ (33)

We call this function $v(x, y, \xi, \eta)$ as the Riemann function or the Riemann

–Green function. Since $L(u)=F$, reduce to

$$[u]_p=[u]_Q-\int_{\tilde{A}}\left[u\left(av-v_y\right)dy-v\left(bu+u_x\right)dx\right]+\iint_{IR}\left(vf\right)dx\,dy \tag{34}$$

This is called Riemann-Green solution for the Cauchy problems described by Eq. (28) when $u\,and\,u_x$ are presented on $\tilde{A}$

The relation gives us the value of u at a point $p\,when\,u\,and\,u_x$ are prescribed on $\tilde{A}$. But $u\,and\,u_x$ are presented on $\tilde{A}$ we obtain

$$[u]_p=[uv]_R-\int_{\tilde{A}}uv\left(a\,dy-b\,dx\right)-\int_{\tilde{A}}\left(uv_x\,dx+vu_y\,dy\right)+\int\int_{\tilde{A}}\left(vF\right)dx\,dy \tag{35}$$

$$[u]_p=\frac{1}{2}\left\{[uv]_Q+[uv]_R\right\}-\int_{\tilde{A}}uv\left(a\,dy-b\,dx\right)-\frac{1}{2}\int_{\tilde{A}}u\left(v_x\,dx-v_y\,dy\right)+\frac{1}{2}\int_{\tilde{A}}v\left(u_x\,dx-u_y\,dy\right)\int\int_{\tilde{A}}\left(vF\right)dx\,dy \tag{36}$$

Thus, we can see that the solution to the Cauchy problem at a point $(\xi.\eta)$ depends only on the Cauchy data on $\tilde{A}$. The knowledge of the Riemann-Green function therefore enable us to solve Eq. with the Cauchy data presented on non-characteristics curve.

Example

Transform the following differential equation to canonical form:

$$u_{xx}+2u_{xy}+4u_{yy}+2u_x+3u_y=0$$

Solution

The discriminant $B^2-4Ac=-12<0$: Hence, the given PDE is elliptic. The characteristics equation are

$$\frac{dy}{dx}=\frac{B-\sqrt{B^2-4AC}}{2A}=+1-i\sqrt{3}$$

$$\frac{dy}{dx}=\frac{B+\sqrt{B^2-4AC}}{2A}=+1+i\sqrt{3}$$

Integration of these equation yields

$$y = +\left(1 - i\sqrt{3}\right)x + c_1, y = +\left(1 + i\sqrt{3}\right)x + c_2$$

Hence, we may take the characteristics equation in the form

$$\xi = y - \left(1 - i\sqrt{3}\right)x\eta = y - \left(1 + i\sqrt{3}\right)x$$

In order, to avoid calculation with complex variables, we introduced the second transformation n

$$\alpha = \frac{\xi + \eta}{2}, \beta = \frac{\xi - \eta}{2i}$$

Therefore,

$$\alpha = y - x, \beta = \sqrt{3x}$$

The canonical form can now be obtained by computing

$$\bar{A} = A\alpha_x^2 + B\alpha_x\alpha_y + C\alpha_y^2 = 3$$

$$\bar{A} = 2\,A\alpha_x\beta_x + B\left(\alpha_x\beta_y + a_y\beta_x\right) + 2\,Ca_y\beta_y = 0$$

$$\bar{C} = A\beta_x^2 + B\beta_x\beta_y + C\beta_y^2 = 3$$

$$\bar{D} = Aa_{xx} + Ba_{xy} + Ca_{yy} + Da_x + Ea_y = 1$$

$$\bar{E} = A\beta_{xx} + B\beta_{xy} + C\beta_{yy} + D\beta_x + E\beta_y = 2\sqrt{3}$$

$$\bar{F} = 0, \bar{G} = 0$$

Thus, the b required canonical form is

$$3u_{aa} + 3u_{\beta\beta} + u_a + 2\sqrt{3u_\beta} = 0$$

$$u_{ua} + u_{\beta\beta} = -\frac{1}{3}\left(u_a + 2\sqrt{3u_\beta}\right)$$

Example

Show that the Green's function for the equation

$$\frac{\partial^2 u}{\partial u\,\partial x} + u = 0$$

$$v(x, y, \xi, \eta) = j_0\sqrt{2(x - \xi)(y - \eta)}$$

where j_0 denotes Bessel's function of the first kind canonical hyperbolic equation we have.

Solution

Comparing with the standard canonical hyperbolic equation we have, it is self-adjoint equation and, therefore, the Green's function v can be obtained form

$$\frac{\partial^2 u}{\partial u\,\partial x}+v=0$$

Subject to

$$\frac{\partial v}{\partial x}=0\,on\,y=\eta$$

$$\frac{\partial v}{\partial x}=0\,on\,x=\xi$$

$$v=1\,at\,x=\xi, y=\eta$$

Let

$$\phi^k=a(x-\xi)(y-\eta)$$

$$\frac{\partial v}{\partial x}=\frac{\partial v}{\partial \phi}\frac{\partial \varphi}{\partial x}$$

But

$$k\phi^{k-1}\frac{\partial \varphi}{\partial x}=a(y-\eta)$$

Therefore,

$$\frac{\partial \varphi}{\partial x}=\frac{a}{k}\phi^{1-k}(y-\eta)$$

Then

$$\frac{\partial v}{\partial x}=\frac{av}{\partial \phi}\frac{a}{k}\phi^{1-k}(y-\eta)$$

$$=\frac{a}{k}\left[\phi^{1-k}\frac{av}{\partial\phi}+(1-k)\phi^{1-k}\frac{\partial v}{\partial\phi}\frac{\partial\phi}{\partial y}(y-\eta)+\phi^{1-k}(y-\eta)\frac{\partial^2 v}{\partial\phi^2}\frac{\partial\phi}{\partial y}\right]$$

However,

$$\frac{a\phi}{\partial y}=\frac{a}{k}\phi^{1-k}(x-\xi)$$

Therefore

$$\frac{\partial^2 u}{\partial u\,\partial x}=\frac{a}{k}\left[\phi^{1-k}\frac{\partial v}{\partial\phi}+(1-k)\phi^{1-k}(x-\xi)(y-\eta)\frac{a}{k}\phi^{1-k}\frac{\partial v}{\partial\phi}+\phi^{1-k}(y-\eta)\frac{\partial^2 v}{\partial\phi^{2-k}}\frac{a}{K}\phi^{1-k}(x-\xi)\right]$$

Hence,

$$\frac{\partial^2 u}{\partial u\,\partial x}+v=0$$

or $$\phi^2 v''+\phi v'+\frac{k^2}{a}\phi^k\, v=0$$

Let $k=2, a=4.$ Then the above equation reduced to

$$\phi^2 v''+\phi v'+\phi^2 v=0=v''+\frac{1}{\phi}c'+v$$ (Bessel's equation)

Its solution is known to be of the form

$$v=j_0(\phi)=j_0(2\sqrt{(x-\xi)(y-\eta)})$$

which is desired Green 's function.

SUMMARY AND DISCUSSION

In this chapter we discussed in detail the basic concepts of Partial Differential Equation, their types, and standard equations for these types of equation.

10.1. Partial Differential Equation

An equation which involves several independent variables (usually denoted by $x, y, z, t, \dots)$ a dependent functions u of these variables and the partial derivatives of the dependent function u with respect to the independent variables such as

$F\left(x, y, z, t \ldots, u_x, u_y, u_z, u_1, \ldots\ldots u_{xx}, u_{yy}, \ldots u_{xy}, \ldots\right) = 0$, is called a partial differential equation. A few well-known examples are:

(i) $u_1 = k\left(u_{xx} + u_{yy} + u_{zz}\right)$ [Linear three-dimensional heat equation]

(ii) $u_{xx} + u_{yy} + u_{zz} = 0$ [Laplace equation in three dimensions]

(iii) $u_{tt} = c^2\left(u_{xx} + u_{yy} + u_{zz}\right)$ [Linear three-dimensional wave equation]

(iv) $u_t + uu_x = \mu u_{xx}$ [Nonlinear one-dimensional Burger equation].

In these examples, u is the dependent function and the subscript denote partial differentiation with respect to these variables.

Definition. The order of the partial differential equation is the order of the highest derivative occurring in the equation. Thus, the above examples are partial differential equation of second order, whereas

$$u_1 = u_{xx} + \sin x$$

is an example for third order partial differential equation.

10.2. Classification of Second Order PDE

The most general linear second order PDE, with one dependent function u on a domain Ù of point $X = \left(x_1, x_2, \ldots\ldots x_n\right), n > 1.$ is

$$\sum_{i,j=1}^{n} a_{ij} u_{xi} x_j + \sum_{i=1}^{n} b_1 u_{x1} + F\left(u\right) = G \qquad (2)$$

The classification of a PDE depends only on the highest order derivatives present.

The classification of PDE is motivated by the classification of the quadratic equation of the form

$$Ax^2 + Bxy + Cy^2 + Dx + Ey + F = 0,$$

which is ellipse, parabolic or hyperbolic.

Thus, we have the following second order linear PDE in two variables $x \, and \, y$

$$Au_{xx} + Bu_{xy} + Cu_{yy} + Du_x + Eu_y + Fu = G$$

where the coefficient $A, B, C, \ldots$ may be functions of $x \, and \, y,$ however for the sake of simplicity we assume them to be constants. Equation can be elliptic, parabolic or hyperbolic at a point (x_0, y_0) according as the discriminant

$$B^2(x_0, y_0) - 4A(x_0, y_0)C(x_0, y_0) = 0$$

is negative, zero or positive. If this is true at all points in a domain Ù, then is said to be elliptic, parabolic in that domain. if the number of independent variables is two or three, transformation can always be found to reduce the given PDE to a canonical form (also called normal form). In general, when the number of independent variables is greater than 3, it is not always possible to find such a transformation except in certain special cases. The idea of reducing the given PDE to a canonical form is that the transformed equation assumes a simple form so that the subsequent analysis of solving the equation is made easy.

REFERENCES

1. C.W. Gear Numerical Initial Value Problems in ordinary Differential Equation, Prentice-Hall, 1971.
2. Encyclopedia of Applied and Computational Mathematics.
3. Hairer, E., Norsett, S. P., & Wanner, G., (1987). *Solving Ordinary Differential Equations I,* Springer-Verlag.
4. Luther, H. A., & Wiles, J. O., Applied Numerical Methods" by Carnahan.
5. Schilling, R., & Harris, S. L., Applied Numerical Methods for Engineers.
6. www.wikipedia.com/differential Equation.
7. Zill, D. G., & Cullen, M. R., Differential equation with boundary value Problem, Brooks/Cole, USA.

INDEX